In addition, various low and mid-bands (e.g., 700 MHz, 800 MHz, 1.8 GHz, 2.1 GHz, 2.3 GHz) are also widely being used in several regions around the world. Besides, 24–28.5-GHz, 31–33-GHz, and 37–40.5-GHz spectra are being studied in Europe, the Republic of Korea, Japan, United States, and India [6]. Millimeter-wave sensing and tracking at 24-GHz ISM band is also being explored all over the world for mm-wave sensing application [7]. An advanced multifunctional RF front end for mm-wave communication and sensing application will require to operate in all these bands simultaneously. It is to be noted that for successful deployment of this system requires high gain multibeam or beam-steering antenna systems as shown in Figure 1.1.

Apart from the recent development of 5G ecosystem, at the time of this writing, a major drive towards the research of 5G beyond and sixth generation (6G) communication system has been initiated. With the development of each generation of communication system, the carrier bandwidth kept increasing from 5 MHz in third generation (3G), 20 MHz in fourth generation (4G) to 100 MHz in 5G, and currently it is targeted to be at least 400 MHz in 6G communication, which will boost the channel capacity of the network [8]. Recently, the International Telecommuncation Union (ITU) Resolution 245 in the World Radiocommunication Conference (WRC)-19 recommended the feasibility study of new spectrum allocation in 6,425–7,025-MHz, 7,025–7,125-MHz, and 10.0–10.5-GHz bands [9]. In 5G spectrum, 600–2,600 MHz was proposed for wide area coverage, 2.5–6 GHz for urban network, and 24–71 GHz for densely populated hotspot coverage. In the upcoming proposal for 6G spectrum, additional bands for new applications, for example, 470–690 MHz (lower band) for extreme wide area IoT, 7–20 GHz (mid-band) for urban area extreme capacity network, and above 92 GHz (high band) for short-range sensing application are discussed [10, 11]. Among these proposed bands, mid-band spectrum may see high potential usage due to good trade-off between capacity and coverage area [12]. Such a high-speed communication channel requires special attention for the design of RF circuit while maintaining a low-profile form factor. The conventional planar technologies exhibit limited performance due to unshielded nature of the line. Hence, a compact, low-profile, shielded transmission line is an urgent need for meeting the requirement of modern wireless communication.

The mm-wave communication in the high band inherits limitation in terms of higher free-space path loss limiting its coverage and also higher sensitivity to obstacles in the communication environment. Therefore, new designs and solutions are required to ensure high-speed communication with billions of devices. The

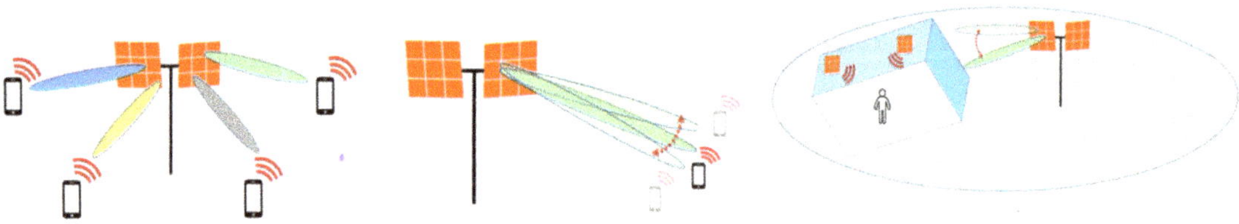

(a) (b) (c)

Figure 1.1 (a) Millimeter-wave multibeam antenna supporting multiuser environment; (b) mm-wave beam-steering antenna for active network planning and tracking; and (c) mm-wave indoor sensing using antennas installed in the wall/rooftop.

modern mm-wave RF front-end systems should be capable of adaptive beamforming to enhance spatial multiplexing by tracking users with individual high gain beams as well as cope up with time variation in the channel. Such a system would enable cochannel interference mitigation, enhancing signal level and power management in wireless communication systems. Therefore, a significant role to successful mm-wave communication lies in tracking and sensing along with communication link. Besides, to ensure satisfactory performance of deep nonline-of-sight communication, a natural solution that may be adopted is dense, small cell deployment to increase coverage to large number of users. Therefore, future RF systems will be modular setup installed in variety of public infrastructure such as street light, traffic light, indoor walls, and tall buildings' rooftop. Therefore, novel, low-cost solutions are required to meet the requirement of modern mm-wave systems maintaining successful coverage with efficient power management. To overcome this excessive path loss, the antennas at these frequencies are required to be high gain to support long-range communication [13]. However, to establish a satisfactory link between all the users with the access point, wider beam coverage is required. Realizing high gain antenna fundamentally limits the beamwidth of the main lobe of the antenna, which restricts its coverage area. Therefore, a high gain antenna solution with multibeam or beam-steering capability is on high demand to implement mm-wave communication links. Also, to overcome the problem of polarization mismatch, these antennas are required to be dual or circular polarized to ensure maximum possible reception of the signal. A possible schematic of future public infrastructure along with modular communication setup is shown in Figure 1.2.

The circuits and systems at mm-wave frequencies become very small, making it suitable to be incorporated in compact handheld devices while supporting higher bandwidth and high data rate. However, the conventional methods for realizing RF circuits using microstrip, coplanar waveguide technologies have limitations for their use in mm-wave frequency bands (e.g., high radiation loss, surface wave loss). Recently, a new technique entitled substrate integrated waveguide (SIW) has emerged as an attractive alternative for use in the mm-wave spectrum [14]. After its inception, several attempts have been made to realize SIW-based antennas for modern communication systems. In this book, a detailed study and analysis to design SIW-based RF front-end subsystems will be discussed in detail.

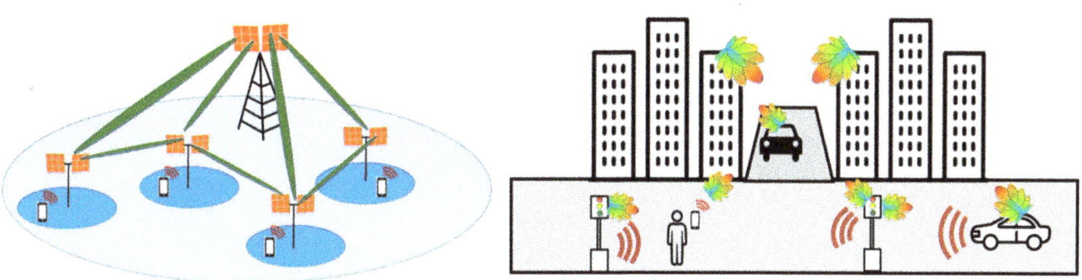

(a) (b)

Figure 1.2 (a) Dense deployment of low-cost mm-wave base station antennas; and (b) mm-wave antenna deployment for outdoor communication and sensing application (e.g., smart traffic management, pedestrian detection).

1.2 Basics of RF Transceiver Systems

As discussed in the previous section, the continuous development in each generation of the communication system has led the revolution in the RF transceiver architecture. To achieve the immense growth of channel capacity, the consecutive development of MIMO technology from 2T2R to 4T4R and finally in 5G, massive MIMO (mMIMO) architecture with a huge number of transmit and receive antennas ranging from 16T16R to 64T64R configuration has been conceptualized. The mMIMO architecture aims to improve the signal to interference and noise ratio in a densely populated communication channel by introducing beamforming algorithm. In such a system, the base station antenna not only exercises time division and frequency division multiplexing; in addition, spatial multiplexing of several beams generated by different subantenna arrays are managed at the back-end processor. In this process, different users get dedicated RF signal using different pencil beams generated from the mMIMO antenna. The simplified architecture of mMIMO transceiver system is shown in Figure 1.3. It is to be noted that the beamforming algorithm may be implemented in the RF module in the case of analog beamforming or in the baseband module in the case of the digital beamforming network. In the case of the analog beamforming network, the baseband signal is first sent to intermediate frequency (IF) module and upconverted to RF where predetermined gain and phase are introduced to deflect the radiation pattern in the desired direction. As shown in Figure 1.3, several parallel RF chains fed by their corresponding baseband channel are implemented to generate simultaneous multiple beams for the transmission of a significant amount of information through the same channel [15]. In the case of a digital beamforming network, the beamforming algorithm generates digital amplitude and phase weight coefficient applied to each of the data streams, which is upconverted to RF and sent through the antenna module [16]. While the analog beamforming provides the simplest architecture, it can only work with a limited number of data streams. However, the digital beamforming architecture can offer huge flexibility in handling digital data streams, but the increased complexity at the digital processing module increases the cost and complexity of the system.

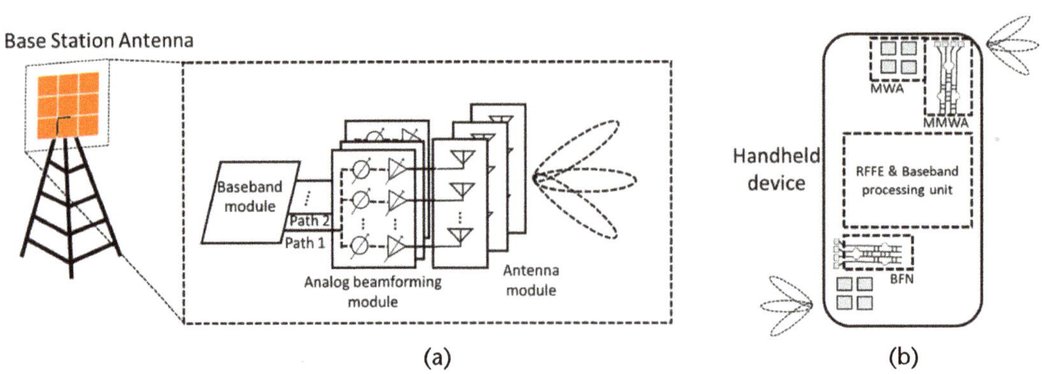

Figure 1.3 (a) Simplified block diagram of mMIMO base station antenna architecture; and (b) antenna module placement in handheld device (MWA: microwave antenna; MMWA: mm-wave antenna; RFFE: RF front end; BFN: beamforming network).

Another popular approach is to have a trade-off between both approaches with the presence of beamforming at both baseband and RF modules, which is known as a hybrid beamforming network [17, 18]. Several attempts have been reported to realize the mMIMO testbed for further studies. A real-life example of 100 antenna testbeds for a massive MIMO operation was built at the University of Lund, Sweden [19]. A 5G testbed using 28-GHz mm-wave phased array module has been implemented in IIT Madras, India, with the collaboration of SAMEER Chennai and NXP Semiconductors [20]. In the case of handheld devices, the architecture is required to be simplified to reduce cost and complexity. In this case, proper placement of microwave and mm-wave antenna module at the side of the device is of the highest priority to ensure sufficient space for the front-end and baseband processing unit and battery unit [21]. In case of an mm-wave module in a handheld device, the multibeam antenna module is realized by analog beamforming network or by using a digital beamforming chip [22, 23]. The use of digital chip typically increases the cost and power consumption of the device and, hence, designers need to trade off between cost, performance, and compactness of the RF front-end module by choosing a proper design architecture. Many of these high-performance RF front-end modules require additional care in circuit design to achieve the stringent specification. The conventional planar technologies (e.g., microstrip and coplanar waveguide (CPW)) are not always the best alternative to meet these requirements, and, hence, a new type of low interference, low-loss transmission line architecture, is very much essential to implement this type of high-performance system. In this book, a detailed study of SIW technology and its design guideline for realization of the basic building blocks of this kind of RF module will be discussed in detail.

1.3 Introduction to SIW Technology

As discussed in the previous section, the introduction of higher-frequency band at millimeter-wave spectrum to develop multigigabit data channels requires a low loss, high shielding transmission line that can exhibit satisfactory performance with low crosstalk, and low radiation loss. The conventional rectangular waveguide is one of the popular choices for such a requirement due to its inherent shielded nature. The bulky waveguide-based communication channel fits the requirement of the RF chain in the base station, although the system becomes bulky and not easy to handle. The maintenance cost of such system is comparatively higher than the conventional printed circuit board (PCB)-based circuits where planar technologies such as microstrip or coplanar waveguide are used. However, at the user equipment (UE) side, the implementation of the transceiver using a conventional waveguide imposes serious limitation in terms of portability and compactness of the design, which does not fit the requirement of a commercial communication setup. To resolve the issue, a new kind of transmission line with a low-profile compact form while meeting the requirement of high-speed communication channel should be explored. A possible solution can be to make the waveguide compact and easy to handle and a low-profile form.

A schematic diagram of conventional waveguide is shown in Figure 1.4 (the dimension of waveguide may be assumed as per conventional American Electronic

Industry Alliance (EIA) standards, e.g., WR16). If we consider the direction of propagation as z, the electric and magnetic fields of the propagating TE_{mn} modes in this waveguide can be expressed as follows [24]:

$$
\begin{aligned}
E_y(x,y,z,t) &= -\frac{\omega\mu}{h^2}\left(\frac{n\pi}{b}\right)H_0\cos\left(\frac{m\pi}{a}x\right)\sin\left(\frac{n\pi}{b}y\right)\sin(\omega t-\beta z)\\
E_y(x,y,z,t) &= \frac{\omega\mu}{h^2}\left(\frac{m\pi}{a}\right)H_0\sin\left(\frac{m\pi}{a}x\right)\cos\left(\frac{n\pi}{b}y\right)\sin(\omega t-\beta z)\\
E_x(x,y,z,t) &= 0\\
H_x(x,y,z,t) &= -\frac{\beta}{h^2}\left(\frac{m\pi}{a}\right)H_0\sin\left(\frac{m\pi}{a}x\right)\cos\left(\frac{n\pi}{b}y\right)\sin(\omega t-\beta z)\\
H_y(x,y,z,t) &= -\frac{\beta}{h^2}\left(\frac{n\pi}{b}\right)H_0\cos\left(\frac{m\pi}{a}x\right)\sin\left(\frac{n\pi}{b}y\right)\sin(\omega t-\beta z)\\
H_z(x,y,z,t) &= H_0\cos\left(\frac{m\pi}{a}x\right)\cos\left(\frac{n\pi}{b}y\right)\cos(\omega t-\beta z)
\end{aligned}
\tag{1.1}
$$

The electric and magnetic field distributions of the dominant TE_{10} mode are shown in Figure 1.4. Here, the concentration of arrows represents the magnitude of the field. Hence, a sinusoidal variation of the electric field along x-axis and no variation along y-axis is observed, which can also be seen from (1.1). The resonant frequency of corresponding modes can be found in (1.2), where a and b are width and height of the waveguide along the x and y axes, respectively.

$$
\begin{aligned}
f_c\big|_{TE_{10}} &= \frac{c}{\sqrt{\varepsilon_r}\,2a}\\
f_c\big|_{TE_{mn}} &= \frac{c}{\sqrt{\varepsilon_r}\,2a}\sqrt{\left(m^2+\left(\frac{na}{b}\right)^2\right)}
\end{aligned}
\tag{1.2}
$$

Considering the dimension of WR-90 (22.86 × 10.16 mm), the resonant frequencies of the TE_{mn} modes can be found in Table 1.2. The consecutive sequence

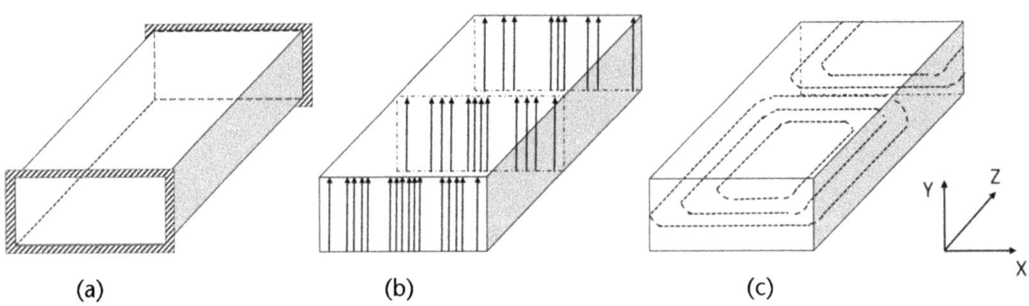

Figure 1.4 (a) Conventional waveguide; (b) electric field of dominant mode; and (c) magnetic field of dominant mode.

Table 1.2 Resonant Frequencies of TE Modes (in Gigahertz)

m	n				
	0	1	2	3	4
0	—	14.625	29.25	43.875	58.5
1	6.5	16.00439	29.96352	44.35387	58.86
2	13	19.56759	32.00879	45.76042	59.92704
3	19.5	24.375	35.15412	48.01318	61.66441
4	26	29.83103	39.13518	51.00015	64.01758
5	32.5	35.63903	43.72428	54.60097	66.9216

of resonant modes in ascending order of resonant frequencies are TE_{10}, TE_{20}, TE_{01}, TE_{11}, TE_{30}, TE_{21}, and so forth.

Now let's examine the effect of reducing the height of the waveguide along the y-axis with a height of 1.14 mm. The corresponding resonant frequencies of TE_{mn} modes are shown in Table 1.3. As we can see, the sequence of resonant modes in ascending order are TE_{10}, TE_{20}, TE_{30}, TE_{40}, TE_{50}, and so forth. Clearly, reducing the height affects the resonant frequencies of the modes for which the second-mode indices (n) are greater than 0. However, if we restrict our mode of operation to the dominant mode and other TE_{m0} modes, specifically if we use a single-mode operation of the waveguide at its dominant mode, we can easily use the reduced height waveguide as well. The electric and magnetic field distributions of the dominant TE_{10} mode in the reduced height waveguide is shown in Figure 1.5, which is like the original waveguide shown in Figure 1.4.

The reduced height waveguide can easily be implemented in a planar, low-profile dielectric board. The commercially available dielectrics have top and bottom copper cladding, which can serve as conductor plate of the waveguide. The only challenge remains in the implementation of a metallic sidewall in the dielectric board. This can be seen in Figure 1.6 where the cross-sectional view of a typical dielectric board is shown in Figure 1.6(a). To realize the metallic sidewall, the most suitable option is to insert a plated through-hole or via, which can resemble the continuous metallic sidewall to a certain extent. The result of the process is shown in Figure 1.6(c), which has top and bottom continuous copper plates and both side rows of metal vias resembling the copper sidewall. As a result, a semi-continuous uniconductor

Table 1.3 Resonant Frequencies of TE Modes for Reduced Height Waveguide (in Gigahertz)

m	n				
	0	1	2	3	4
0	—	130	260	390	520
1	6.5	130.1624	260.0812	390.0542	520.0406
2	13	130.6484	260.3248	390.2166	520.1625
3	19.5	131.4544	260.7302	390.4872	520.3655
4	26	132.5745	261.2968	390.8657	520.6496
5	32.5	134.0009	262.0234	391.3518	521.0146

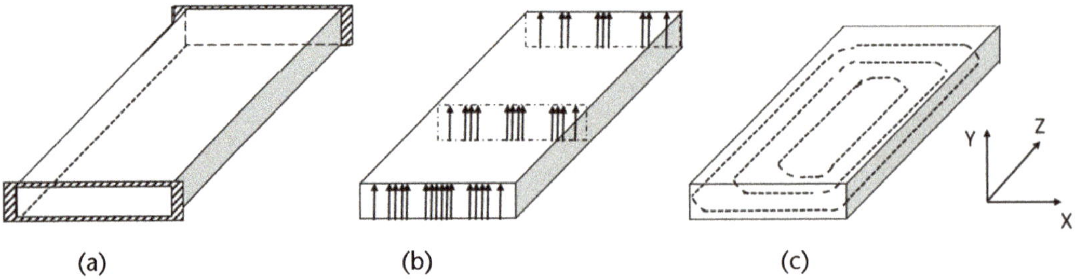

Figure 1.5 (a) Reduced height waveguide; (b) dominant mode electric field; and (c) dominant mode magnetic field.

transmission line can be implemented in a planar dielectric board, which can resemble a dielectric field waveguide. The first report of such structure can be found in [25], which mentions a waveguide line. Later, several similar designs were reported with the name of post wall waveguide (PWWG) [26], laminated waveguide [27], and so forth. Finally, a series of similar structures were reported where the technology was called a substrate integrated circuit (SIC) [4]. Specifically, the waveguide structure as shown in Figure 1.6 was named as substrate integrated waveguide (SIW). This name became popular among researchers all over the world and hence we will use the same name throughout the book. However, it must be noted that

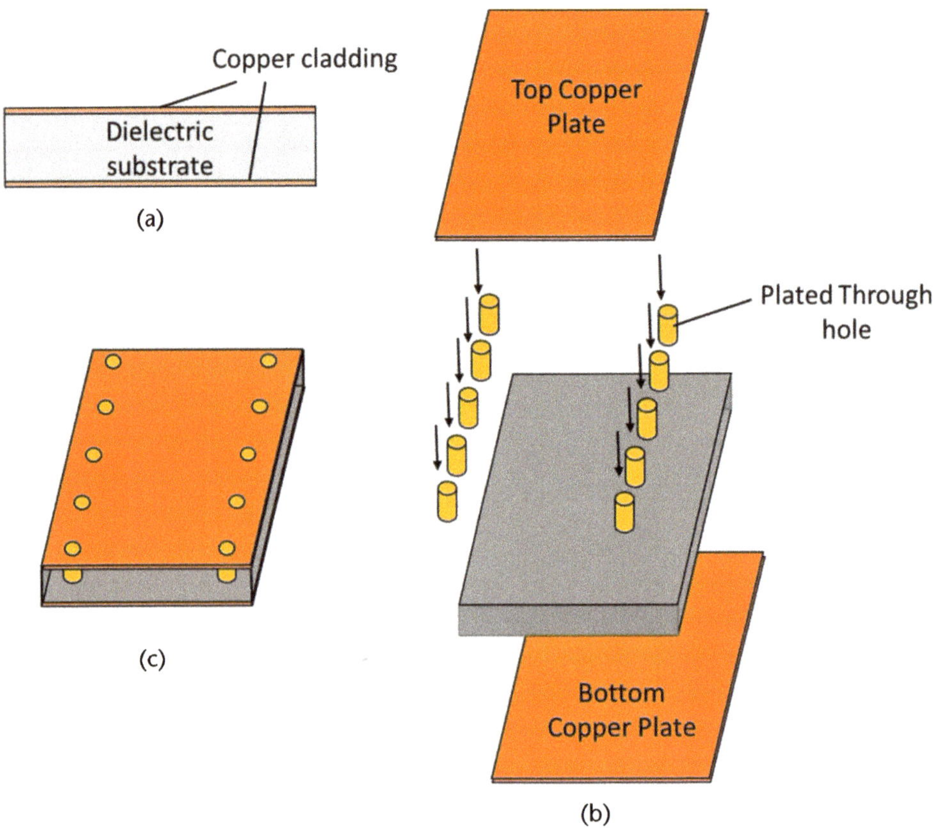

Figure 1.6 (a) Cross-section view of dielectric substrate; (b) layer-wise structure of SIW; and (c) SIW section.

all these concepts of PWWG, laminated waveguide, and SIW refers to a similar kind of implementation of waveguide-based structures in dielectric substrate in a low-profile, planar form. Another important point to be noted is that the reduction of the height of the waveguide will also have an effect on the resonant frequencies of the TM_{mn} mode. However, as we will see in later chapters, the TE_{0n}, $TE_{mn\,(n>0)}$, and TM_{mn} modes does not exist in SIW structures and hence, for simplicity, the corresponding discussion is excluded in this section. The readers are encouraged to explore the effect on TM_{mn} modes on their own.

1.4 Types of SIW Structures

The SIW structure can be represented as a planar form of dielectric loaded rectangular waveguide (RWG). It inherits many of the properties of conventional RWG such as good electromagnetic shielding, low loss, and high Q. The planar form of the structure opens a new possibility to implement a special kind of miniaturized pseudo-waveguide structure that can be judiciously used to special-purpose circuits. In this section, we will discuss some of these special types of SIW structures that are very commonly used as miniaturization technique for these kinds of circuits.

1.4.1 Half-Mode SIW

The field distribution in SIW is similar to a dielectric loaded RWG, which can be seen in Figure 1.7(a). Now, if we closely study the field distribution, it can be seen that a perfect open (perfect magnetic conductor (PMC)) symmetry plane can be drawn along the center of the waveguide, as shown in Figure 1.7(a). The field distribution in left and right sides of this imaginary plane will follow the boundary condition of the PMC imaginary wall. By applying the image theory, we can then bisect the waveguide into half of it with the PMC boundary at the bisected end. The same is shown in Figure 1.7(b). Such a kind of structure is called a half-mode SIW (HMSIW) [28]. By following this approach, one can reduce the SIW circuit footprint by almost half and the rest of the space can be utilized to accommodate other circuitry. The HMSIW top copper plate is spread almost half of the width of the original SIW while the bottom one is usually kept extended to a certain extent. One of the sidewall has metallic via walls while other side is kept open. The electric field distribution of the dominant mode of the HMSIW is shown in Figure 1.7(c). Clearly, such a dielectric aperture window at the side wall of the HMSIW does not behave as perfect open and a fringing field is expected to arise at this end, as shown in Figure 1.7(c). The fringing field affects the cutoff frequency of the HMSIW structure. As a result, to get the desired cutoff frequency, the desired width of HMSIW is calculated with a correction factor as shown in (1.3) [29].

$$
\begin{aligned}
W_{\mathrm{HMSIW}} &= \frac{W_{\mathrm{SIW}}}{2} + \Delta W \\
\frac{\Delta W}{h} &= \left(0.05 + \frac{0.30}{\varepsilon_R}\right) X \ln\left(0.79\frac{W_{\mathrm{HMSIW}}^2}{h^3} + \frac{104W_{\mathrm{HMSIW}} - 261}{h^2} + \frac{38}{h} + 2.77\right)
\end{aligned}
\qquad (1.3)
$$

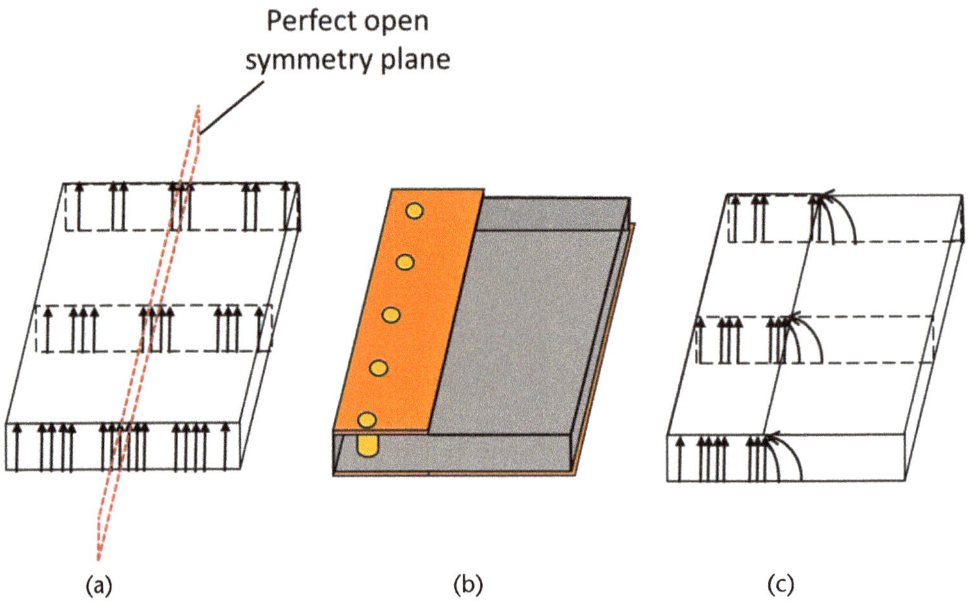

Figure 1.7 (a) Full waveguide with a perfect open symmetry plane; (b) half-mode SIW structure; and (c) field distribution in HMSIW.

In continuation of the similar working principle, in case of a SIW-based cavity resonator, the cavity can be bisected into the HMSIW cavity and further bisected along another symmetry wall to realize quarter-mode SIW (QMSIW) cavity resonator. Such miniaturization technique has been widely used in SIW-based systems. The detailed discussion of QMSIW will be done in the next chapter.

1.4.2 Ridge SIW

The bandwidth of the mono-mode operation of RWG is generally narrow and one of the effective ways to extend the bandwidth of a single-mode operation is using the ridge in one of the walls of waveguide. In case of conventional waveguide, the ridge can be implemented in top or bottom wall as well as in sidewall. However, in the case of SIW, there is a limitation in terms of fabrication since the ridge in the side wall is difficult to implement. In this case, a ridge in top wall or bottom conducting wall is typically implemented using the multilayer PCB fabrication technique [30]. The cross-sectional view of the ridge waveguide and its implementation in SIW technology are shown in Figure 1.8. To implement the ridge, four-layer PCB is typically used. A conducting layer in the middle plane with a ridge width is implemented, which is connected to either top or bottom layer using one or two rows of metallic via, as shown in Figure 1.8(b, c). The electric field distribution of the ridge SIW (RSIW) is shown in Figure 1.8(d). The ridge SIW is also considered as one of the effective miniaturization techniques as it reduces the lateral footprint at the expense of fabrication complexity.

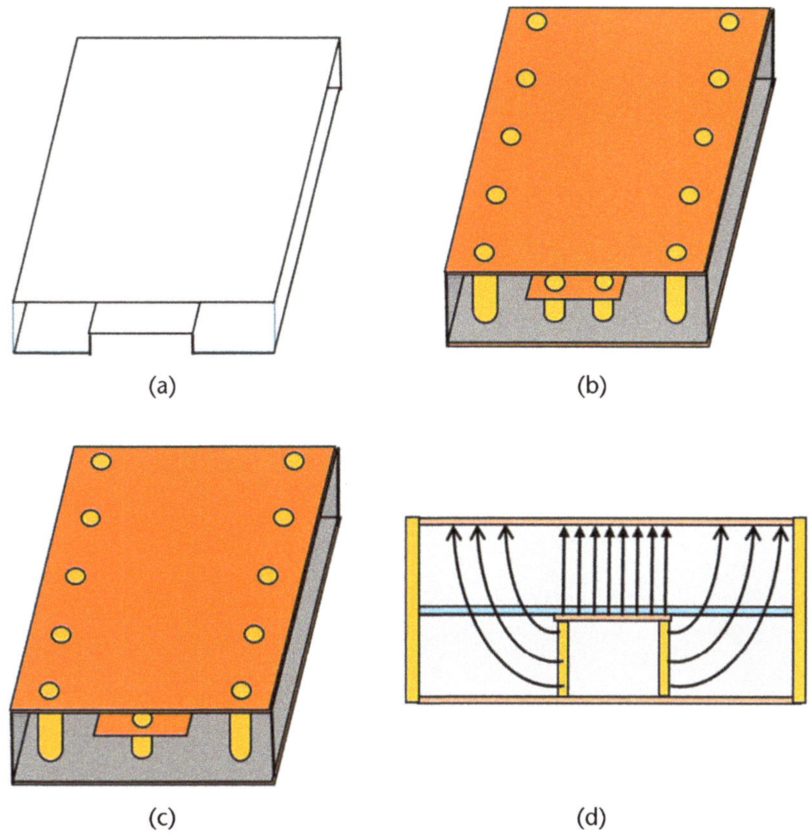

Figure 1.8 (a) Cross-section of ridge waveguide; (b) ridge SIW; (c) alternate implementation of ridge SIW; and (d) field distribution in ridge SIW.

1.4.3 Folded SIW

Another miniaturization technique of SIW technology using the multilayer fabrication technique is introducing a septum in the SIW structure as shown in Figure 1.9. Such a structure is called folded SIW (FSIW) [31]. The field distribution of the structure is shown in Figure 1.9(c) where the maximum of the field exists at the open end of the septum. In effect, the FSIW almost makes the effective width of

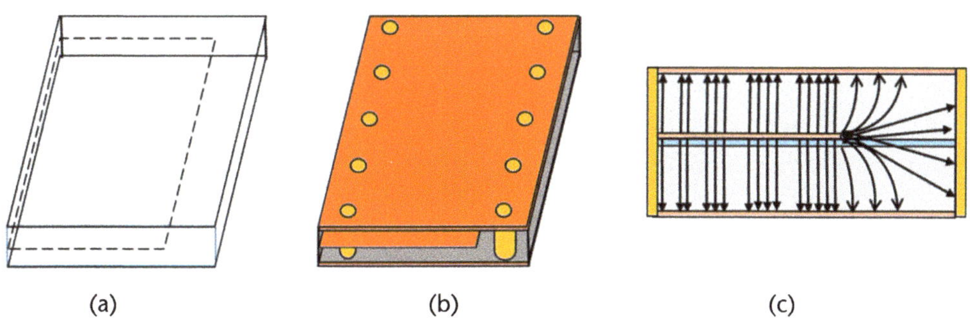

Figure 1.9 (a) Cross-section of folded waveguide; (b) schematic of folded SIW; and (c) field distribution of folded SIW.

the FSIW with a factor of 0.5. As a result, miniaturization in the lateral footprint is observed. The major difference with half-mode SIW and FSIW is that the latter one is a completely shielded structure. However, the fabrication complexity of the design is higher than the former one.

1.5 Overview of Analysis Techniques for SIW

Various numerical techniques are adopted to realize the SIW structures. Since in the SIW structure, the metallic side walls of the conventional waveguide are replaced by a pair of rows of metallic vias, which lead to complex periodic guided wave structure; therefore, accurate and efficient methods are highly required to analyze such a guided wave structure [32]. In the SIW structure, it is very important to know details about propagation characteristics of various complex-mode, bandstop phenomena of the periodic SIW structure and the possible leakage from the SIW structure with the variation of spacing of the periodic gap. To investigate such a basic characteristic effect on the performance of the SIW structure, various numerical techniques are carried out.

First, the finite-difference frequency-domain (FDFD), full-wave electromagnetic numerical method was proposed to model the guided-wave properties of the arbitrary complex open periodic SIW structures. The leakage through the periodic gap introduces wave attenuation. Bilateral unbounded periodic SIW structures with absorbing boundary conditions and perfectly matched layer (PML) are analyzed with the help of the Floquet mode in which the problem is converted into a generalized matrix eigenvalue problem and then transformed to a standard matrix eigenvalue problem and solved with efficient subroutines. FDFD analysis can also be suitable for complex geometries and anisotropic substrate. Numerical multimode calibration technique is another mathematical technique that is integrated with commercial full-wave EM simulator exhibiting TE_{m0} mode that only exist in the SIW structures [33]. The complex propagation constant of the SIW transmission line can be obtained by using a multimode calibration technique where multiple SIW transmission lines are used for analysis. Similar to FDFD, this technique can also be used for accurate analysis of various guided wave properties such as dispersion characteristics, leakage loss, and bandstop properties. Again, the different parameters of SIW-based arbitrary complex circuits and structures of microwave and millimeter-wave components such as filters, resonators, and antennas are accurately determined by combing domain decomposition FDTD (DD-FDTD) and numerical through-line (TL) calibration technique. This technique saves the computation time and memory space with good accuracy of the numerical simulation [34]. The Method of Moments (MoM) and transverse resonance techniques are combined together to determine complex propagation constant of SIW structure accurately. In this case, the rows of metallic vias are modeled using surface impedance concept [35]. Further, this method is applied to introduce the design rules based on physical parameters such as the cylinder diameter and the period length, which depends on desired cutoff frequency.

To realize the SIW-based network with the help of design rules, the appropriate parameters and operational regions should be carefully chosen. The method of moments analysis is also adopted to realize the SIW-based slot array antenna where slots act as a radiating element modeled as a slotted parallel plate waveguide and each slot is considered as equivalent magnetic current sources [36]. The electromagnetic (EM) wave due to radiating slots are computed using the MoM technique. The field into the SIW is expressed by sum of individual contribution of field scattered by metallic vias and radiating slot. This semi-analytical, resource-efficient method provides an appropriate result applied to the analysis of structures of large extension. Another interesting full-wave numerical technique, namely, method of lines (MoL), is also used to analyze the propagation characteristics of SIWs. Here, the computing efficiency is improved by deriving the generalized matrix of the eigenvalue equation. The accuracy of propagation constant is further improved with the help of the Z-transform absorbing boundary condition (Z-ABC). This numerical technique is also helpful for designing SIW-based various microwave and millimeter-wave integrated circuits [37]. The SIW-based discontinuities are also accurately modeled by the Boundary Integral-Resonant Mode Expansion (BI-RME) method [38]. By this method, an accurate modeling of equivalent circuits of SIW-based structures can be calculated with high accuracy. SIW-to-substrate mounted waveguide (SMW) transition is implemented using the mode-matching technique [39]. Interested readers are encouraged to refer to the detailed discussion in the corresponding research papers to find out more details about these individual methods.

1.6 SIW Fabrication

As discussed in previous sections, the substrate integrated waveguide is conceptualized to realize waveguide-based structure using printed circuit (PCB) technology. The fabrication of such circuits require extra caution, as it involves the handling of minute dimensional tolerances, and a huge number of plated through-holes. In addition, as discussed in Section 1.4, some miniaturization techniques require a multilayer PCB fabrication process. In this section, we will discuss briefly the steps involved in the fabrication of the SIW structure.

The basic steps of the PCB fabrication process are shown in Figure 1.10 [40]. As discussed in the previous section, the SIW circuits are designed using a commercially available electromagnetic solver or by using computational methods. Once the design is finalized, GERBER file and Drill file corresponding to the final design are required to be generated. Typically, the commercially available software provides the option to generate these files. In the next step, the generated fabrication files are communicated to PCB fabrication service provider.

As shown in Figure 1.10, the PCB fabrication service provider uses the Drill file separately for the drilling of holes and for a layer-to-layer to connection. The maximum and minimum hole sizes are typically specified by the service provider and the design needs to meet these dimensional criteria. Typical specification of size of drill holes are within the range of 0.2 to 3.0 mm. However, these specifications may vary from vendor to vendor depending on their available fabrication facility. The readers are encouraged to consult the service provider before proceeding to fabrication. If the design includes dimensions outside these range, reiteration in the design steps are required. After the drilling, the boards are brushed using a brushing machine

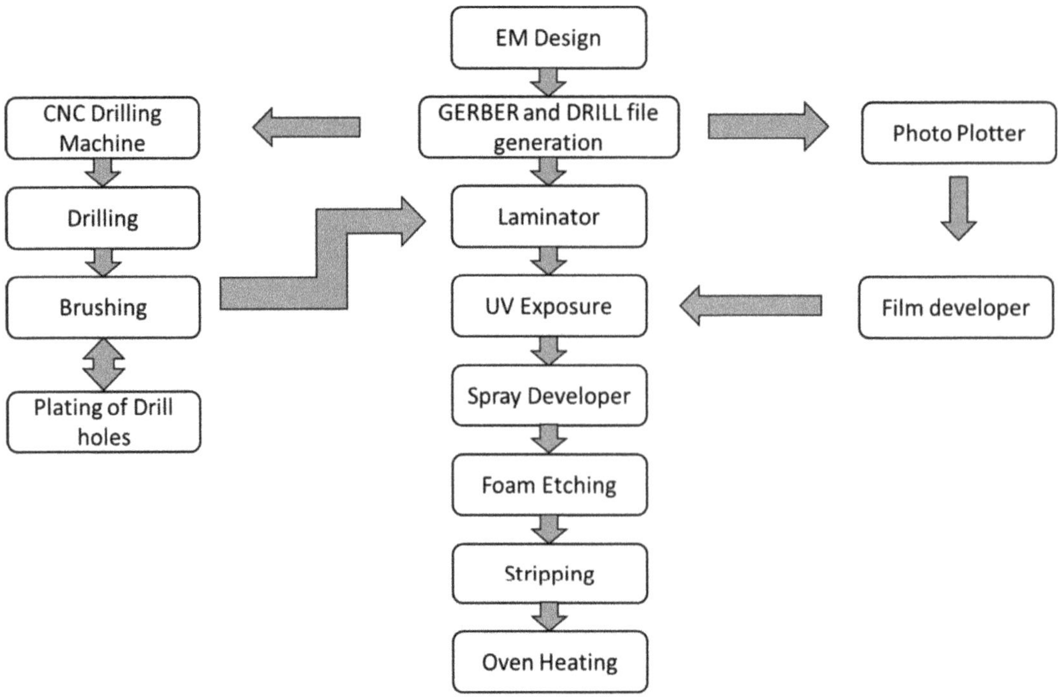

Figure 1.10 Step-by-step guide for PCB fabrication.

to remove any stray particles in the board. Next, the drilled vias are required to be plated with copper for connection between the top and bottom layers. The simplified fabrication steps of SIW circuits are shown in Figure 1.11. The plating of drill holes or vias are done by an electrolysis process. The board is dipped in a copper-based chemical solution and is stirred mechanically for a longer duration. Due to the principle of electrolysis, copper is deposited slowly in the holes, which ensures connection between different layers. However, before the copper deposition, the board is processed through degreasing, rinsing, catalyzing, activating, and pickling solutions to ensure removal of contaminants from the board surfaces. After the plating of holes, the etching process begins to generate the patterned copper layer at the top and bottom surfaces of the SIW board. The first step of the etching process is lamination of board with the photoresist (PR) layer. Here, positive or negative PR can be used. Caution is required to be taken during generation of mask film, which needs to be altered based on the use of the type of PR. The mask can be developed using the GERBER file using the photo plotter machine.

After the lamination and mask film generation, the laminated board along with the mask film is placed on an ultraviolet (UV) exposure machine. The opaque region of the mask prevents UV exposure and as a result, the variation of exposure to PR is implemented as per the pattern printed on the mask. If we use negative PR and negative film, the exposed portion of the PR layer becomes etch-resistant. Next, the board is sent to the spraying unit where the board is treated with alkaline solution to soften the nonexposed portion of the PR. Now, the board is ready for chemical etching process where the board is placed in a chemical foaming fluid flow, which etches out the nonexposed PR and the copper underneath the board. Typical minimum dimensions that can be etched out are in the order of 0.2-mm track width

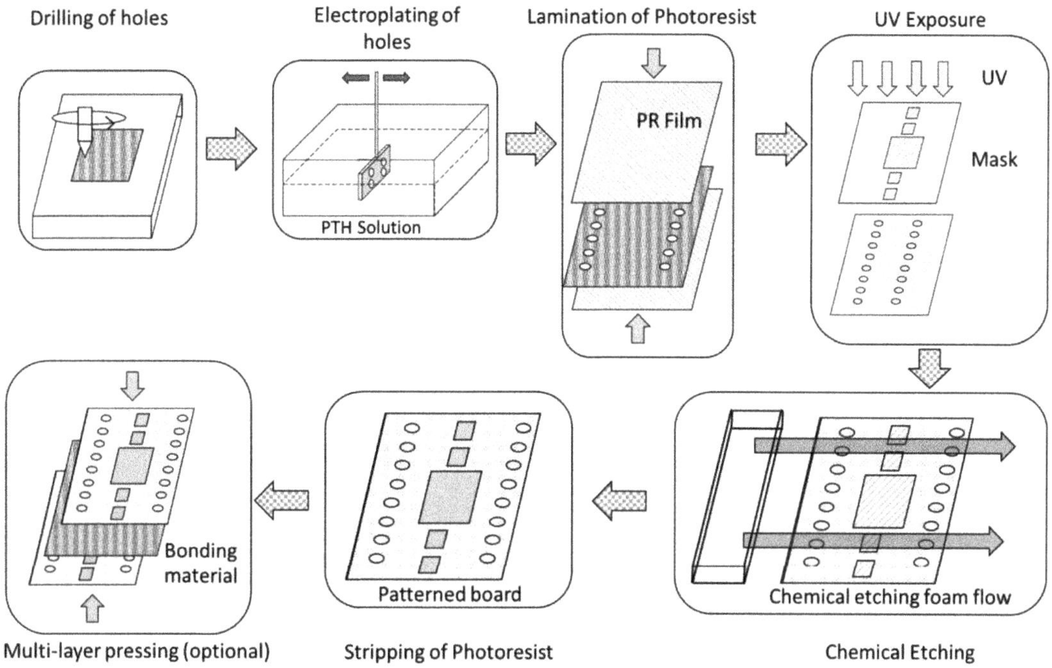

Figure 1.11 Simplified schematic of SIW fabrication.

and 0.2-mm spacing between successive tracks. However, these specifications may vary from vendor to vendor depending on their available facility. After the etching process, we get the board with a desired layout pattern of copper layer with an etch-resistant layer of PR. This additional layer is removed in a stripping cuvette machine with the help of another alkaline solution. Finally, the board is dried using an oven.

In the case of multilayer SIW fabrication, the previously mentioned steps are required to be followed for each layer of circuit. Finally, at the last step, these boards are placed one above another along with a number of bonding material or prepreg layers in between the boards. Care must be taken during the alignment of different layers for successful operation of the final circuit. Once the layers are aligned, the complete setup is placed in a multipress machine, which applies regulated pressure and temperature to bond different layers of the boards into a single circuit. The final circuit is cooled and can be taken out from the machine for finishing the fabrication process. Sometimes vendors may ask for additional information about the solder masking, legend printing, and so forth. These requirements vary based on the application of the finished board and readers are encouraged to discuss with vendors before giving the assignment.

References

[1] Westberg, E., et al., "5G Infrastructure RF Solutions: Challenges and Opportunities," *IEEE Microwave Magazine*, Vol. 20, No. 12, 2019, pp. 51–58.

[2] Flamini, R., et al., "Towards a Heterogeneous Smart Electromagnetic Environment for Millimeter-Wave Communications: An Industrial Viewpoint," *IEEE Transactions on Antennas and Propagation*, 2002.

[3] Azar, Y., "28 GHz Propagation Measurements for Outdoor Cellular Communications Using Steerable Beam Antennas," *2013 IEEE International Conference on Communications (ICC)*, New York, 2013.

[4] Wu, K., D. Deslandes, and Y. Cassivi, "The Substrate Integrated Circuits—A New Concept for High-Frequency Electronics and Optoelectronics," *TELSIKS*, November 2003.

[5] Meredith, J. M., "NR; User Equipment (UE) Radio Transmission and Reception; Part 2: Range 2 Standalone," 3GPP TS 38.101-2 V15.4.0, Release 15, December 2018.

[6] Hong, H., "Solving the 5G Mobile Antenna Puzzle: Assessing Future Directions for the 5G Mobile Antenna Paradigm Shift.," *IEEE Microwave Magazine*, Vol. 18, No. 7, 2017, pp. 86–102.

[7] Han, L., and K. Wu, "24-GHz Integrated Radio and Radar System Capable of Time-Agile Wireless Communication and Sensing," *IEEE Transactions on Microwave Theory and Techniques*, Vol. 60, No. 3, 2012, pp. 619–631.

[8] Holma, H., and H. Viswanathan, "The 6G Era's Enormous Capacity Demands Will Require New Spectrum and Extreme Massive MIMO," Nokia BELL Labs, October 6, 2021, https://www.bell-labs.com/institute/blog/6g-eras-enormous-capacity-demands-will-require-new-spectrum-and-extreme-massive-mimo/#gref.

[9] "Studies on Frequency-Related Matters for the Terrestrial Component of International Mobile Telecommunications Identification in the Frequency Bands 3 300-3 400 MHz, 3 600-3 800 MHz, 6 425-7 025 MHz, 7 025-7 125 MHz and 10.0-10.5 GHz," WRC-19, RESOLUTION 245, October 28–November 22, 2019.

[10] https://group.ntt/en/newsrelease/2023/02/27/230227a.html.

[11] https://www.rcrwireless.com/20220308/5g/advanced-mimo-technologies-usher-in-new-wide-area-5g-capabilities.

[12] "6GTaskforceReport: 6GSpectrum," chrome-extension://efaidnbmnnnibpcajpcglclefindmkaj/https://dot.gov.in/sites/default/files/Taskforce%20Report%20-%206G%20Spectrum.pdf?download=1.

[13] Ershadi, S., et al., "Wideband Subarray Design for 5G and Antenna Arrays," *2016 URSI Asia-Pacific Radio Science Conference (URSI AP-RASC)*, 2016.

[14] Gatti, F., et al., "A Novel Substrate Integrated Coaxial Line (SICL) for Wide-Band Applications," *2006 European Microwave Conference*, 2006.

[15] Higham, E., "Semiconductor Trends in Sub-6 GHz 5G Networks," *Microwave Journal*, 2020.

[16] Yu, C., et al., "Full-Angle Digital Predistortion of 5G," *IEEE Transactions on Microwave Theory and Techniques*, Vol. 67, No. 7, 2019, pp. 2847–2860.

[17] Luo, Q., et al., "Single-Receiver Over-the-Air Digital Predistortion for Massive MIMO Transmitters with Antenna Crosstalk," *IEEE Transactions on Microwave Theory and Techniques*, Vol. 68, No. 1, 2020, pp. 301–315.

[18] Delos, P., S. Ringwood, and M. Jones, "Hybrid Beamforming Receiver Dynamic Range Theory to Practice," Analog Device, June 13, 2022.

[19] Malkowsky, S., et al., "The World's First Real-Time Testbed for Massive MIMO: Design, Implementation, and Validation," *IEEE Access*, Vol. 5, 2017, pp. 9073–9088.

[20] Prasad, S., et al., "Mm-wave Multibeam Phased Array Antenna for 5G Applications," *Journal of Electromagnetic Waves and Applications*, Vol. 35, No. 13, 2021, pp. 1802–1814.

[21] Gholb, Y. E., and N. El Amrani El Idrissi, "5G Mobile Antennas: MIMO Implementation," *2019 International Conference on Wireless Technologies, Embedded and Intelligent Systems (WITS)*, Morocco, 2019.

[22] Ikram, M., et al., "Integrated Frequency-Reconfigurable Slot Antenna and Connected Slot Antenna Array for 4G and 5G Mobile Handsets," *IEEE Transactions on Antennas and Propagation*, Vol. 67, No. 12, December 2019, pp. 7225–7233.

[23] SalarRahimi, M., et al., "A Low-Cost Analog Beamforming Antenna for 5G mm-Wave Handset Applications," *13th European Conference on Antennas and Propagation (EuCAP 2019)*, 2019.

[24] Cheng, D. K., *Field and Wave Electromagnetics*, Boston, MA: Pearson, 1989.

[25] Shigeki, F., "Waveguide Line," Japan Patent 06-053 711, February 25, 1994.

[26] Hirokawa, J., and M. Ando, "Single-Layer Feed Waveguide Consisting of Posts for Plane TEM Wave Excitation in Parallel Plates," *IEEE Transactions on Antennas and Propagation*, Vol. 46, 1998, pp. 625–630.

[27] Uchimura, H., T. Takenoshita, and M. Fujii, "Development of the Laminated Waveguide," *IEEE Transactions on Microwave Theory and Techniques*, Vol. 46, No. 12, 1998, pp. 2438–2443.

[28] Hong, W., et al., "Half Mode Substrate Integrated Waveguide: A New Guided Wave Structure for Microwave and Millimeter Wave Application," *Joint 31st International Conference on Infrared Millimeter Waves and 14th International Conference on Terahertz Electronics*, Shanghai, China, 2006.

[29] Lai, Q., et al., "Characterization of the Propagation Properties of the Half-Mode Substrate Integrated Waveguide," *IEEE Transactions on Microwave Theory and Techniques*, Vol. 57, No. 8, 2009, pp. 1996–2004.

[30] Bozzi, M., S. A. Winkler, and K. Wu, "Broadband and Compact Ridge Substrate Integrated," *IET Microwaves, Antennas & Propagation*, Vol. 4, No. 11, 2010, pp. 1965–1973.

[31] Grigoropoulos, N., B. Sanz-Izquierdo and P. R. Young, "Substrate Integrated Folded Waveguides," *IEEE Microwave and Wireless Components Letters*, Vol. 15, No. 12, 2005, pp. 829–831.

[32] Xu, F., et al., "Finite-Difference Frequency-Domain Algorithm for Modeling Guided-Wave Properties of Substrate Integrated Waveguide," *IEEE Transactions on Microwave Theory and Techniques*, Vol. 51, No. 11, pp. 2221–2227.

[33] Xu, F., and K. Wu, "Guided-Wave and Leakage Characteristics of Substrate Integrated Waveguide," *IEEE Transactions on Microwave Theory and Techniques*, Vol. 53, No. 1, 2005, pp. 66–73.

[34] Xu, F., K. Wu, and W. Hong, "Domain Decomposition FDTD Algorithm Combined with Numerical TL Calibration Technique and Its Application in Parameter Extraction of Substrate Integrated Circuits," *IEEE Transactions on Microwave Theory and Techniques*, Vol. 54, No. 1, 2006, pp. 329–338.

[35] Deslandes, D., and K. Wu, "Accurate Modeling, Wave Mechanisms, and Design Considerations of a Substrate Integrated Waveguide," *IEEE Transactions on Microwave Theory and Techniques*, Vol. 54, No. 6, 2006, pp. 2516–2526.

[36] Arnieri, E., and G. Amendola, "Method of Moments Analysis of Slotted Substrate Integrated Waveguide Arrays," *IEEE Transactions on Antennas and Propagation*, Vol. 59, No. 4, 2011, pp. 1148–1154.

[37] Yan, L., et al., "Investigations on the Propagation Characteristics of the Substrate Integrated Waveguide Based on the Method of Lines," *Proc. Inst. Elect. Eng.—Microw. Antennas Propag.*, Vol. 15, No. 2, 2005, pp. 35–42.

[38] Bozzi, M., and L. Perregrini, "Numerical Modeling and Design of Substrate Integrated Waveguide (SIW) Components," *2012 International Conference on Mathematical Methods in Electromagnetic Theory*, Kharkiv, Ukraine, 2012.

[39] Schorer, J., and J. Bornemann, "A Mode-Matching Technique for the Analysis of Waveguide-on-Substrate Components," *IEEE MTT-S International Conference on Numerical Electromagnetic and Multiphysics Modeling and Optimization (NEMO)*, Ottawa, ON, Canada, 2015.

[40] "PCB Design Guide Sheet," Indian Institute of Technology Kanpur, Kanpur, India, 2009.

CHAPTER 2

The Design and Characteristics of SIW

In Chapter 1, the SIW and its different types were introduced. The comparative study of smaller-height rectangular waveguide and SIW was discussed in detail. In this chapter, we will focus on designing the SIW structure for the desired operating band. Also, detailed design steps to realize the SIW-based cavity to resemble the conventional waveguide cavity in a planar form will be discussed in detail.

2.1 Working Principle

The SIW is interpreted as a dielectric-filled rectangular waveguide implemented in a planar dielectric substrate. Therefore, to design the SIW for the desired operating frequency, we will follow the similar approach as it is followed for conventional rectangular waveguide. To design a waveguide at the desired operating band, the operating mode of the waveguide is also an important parameter to be considered. Conventionally, all waveguide structures are designed to operate at the dominant mode only (single-mode/mono-mode operation) to simplify the excitation and termination circuit at the input and output ends. To start with, we will study the design of the SIW for a dominant-mode operation at the desired frequency. Later, we will investigate the design steps for the operation of the SIW structures in a higher-order mode. As described in Section 1.3, the relation of the cutoff frequency and the width of the waveguide is related by (1.2).

In the case of a conventional rectangular waveguide, the sequence of cutoff frequencies of subsequent modes in ascending order depends on the width (a) and height (b) [1]. To understand this, a mode chart for the typical WR-90 rectangular waveguide is shown in Figure 2.1. In the chart, the width (a) to height (b) ratio is varied gradually from 1 to 3. It can be seen that while the dominant-mode (TE_{10}) resonant frequency remains the same for all the cases, the next higher-order mode resonant frequency can be varied if we change the width to height ratio. It is clear from Figure 2.1 that if we choose the ratio greater than 2, the maximum mono-mode bandwidth can be achieved as $\Delta f = f_{cTE10}$ ($\Delta f = f_H - f_L$; $f_H = f_{cTE20}$; $f_L = f_{cTE10}$) since $f_{cTE20} < f_{cTE01}$.

In the case of the SIW, the height of the guide is chosen from the list of commercially available dielectric substrate, which is typically much smaller than the width of the waveguide and hence the maximum mono-mode bandwidth is maintained in such structure. Next, to determine the width of the SIW, we need to know the desired cutoff frequency based on the intended application. However, to ensure satisfactory performance of the guide (negligible insertion loss and return loss as

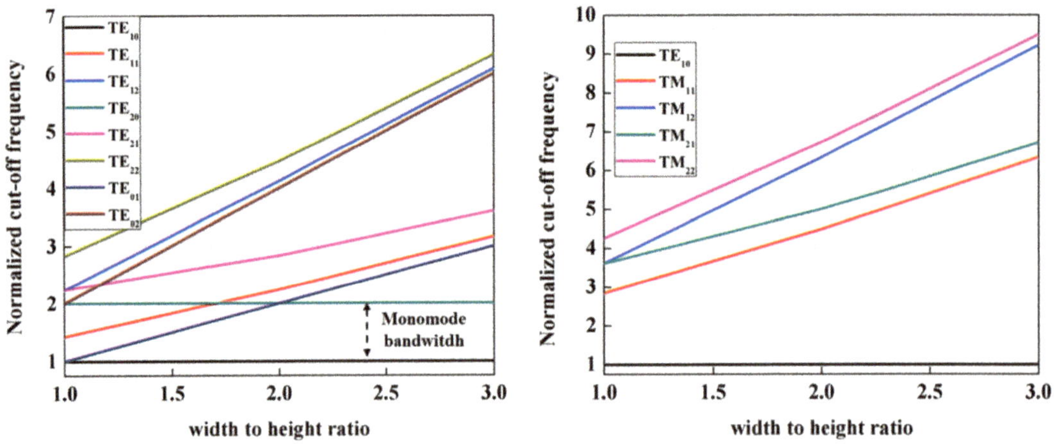

(a) (b)

Figure 2.1 Variation of normalized cutoff frequency of different modes with the width-to-height ratio of the waveguide: (a) TE modes; and (b) TM modes.

minimum as possible), the guide needs to be operated almost at 1.2 times of the cutoff frequency. This can be better understood from the response of a SIW transmission line operating at X-band (8 to 12 GHz). We can start with the dimension of the WR-90 waveguide and follow the above-mentioned procedure to have the dimension of the SIW section.

However, as shown in Figure 2.2, although the transmission of signal in the guide starts at the cutoff frequency, the desired performance of the guide (negligible insertion loss and return loss as minimum as possible) is achieved almost at 1.2 to 1.5 times of the cutoff frequency. To account for this effect, a rule of thumb can be followed that the center frequency of the desired operating frequency band is to be kept at 1.5 times of the cutoff frequency of the guide [2].

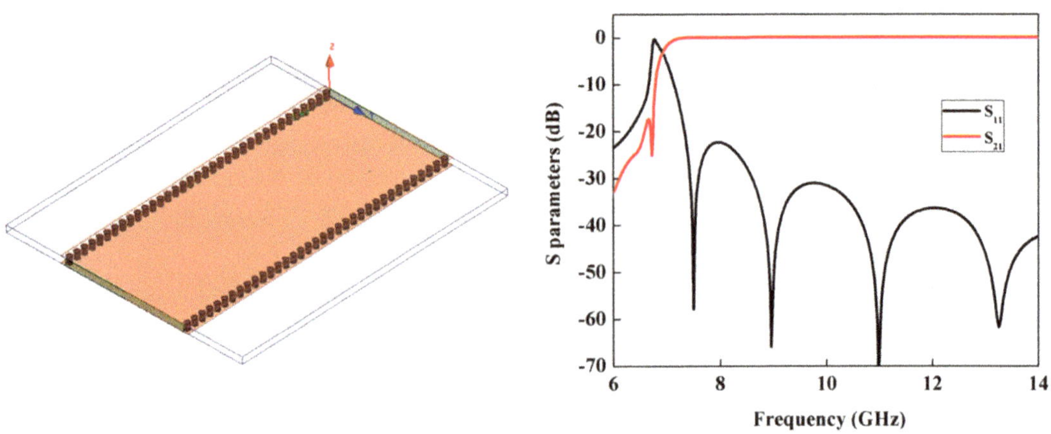

(a) (b)

Figure 2.2 (a) SIW section; and (b) response of SIW section (S-parameter response) (width = 15 mm; dia = 0.6 mm; pitch = 1 mm; height = 0.8 mm; dielectric constant of medium = 2.2; loss tangent = 0.0009).

Once the cutoff frequency is determined, the width of the guide can be determined from (1.2). Next, we need to put the rows of metallic vias along the two side walls of the guide as shown in Figure 2.2. Therefore, the diameter and pitch of the vias need to be determined. As explained in Section 1.3, the rows of metallic vias are used to realize the metallic side wall of the SIW in the dielectric substrate. Essentially, a sequence of a metal post and open aperture in alternate fashion are created, which effectively should be sufficient to guide the EM energy within the two via walls and, ideally, no leakage of EM waves should happen outside the wall like the conventional waveguide with continuous metal sidewall. The leakage of the EM wave can be reduced to a significant extent with a smaller pitch of the via rows. Therefore, the lowest limit of the pitch of the via is equal to the diameter of the via, which resembles a continuous side wall. Any value of the pitch less than this refers to overlapping vias, which are unrealizable. However, it needs to be ensured that the vias are not too close to each other, which will degrade the mechanical stability of the substrate. Therefore, as a rule of thumb, the minimum pitch of the vias that is allowed to be maintained is 20 vias per wavelength so that the substrate remains mechanically rigid and stable. In addition, the pitch of the via rows are chosen among the range of possible options typically provided by the fabrication process. Therefore, we need to select a maximum allowable pitch such that the open aperture between two consecutive metallic vias are not large enough to leak the EM wave. This can be better understood with an example as shown in Figure 2.3. Here, the pitch of the vias is kept approximately twice the diameter of the side-wall vias of the waveguide. As a result, we can see that the EM wave cannot be contained within the two-via wall and the wave starts leaking through the aperture between the two consecutive vias. Such leakage should be avoided for successful operation of the SIW.

To ensure proper guiding of EM waves, as a rule of thumb, the pitch of the via should be kept to less than twice the value of the via diameter while the via diameter should be much smaller than the guide wavelength. The smaller diameter ensures a lesser pitch value as well as less reflection or scattering from an individual via wall as mentioned in Section 1.5.

Additionally, the sequential placement of rows of vias resembles a periodic loading structure in the guide. As is well known, the periodically loaded line typically

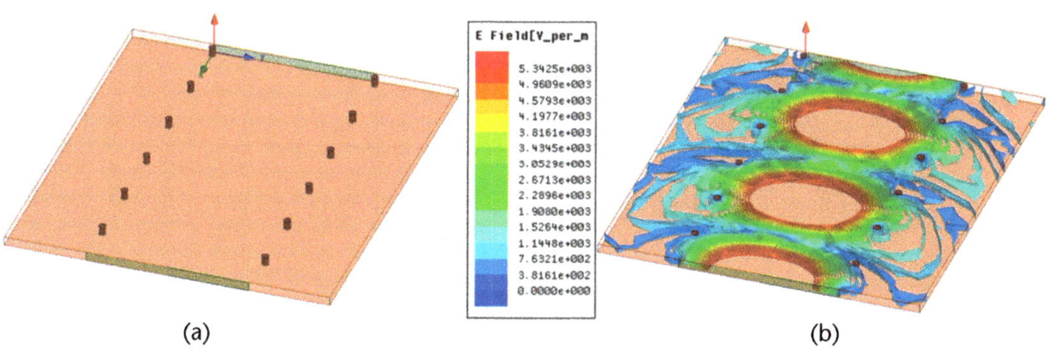

Figure 2.3 (a) SIW structure with large pitch of vias; and (b) magnitude of electric field for the dominant mode propagating through the line (width = 15 mm; dia = 0.6 mm; pitch = 2 mm; height = 0.8 mm; dielectric constant of medium = 2.2; loss tangent = 0.0009).

exhibits passband and stopband characteristics [3]. The typical bandstop response will occur when the spacing between two consecutive unit cells becomes comparable to the half-wavelength at the operating frequency. Therefore, for a bandstop response to occur in the SIW, the following condition needs to be followed:

$$s = n * \frac{\lambda_g}{2}$$
$$\beta_g * s = n\pi \tag{2.1}$$

where s is the pitch of the side-wall vias and λ_g is the guided wavelength in the guide.

This can be better understood by the help of an example as shown in Figure 2.4. Here the operating frequency of the SIW is chosen at a higher-frequency band where the pitch of the via will be comparable to the operating wavelength. As shown in Figure 2.4(b), there is a bandstop phenomenon occurring at 83 GHz where (2.1) is satisfied. Hence, it should be ensured that the stopband should be kept outside the desired operating band ($1.2f_{cTE10} < f < 1.8\ f_{cTE10}$) of the waveguide. By examining (2.1), it can be said that the bandstop response can be pushed away further from the operating frequency range of the line if we choose a smaller pitch (s). In (2.1), if we choose $n = 1$, we will find the relation as (2.2).

$$s = \frac{\lambda_g}{2}$$
$$\lambda_g = \frac{\lambda}{\sqrt{1-\left(\frac{\lambda}{\lambda_c}\right)^2}} \tag{2.2}$$

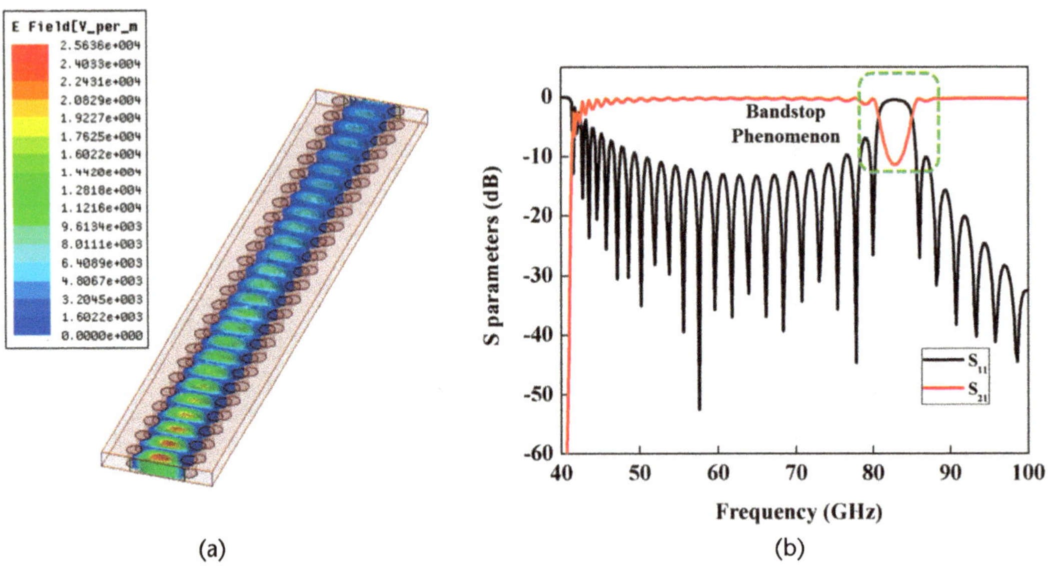

Figure 2.4 (a) Magnitude of electric field; and (b) S-parameter response of the SIW operating at the bandgap region (width = 3 mm; dia = 0.8 mm; pitch = 1.4 mm; height = 0.8 mm; dielectric constant of medium = 2.2; loss tangent = 0.0009).

where λ is the free-space wavelength, λ_c is the cutoff wavelength, and λ_g is the guided wavelength in the guide. As a rule of thumb, we can choose pitch (*s*) less than one-fourth of the cutoff wavelength, which will ensure that the stopband will be pushed away towards higher-frequency side.

2.1.1 The Modes of the SIW

As explained earlier, the rows of metallic vias in the SIW mimic the metallic side wall of a rectangular waveguide to shield the EM wave within the structure. The objective is to maintain field distribution along with associated surface current distribution in the outer single conductor of the SIW to ensure the propagation of the desired mode. However, because of the presence of the series of open aperture between consecutive vias in the side wall, the corresponding boundary condition becomes different as compared to the conventional waveguide. We can understand this better with the help of surface current distribution in the side wall. In the case of conventional waveguide, the surface current at the side wall can be determined from the dielectric-conductor boundary condition as follows:

$$J_s = \hat{a}_n \times H \tag{2.3}$$

Referring to Figure 1.4, the normal direction to the surface for the two opposite side walls would be $\pm\mathbf{a_x}$. Hence, we can derive the equation for the surface current at the side wall of the waveguide starting from the magnetic field equation for the waveguide as mentioned in (1.1).

$$\begin{aligned}
H_y(x,y,z,t) &= -\frac{\beta}{h^2}\left(\frac{n\pi}{b}\right)H_0\cos\left(\frac{m\pi}{a}x\right)\sin\left(\frac{n\pi}{b}y\right)\sin(\omega t - \beta z)\\
H_z(x,y,z,t) &= H_0\cos\left(\frac{m\pi}{a}x\right)\cos\left(\frac{n\pi}{b}y\right)\cos(\omega t - \beta z)\\
J_s &= \hat{a}_x \times \left(\hat{a}_y H_y + \hat{a}_z H_z\right)\\
J_{sz} &= -\frac{\beta}{h^2}\left(\frac{n\pi}{b}\right)H_0\cos\left(\frac{m\pi}{a}x\right)\sin\left(\frac{n\pi}{b}y\right)\sin(\omega t - \beta z)\\
J_{sy} &= -H_0\cos\left(\frac{m\pi}{a}x\right)\cos\left(\frac{n\pi}{b}y\right)\cos(\omega t - \beta z)
\end{aligned} \tag{2.4}$$

Now, if we examine the SIW structure as shown in Figure 2.1, the **y** component of the surface current is parallel to the open aperture existing between two consecutive vias, while for the **z** component, it is perpendicular to it. Therefore, in the case of SIW, maintaining the **y** component of the surface current at the via wall is much easier with negligible disturbance compared to the **z** component. It is evident that the **z** component of the surface current will be heavily disturbed by the presence of the open aperture, which eventually will result in radiation of the EM wave from the side wall.

It can be seen that the cutoff frequency of SIW is higher than the rectangular waveguide for the same dimension of width, which means the effective electrical width of the SIW is smaller than the actual physical width of the guide. This phenomenon is studied in detail in [2, 4]. This can be understood from the Figure 2.8 where it is evident that the effective width of the SIW is constantly varying between W and $W - d$ considering closely spaced via holes.

Therefore, the effective width of the guide will be an approximate average of the two values. However, it has been extensively studied in [4] and it was shown that the effective width of the SIW depends on physical width (W), diameter (d) and pitch (s) of the via wall. Several closed-form expressions have been reported by various researchers [5–7]. However, following two expressions have been popularly used by most of the researchers to get accurate estimation of effective width of the SIW [4]:

$$W_{eff} = W - \frac{d^2}{0.95s} \tag{2.5}$$

$$W_{eff} = W - 1.08\frac{d^2}{s} + 0.1\frac{d^2}{W} \tag{2.6}$$

In this book, both equations will be used for designing SIW components in subsequent chapters.

2.1.2 Losses in the SIW

As discussed before, the guided wave in the SIW follow similar field distribution as a rectangular waveguide. Therefore, the fundamental reasons for losses in the SIW remain similar to that of the rectangular waveguide. There are three major losses

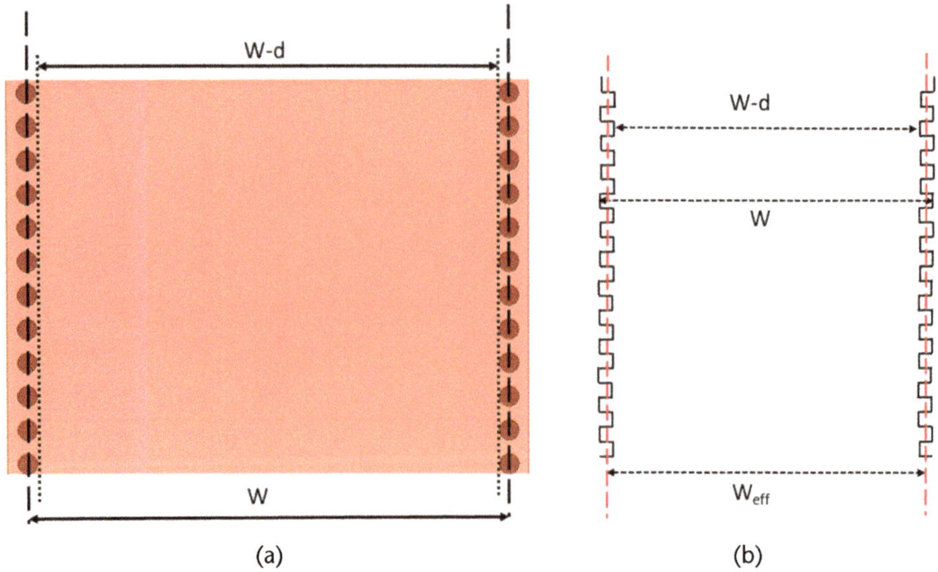

Figure 2.8 Effective width of the SIW.

that we need to consider for the SIW structure, namely, conductor loss, dielectric loss, and radiation loss.

In the case of the conventional rectangular waveguide, the metallic wall creating four boundaries along each side introduces loss due to finite conductivity. The conductor loss in waveguide for dominant TE_{10} mode can be determined as follows:

$$\alpha_c = \frac{R_s}{\eta b} \frac{\left[1 + \frac{2b}{a}\left(\frac{f_c}{f}\right)^2\right]}{\sqrt{1 - \left(\frac{f_c}{f}\right)^2}} \tag{2.7}$$

Similarly, in the SIW, the conductivity of the metal used in the top and bottom cladding of the substrate and the plating of via holes will determine the amount of conductor loss expected to form the line. By using (2.5) or (2.6) in (2.7), the effective conductor loss in the SIW for a dominant TE_{10} mode can be determined as follows [2]:

$$\alpha_c = \frac{R_s}{\eta W_{\text{eff}}} \frac{\left[\frac{W_{\text{eff}}}{h} + 2\left(\frac{k_c}{k_0}\right)^2\right]}{\sqrt{1 - \left(\frac{k_c}{k_0}\right)^2}} \tag{2.8}$$

$$R_s = \sqrt{\frac{\omega \mu_0}{2\sigma}} \tag{2.9}$$

In Figure 2.9, a comparative study of loss in the SIW due to different conductive metal cladding over the dielectric has been shown. As indicated in (2.8) and (2.9), the increase in conductivity introduces conduction loss in the guide. It is also indicated in (2.9) that the loss depends on the frequency of operation and, hence, significant loss is expected at a higher operating frequency.

In the SIW, the top and bottom conductors are supported by dielectric substrate sandwiched between the two metal plates. Hence, it resembles a conventional dielectric-filled waveguide. Therefore, the dielectric loss due to the finite loss tangent value of any commercially available dielectric substrate is also evident in the case of the SIW, which can be designated as the dielectric loss of the SIW. The corresponding expression will be similar to that of a dielectric-filled waveguide:

$$\alpha_d = \frac{1}{2}\tan\delta \frac{2\pi}{\lambda\sqrt{1 - \left(\frac{f_c}{f}\right)^2}} = \frac{1}{2}\tan\delta \frac{k_0^2}{k_z} \tag{2.10}$$

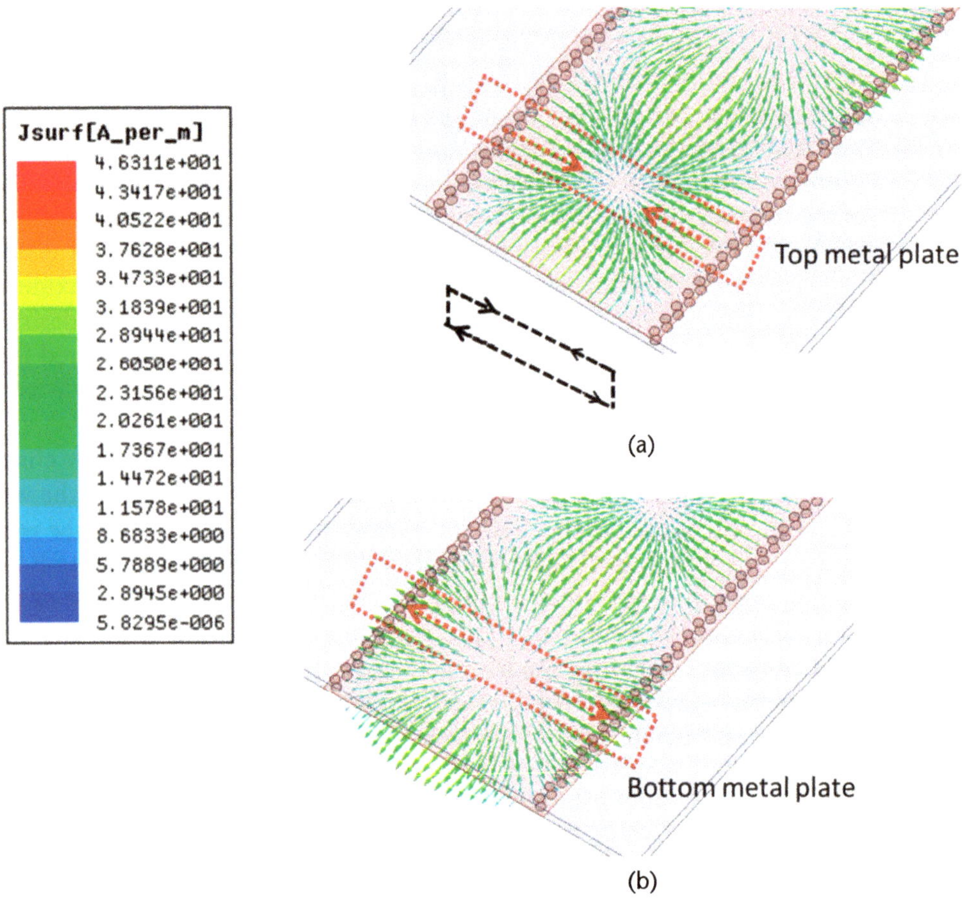

Figure 2.11 Surface current distribution in the top and bottom metal plates in the SIW.

included for successful excitation of the guide. In this book, two examples of such circuits will be discussed.

2.2.1 Coaxial-Fed SIW

To feed the SIW with a coaxial line and to excite the dominant TE_{10} mode in the guide, we will follow similar approaches as in the case of the conventional waveguide. Since the height of the substrate is small, the voltage probe approach is difficult to succeed and, hence, we need to make an arrangement so that a current probe is created at one end of the guide [8]. As shown in Figure 2.11, the current going from the top plate thorough the side via to the bottom plate makes a perfect current loop perpendicular to the magnetic field of the dominant mode. Hence, we need to use the coaxial line to generate the desired current in the guide. An example of the design of current probe feeding is shown in Figure 2.12. Here, the central conductor of the coaxial line is connected to the top metal plate and the outer conductor is soldered to the bottom metal plate, which essentially creates 180° out-of-phase surface current distribution at the top and bottom of the guide. To put the central conductor up to the top plate, an etched hole in the bottom plate and through via was implemented. The coaxial connector is placed along the center line of the SIW section to excite the dominant TE_{10} mode. The arrangement will create the desired alternating current loop operating at the source frequency, which will generate the desired magnetic field and the alternating magnetic field will help to generate the desired electric field of the dominant mode of the guide. The generated EM wave will start propagating in either direction of the guide. To manage this, one side of the guide is short circuited at one-fourth of the guided wavelength at center frequency (10 GHz) from the probe position.

The circuit can be analyzed using an equivalent circuit model as shown in Figure 2.13. The SIW is fed by a 50Ω coaxial connector at port 1. The probe going through the guide can be modeled as inductance (L_probe) and the gap at the bottom and top metal plates can be modeled as capacitance (C_upper and C_lower) creating a π-equivalent circuit. The energy transfer from the central conductor to the guide by means of current probe method can be modeled as transformer turns ratio (n:1). The effect of reactance due to the excitation probe at the transition is compensated by the proper choice of the C_upper value, which is determined by varying the capacitive gap g_r at the top metallic surface. The values of the equivalent inductance and capacitance of the excitation probe of the final optimized design are calculated from Y-parameters deduced from two-port analysis of the probe itself and are shown in Figure 2.13.

Now at reference plane TT', the port impedance seen by the excitation probe, as shown in Figures 2.12(b) and 2.13, has much higher value than that of the SIW line at plane PP'. The impedance of the SIW section at PP' can be calculated by computing the Z_{PI} value (power-current definition). A relation between port impedance Z_{PI} and wave impedance Z_W of equivalent waveguide of the SIW section is given as [9, 10]

$$Z_{PI} = \frac{\pi^2}{8}\frac{h}{W_{eff}} Z_W \tag{2.12}$$

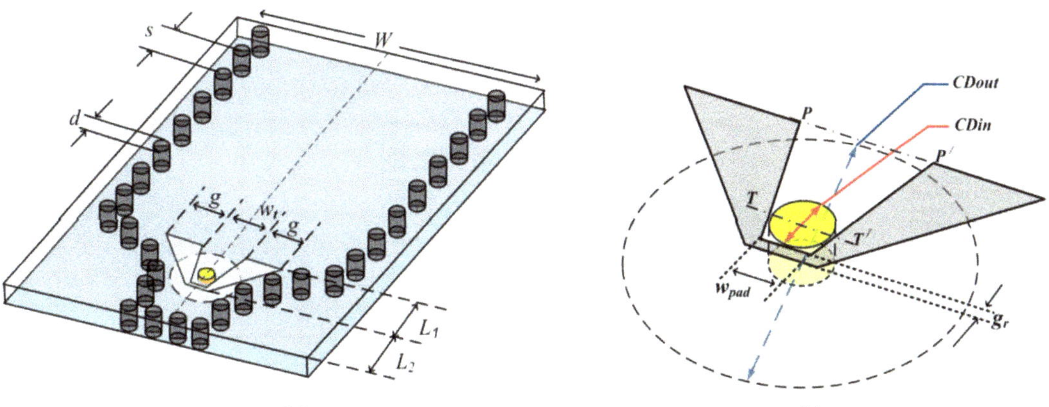

Figure 2.12 (a) Coaxial-fed SIW; and (b) enlarged view (W = 16.16 mm; W_{pad} = 1.5 mm; d = 1.5 mm; g_r = 0.5 mm; s = 2.42 mm; CD_{in} = 1 mm; W_t = 5 mm; CD_{out} = 5 mm; g = 3.7 mm; L_2 = 4.45 mm; L_1 = 4.45 mm). (© 2013 IEEE [8].)

been the most popular choice among the designers. Several empirical formulas for the tapered section have been proposed [12–14]. However, the final design in most of the cases requires little optimization with the help of computational software to get the best result.

At higher frequencies above 60 GHz, microstrip line-based open structures may exhibit radiation loss or surface wave loss. Therefore, at these frequencies, other type of transitions such as CPW feeding and waveguide-to-SIW transition are preferred based on the applications. The readers are encouraged to explore different transition options available in the literature [15–17].

2.3 Design of the SIW Cavity

The conventional rectangular waveguide can be short-circuited to both ends to become a shielded structure with conducting wall on all its six sides. Such a structure behaves as a resonator or cavity. An example of such a structure is shown in Figure 2.17. If we introduce an EM field inside the cavity, it starts propagating until it reaches at the both ends where full reflection occurs due to the dielectric-conductor boundary. The reflected wave from both sides ($+X$ and $-X$) will coincide and create standing wave pattern. Therefore, in a cavity, the EM wave will start resonating and the energy will be stored. No wave propagation will happen in such a structure. Such a circuit has profound use as resonator since the quality factor is very high due to its shielded environment.

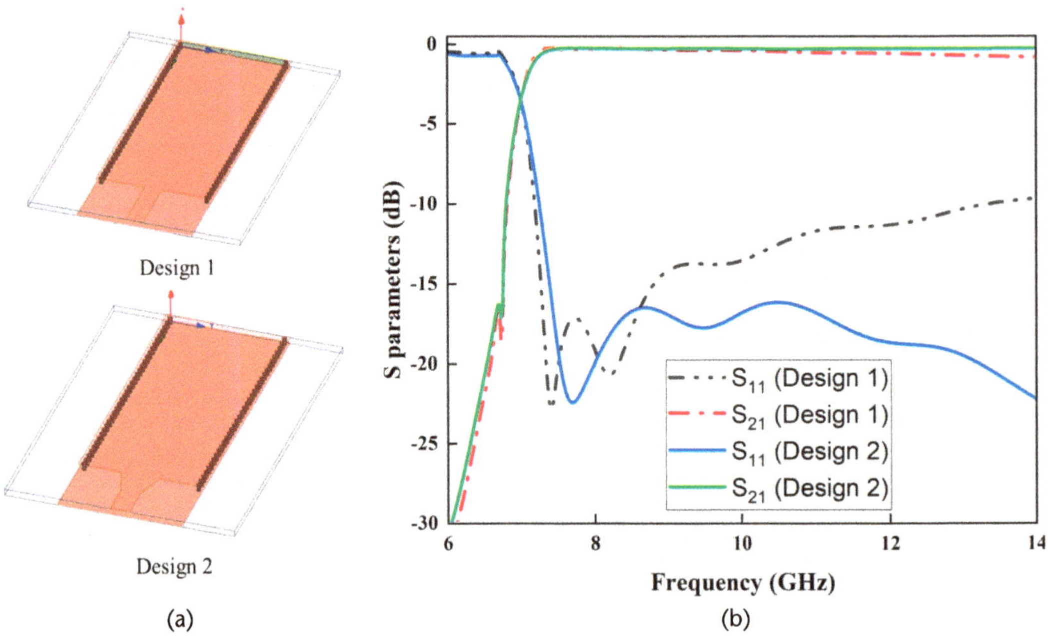

Figure 2.16 Microstrip-to-SIW transition: (a) design examples; and (b) performance comparison (width = 15 mm; dia = 0.6 mm; pitch = 1 mm; height = 0.8 mm; dielectric constant of medium = 2.2; loss tangent = 0.0009; $M_{\text{line width}}$ = 2.46 mm; $M_{\text{line taper width}}$ = 5 mm; $M_{\text{line taper length}}$ = 4 mm).

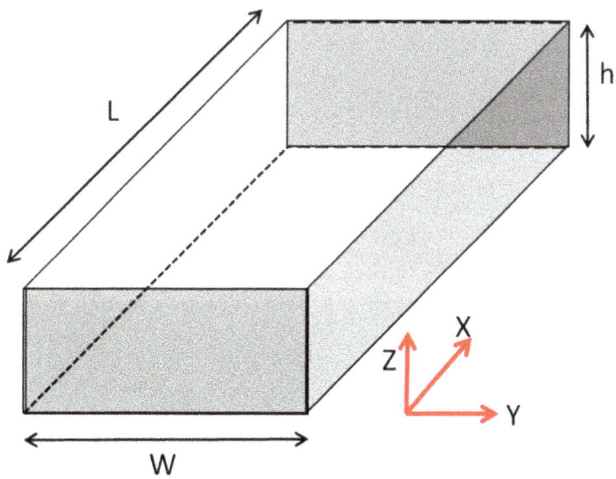

Figure 2.17 Rectangular waveguide cavity.

The resonance modes and its harmonics can be identified with the help of TE or TM modes, as both of them exist in a conventional rectangular waveguide. However, identifying TE or TM modes in a rectangular waveguide cavity is not unique, as we are free to choose the direction of propagation for reference since there is no longitudinal wave propagation inside the cavity.

In a similar manner, the SIW guide can be short-circuited at both ends to make the SIW-based cavity resonator [18]. The circuit will behave similarly to the conventional rectangular waveguide cavity. An example of an SIW cavity resonator is shown in Figure 2.18. Here, the cross-section (YZ plane) of the cavity resonator is a rectangle just like the conventional waveguide cavity. In this case, we assume an additional boundary condition of dielectric to metal interface at the front and back faces of the structure. The field distribution in such a structure can be solved in a similar manner as in the case of the rectangular waveguide. However, the presence of the via and the open aperture in an alternate fashion will alter the effective dimension of the cavity, which, in effect, changes the resonating frequency of each resonating mode. Following a similar argument as in the case of the SIW equivalent width, the length and width of the SIW cavity can be expressed as follows:

$$L_{\mathrm{Ceff}} = L_{\mathrm{C}} - \frac{d^2}{0.95s} \tag{2.13}$$

$$W_{\mathrm{Ceff}} = W_{\mathrm{C}} - \frac{d^2}{0.95s} \tag{2.14}$$

We can fix mode indices as m, n, and p along the X, Y, and Z axes, respectively. By solving the boundary condition with the equivalent dimensions of the cavity and assuming a direction of propagation along X, the TE^{X}_{mnp} field distribution in the cavity can be expressed as follows:

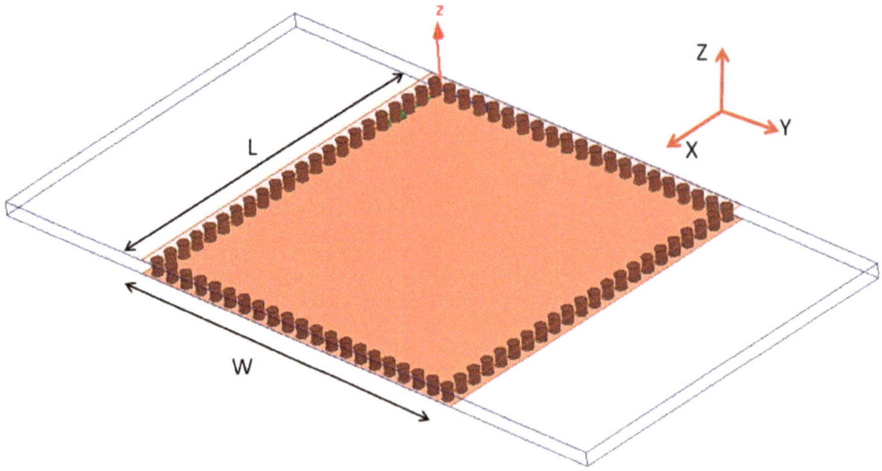

Figure 2.18 SIW rectangular cavity.

$$
\begin{aligned}
E_x(x,y,z,t) &= 0 \\
E_y(x,y,z,t) &= \frac{A_{mnp}}{\varepsilon}\left(\frac{p\pi}{h}z\right)\sin\left(\frac{m\pi}{L_{\text{Ceff}}}x\right)\cos\left(\frac{n\pi}{W_{\text{Ceff}}}y\right)\sin\left(\frac{p\pi}{h}z\right) \\
E_z(x,y,z,t) &= -\frac{A_{mnp}}{\varepsilon}\left(\frac{n\pi}{W_{\text{Ceff}}}\right)\sin\left(\frac{m\pi}{L_{\text{Ceff}}}x\right)\sin\left(\frac{n\pi}{W_{\text{Ceff}}}y\right)\cos\left(\frac{p\pi}{h}z\right) \\
H_x(x,y,z,t) &= -\frac{j}{\omega\mu\varepsilon}\left(\left(\frac{p\pi}{h}\right)^2+\left(\frac{n\pi}{W_{Ceff}}\right)^2\right)A_{mnp}\sin\left(\frac{m\pi}{a}x\right)\cos\left(\frac{n\pi}{b}y\right)\cos\left(\frac{p\pi}{h}z\right) \\
H_y(x,y,z,t) &= \frac{j}{\omega\mu\varepsilon}\left(\frac{n\pi}{W_{\text{Ceff}}}\right)\left(\frac{m\pi}{L_{\text{Ceff}}}\right)A_{mnp}\cos\left(\frac{m\pi}{L_{\text{Ceff}}}x\right)\sin\left(\frac{n\pi}{W_{\text{Ceff}}}y\right)\cos\left(\frac{p\pi}{h}z\right) \\
H_z(x,y,z,t) &= \frac{j}{\omega\mu\varepsilon}\left(\frac{m\pi}{L_{\text{Ceff}}}\right)\left(\frac{p\pi}{h}\right)A_{mnp}\sin\left(\frac{m\pi}{L_{\text{Ceff}}}x\right)\cos\left(\frac{n\pi}{W_{\text{Ceff}}}y\right)\sin\left(\frac{p\pi}{h}z\right)
\end{aligned}
\tag{2.15}
$$

Since the height of the substrate is very small compared to the operating wavelength of the cavity, we can assume no standing-wave pattern along the height of the substrate. Therefore, for the SIW cavity, there is $p = 0$. Therefore, the resonant frequency of the particular $mn0$ mode of the SIW cavity can be defined with the help of (2.13) and (2.14) as follows:

$$
f_{mn0} = \frac{c}{2\sqrt{\varepsilon_r}}\sqrt{\left(\frac{m}{L_{\text{Ceff}}}\right)^2+\left(\frac{n}{W_{\text{Ceff}}}\right)^2} \tag{2.16}
$$

From (2.15), it can be seen that the TE^X_{mn0} mode will have following field components:

$$\begin{aligned}
E_x(x,y,z,t) &= 0 \\
E_y(x,y,z,t) &= 0 \\
E_z(x,y,z,t) &= -\frac{A_{mnp}}{\varepsilon}\left(\frac{n\pi}{W_{\text{Ceff}}}\right)\sin\left(\frac{m\pi}{L_{\text{Ceff}}}x\right)\sin\left(\frac{n\pi}{W_{\text{Ceff}}}y\right) \\
H_x(x,y,z,t) &= -\frac{j}{\omega\mu\varepsilon}\left(\left(\frac{n\pi}{W_{\text{Ceff}}}\right)^2\right)A_{mnp}\sin\left(\frac{m\pi}{a}x\right)\cos\left(\frac{n\pi}{b}y\right) \\
H_y(x,y,z,t) &= \frac{j}{\omega\mu\varepsilon}\left(\frac{n\pi}{W_{\text{Ceff}}}\right)\left(\frac{m\pi}{L_{\text{Ceff}}}\right)A_{mnp}\cos\left(\frac{m\pi}{L_{\text{Ceff}}}x\right)\sin\left(\frac{n\pi}{W_{\text{Ceff}}}y\right) \\
H_z(x,y,z,t) &= 0
\end{aligned} \tag{2.17}$$

If we follow closely, we can choose the direction of propagation in Figure 2.18 as Z. In that case, we can call the mode the $\text{TM}^Z{}_{mn0}$ mode, which will lead to same set of field components as in (2.17). This is clearly up to the reader to choose the name of the mode in the cavity. However, conventionally in the literature, the SIW rectangular cavity modes have been named as the TE_{mn0} mode and, hence, the same will be followed in this book. A list of consecutive modes resonating in the SIW rectangular cavity is shown in Figure 2.19. These modes are obtained from the eigenmode analysis of the structure.

By referring to (2.16), it is evident that by changing the length-to-width ratio of the cavity, the sequence of the resonant frequencies of successive mode in an SIW

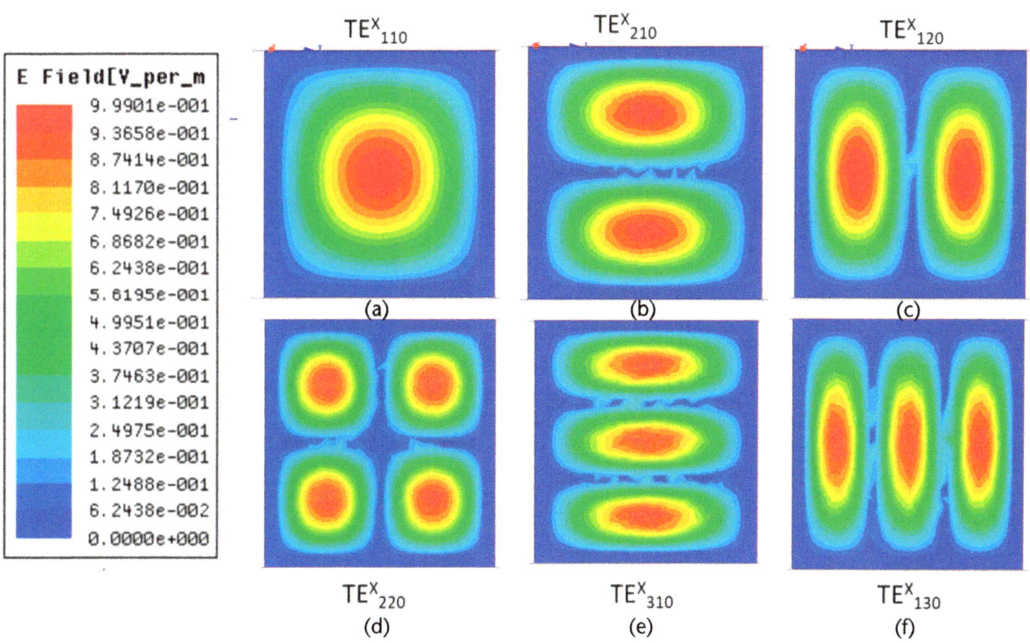

Figure 2.19 Modes in the SIW rectangular cavity (length = 22 mm, width = 20 mm, height = 0.8 mm, dia = 0.6 mm; pitch = 1 mm; height = 0.8 mm; dielectric constant of medium = 2.2; loss tangent = 0.0009).

cavity can be altered as per our design criteria. In Figure 2.20, a mode chart graph for a rectangular SIW cavity is shown. Such graphs become useful while designing components using the SIW cavity because we want to utilize certain resonating modes while suppressing all other modes.

In the case of the conventional waveguide-based resonator, we can choose any other shape of the cross-section besides rectangular such as circular or elliptical, giving us an additional degree of freedom. In the case of the SIW, the height of the substrate is constrained; hence, the implementation of a circular waveguide cavity becomes difficult if we restrict our direction of the reference as X. However, we can definitely rearrange the via side wall in a circular fashion, which, along with the top and bottom metal plates, can work as a circular waveguide cavity. An example of the SIW circular cavity resonator is shown in Figure 2.21. The difference of the analysis of such a structure with its previous example would be in the choice of reference of direction, which is Z instead of X (circular cross-section along XY). Hence, conventionally, the circular SIW cavity modes are termed as TM^{Z}_{mn0} mode, which will be followed in this book. Also in this case, the effective radius of the SIW cavity will be different from the physical radius of the cavity and it will be function of the diameter and the pitch of the via wall in a fashion similar to (2.13), which affects the resonant frequency of the modes of the cavity. A list of consecutive modes resonating in SIW circular cavity is shown in Figure 2.22.

The modes of the SIW circular cavity have been studied in detail by many researchers [19, 20]. It can be seen in Figure 2.21 that multiple degenerate orthogonal modes are possible, which are represented as Modes A and B. These modes can be efficiently utilized to realize a desired filter response, which will be discussed in a later chapter.

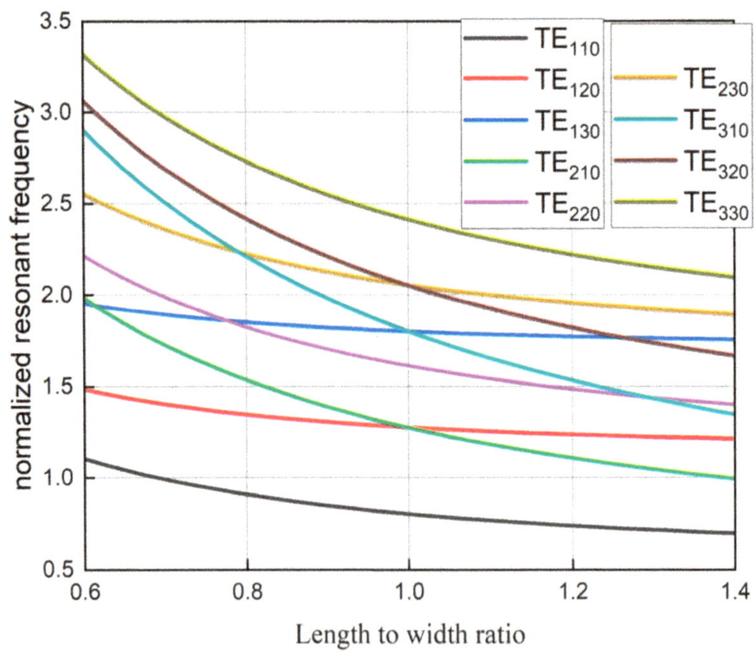

Figure 2.20 Mode chart with a length-to-width ratio of the rectangular SIW cavity.

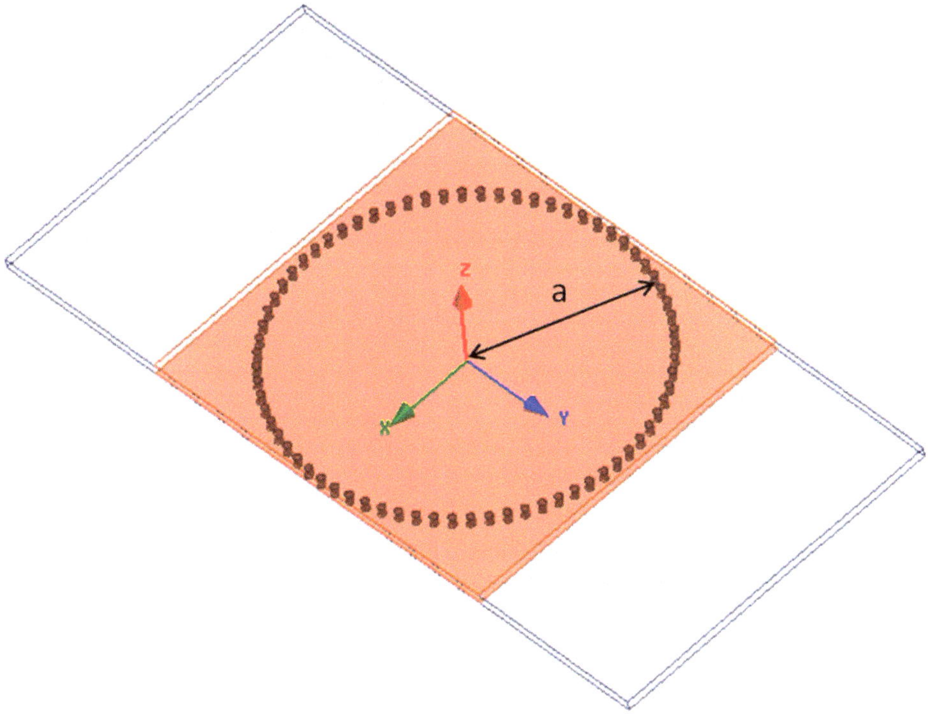

Figure 2.21 SIW circular cavity.

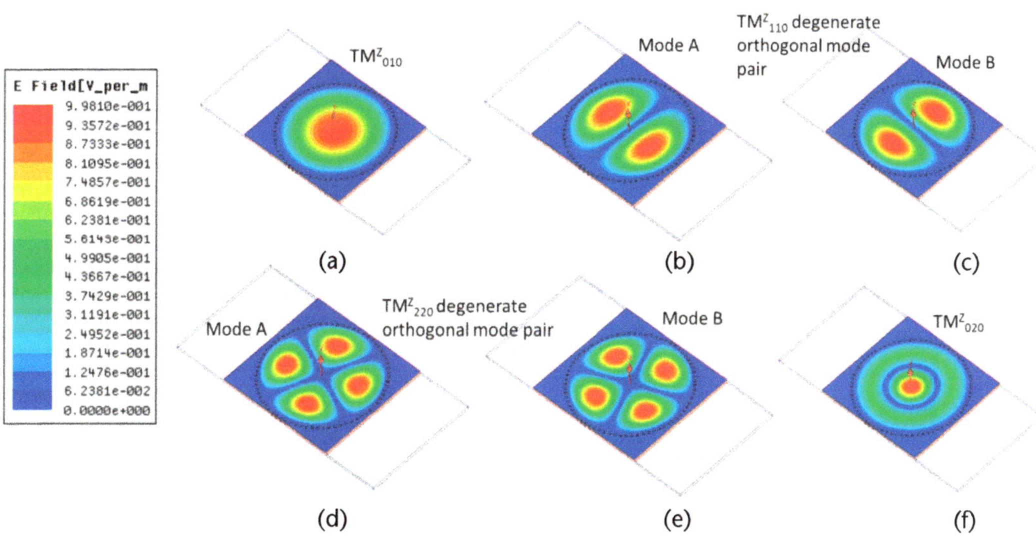

Figure 2.22 Modes in the SIW circular cavity (cavity radius = 14 mm, height = 0.8 mm, dia = 0.6 mm; pitch = 1 mm; height = 0.8 mm; dielectric constant of medium = 2.2; loss tangent = 0.0009).

2.3.1 Excitation of the SIW Cavity

To excite the SIW cavity, the current probe method similar to the SIW guide can be utilized. A 50Ω microstrip line can be directly connected to the top metal plate of the cavity to generate the desired surface current distribution in the cavity, which to be a unique choice of planar resonant circuit with better performance, which can find application in the antenna, filter, or active circuits.

References

[1] Balanis, C. A., *Advanced Engineering Electromagnetics*, New York: Wiley, 1989.

[2] Deslandes, D., and K. Wu, "Accurate Modeling, Wave Mechanisms, and Design Considerations of a Substrate Integrated Waveguide," *IEEE Transactions on Microwave Theory and Techniques*, 2006, pp. 2516–2526.

[3] Pozar, D. M., *Microwave Engineering*, New York: Wiley, 2008.

[4] Xu, F., and K. Wu, "Guided-Wave and Leakage Characteristics of Substrate Integrated Waveguide," *IEEE Transactions on Microwave Theory and Techniques*, Vol. 53, No. 1, 2005, pp. 66–73.

[5] Che, W., et al., "Equivalence Between Waveguides with Side Walls of Cylinders (SIRW) and of Regular Solid Sheets," *Asia-Pacific Microwave Conference Proceedings*, Suzhou, 2005.

[6] Tseng, C. -H., and T. -H. Chu, "Measurement of Frequency-Dependent Equivalent Width of Substrate Integrated Waveguide," *IEEE Transactions on Microwave Theory and Techniques*, Vol. 54, No. 4, 2006, pp. 1431–1437.

[7] Salehi, M., and E. Mehrshahi, "A Closed-Form Formula for Dispersion Characteristics of Fundamental SIW Mode," *IEEE Microwave and Wireless Components Letters*, Vol. 21, No. 1, 2011, pp. 4–6.

[8] Mukherjee, S., et al., "Design of a Broadband Coaxial to Substrate Integrated Waveguide (SIW) Transition," *Asia-Pacific Microwave Conference Proceedings (APMC)*, Seoul, South Korea, 2013.

[9] Yau, C. -K., et al., "Design of 30GHz Transition Between Microstrip Line and Substrate Integrated Waveguide," *Asia-Pacific Microwave Conference*, Bangkok, Thailand, 2007.

[10] Ju, B. Y., "Tunable Ferrite Phase Shifters Using Substrate Integrated Waveguide Techniques," Master of Sciences Dissertation, Université de Montreal, December 2010.

[11] Deslandes, D., and K. Wu, "Integrated Microstrip and Rectangular Waveguide in Planar Form," *IEEE Microwave and Wireless Components Letters*, Vol. 11, No. 2, 2001, pp. 68–70.

[12] Deslandes, D., "Design Equations for Tapered Microstrip-to-Substrate Integrated Waveguide Transitions," *IEEE MTT-S International Microwave Symposium*, 2010.

[13] Miralles, E., et al., "Improvement for the Design Equations for Tapered Microstrip-to-Substrate Integrated Waveguide Transitions," *International Conference on Electromagnetics in Advanced Applications*, 2011.

[14] Kordiboroujeni, Z., and J. Bornemann, "New Wideband Transition from Microstrip Line to Substrate Integrated Waveguide," *IEEE Transactions on Microwave Theory and Techniques*, Vol. 62, No. 12, 2014, pp. 2983–2989.

[15] Suntives, A., and R. Abhari, "Transition Structures for 3-D Integration of Substrate Integrated Waveguide Interconnects," *IEEE Microwave and Wireless Components Letters*, Vol. 17, No. 10, 2007, pp. 697–699.

[16] Mohamed, I., and A. Sebak, "Broadband Transition of Substrate-Integrated Waveguide-to-Air-Filled Rectangular Waveguide," *IEEE Microwave and Wireless Components Letters*, Vol. 28, No. 11, 2018, pp. 966–968.

[17] Deslandes, D., and K. Wu, "Analysis and Design of Current Probe Transition from Grounded Coplanar to Substrate Integrated Rectangular Waveguides," *IEEE Transactions on Microwave Theory and Techniques*, Vol. 53, No. 8, 2005, pp. 2487–2494.

[18] Cassivi, Y., et al., "Low-Cost and High-Q Millimeter-Wave Resonator Using Substrate Integrated Waveguide Technique," *32nd European Microwave Conference*, Milan, Italy, 2002.

[19] Liu, Q. Z., et al., "High-Selective Triple-Mode SIW Bandpass Filter Using Higher-Order Resonant Modes," *Electron. Lett.*, Vol. 56, 2020, pp. 37–39.

[20] Yang, X. -L., et al., "Q-Band Dual-Mode Substrate Integrated Waveguide Filter and Diplexer with Circular Cavity," *Int. J. RF Microw. Comput. Aided Eng.*, Vol. 31, 2021.

[21] Grine, F., et al., "High-Q Substrate Integrated Waveguide Resonator Filter with Dielectric Loading," *IEEE Access*, Vol. 5, 2017, pp. 12526–12532.

system. The design can be further enhanced if we replace the reflector with a cavity with a height of a quarter-wavelength as shown in Figure 3.1. Such a setup is called a cavity-backed slot antenna. Conventionally cavity-backed slot antennas are excited by the coaxial probe as shown in Figure 3.1.

The same setup may be realized in a planar substrate with the help of SIW technology. A SIW cavity can be designed in a planar substrate and a slot can be etched in the top or bottom metal plate, as shown in Figure 3.2. The most common method to excite the slot antenna placed at the top or bottom metal plate is done through exciting the SIW rectangular cavity TE^{y}_{110} or TE^{y}_{210} mode. To understand the working principle of the excitation of the slot antenna, the concept of equivalent magnetic current (M_s) can be used.

The slot antenna can be excited by introducing magnetic current component along the slot as shown in which are given by

$$M_s = -2\hat{n} \times E_s \tag{3.1}$$

where the E_s is the electric field across the slot. Therefore, in effect, we need to generate potential difference along the opposite sides of the slot such that E_s is generated. It can be accomplished by placing the slot at proper position in the top metal plate of the SIW cavity, which is excited in the desired mode. This can be better understood by investigating the field distribution of the SIW cavity modes.

As shown in Figure 3.2, if we place the slot very close to one side of the SIW cavity and excite the same using TE^{y}_{110} mode, the magnitude of the electric field at both sides of the slot are at different levels. This helps to create a potential difference across the slot which in effect creates the desired E_s and accordingly the M_s. As a result, the slot antenna is excited and electromagnetic radiation is generated.

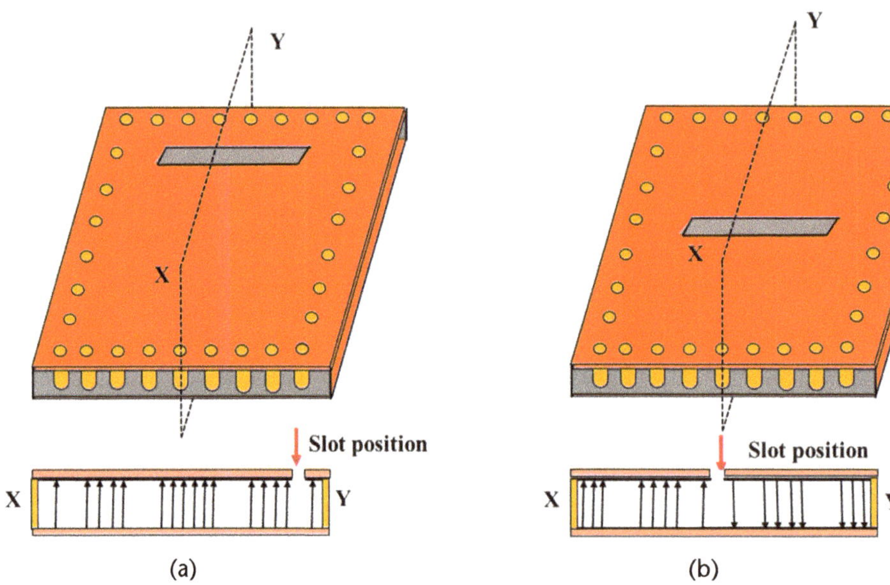

Figure 3.2 SIW cavity backed slot antenna (E-field shown in the figure): (a) TE^{y}_{110} mode excitation; and (b) TE^{y}_{210} mode excitation.

The radiation in the back side of the slot is reflected back by the cavity and a unidirectional radiation pattern is achieved. However, due to the small height of the cavity, the superposition of the reflected field with a forward radiated field may not be completely in phase, and, hence, comparatively smaller gain is achieved with respect to that of a quarter-wavelength height cavity.

The TE^{y}_{210} SIW cavity mode can also be used to excite slot antenna placed on the top metallic plate of the cavity. In this case, the slot is placed at the middle plane of the cavity where there is a null in the electric field or where there is maximum surface current distribution. The placement of the slot at the middle ensures that two opposite polarities' electric field exists at the opposite side of the slot, which essentially ensures the generation of the desired excitation electric field E_s along the slot. The excitation of slot antenna can also be explained from the fact that the placement of the slot at the middle of the cavity disturbs the maximum amount of the surface current, which helps to get the desired excitation field across the slot.

To feed the SIW cavity-backed slot, several possible approaches (e.g., microstrip line, GCPW line, or SIW inductive coupling window followed by SIW feed line section) may be used. The feeding of the SIW cavity by the microstrip line was discussed in Section 2.3. The same strategy with a small modification can be used to feed the SIW cavity-backed slot antenna. As explained earlier, the feeding mechanism includes the microstrip feed line to excite the desired mode of the cavity, which will excite the slot antenna radiation. The working principle can be better understood by the help of an equivalent circuit model. As shown in Figure 3.3, the feed line couples incident EM energy to the SIW resonator and the amount of coupling (n_1) is determined by the inset and the offset of the feed line. The energy from SIW resonator is coupled to slot antenna which can be modelled as lossy resonator. Here, the coupling (n_2) can be controlled by the proper placement of the slot in the top metallic plate of the antenna. Altogether, the coupling at two different stages determines the input impedance seen at the input port of the antenna, which needs to be matched to characteristic impedance (50Ω) of the line.

The working principle can be validated by a design example, as shown in Figure 3.4 [1]. Here, a slot is etched at the top plate of the SIW cavity. The cavity is fed by the microstrip line and, to get impedance matching, the offset and inset of the feed line are employed. This can be interpreted as change in the coupling (n_1) in the equivalent circuit model. The same can be verified by the Smith chart plot shown in Figure 3.5 where the solid line denotes a constant d_{shift} curve and the dotted line denotes a constant l_{in} curve. For this design, the position of the slot is fixed at $d_s = 8.2$ mm, which can be attributed to the fixed value of coupling (n_2) between cavity resonator and slot antenna. The input impedance seen at the feed point can

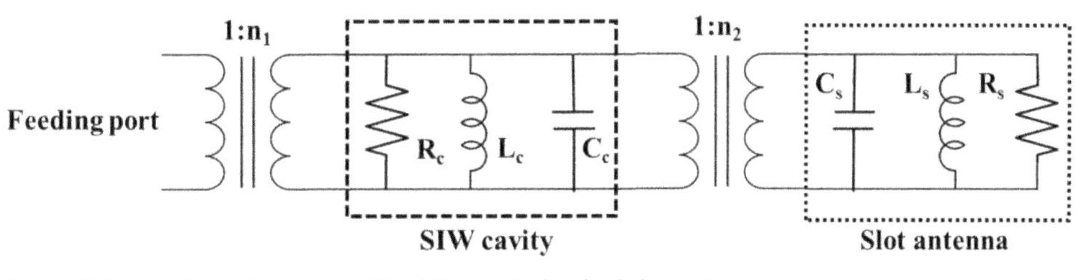

Figure 3.3 Equivalent model of the SIW cavity-backed slot antenna.

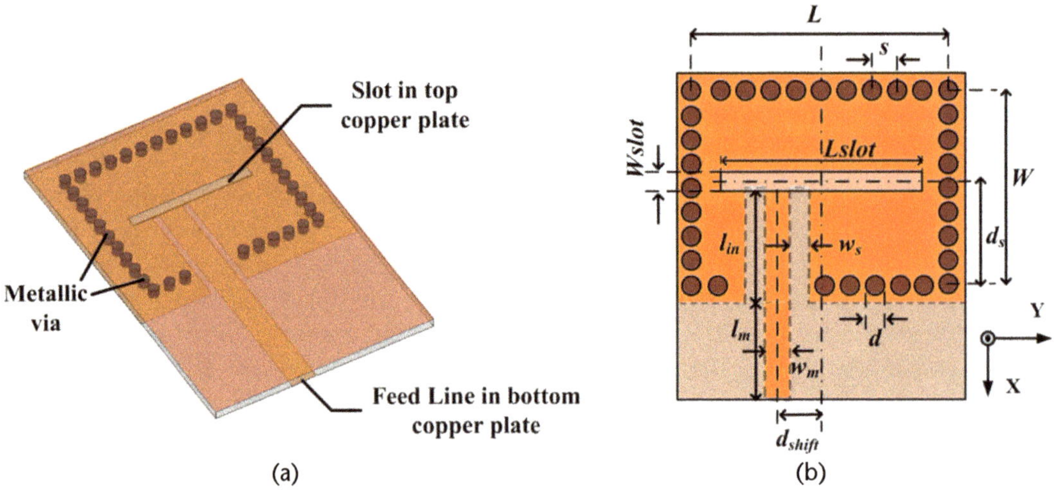

Figure 3.4 Design example of the SIW cavity-backed slot antenna (L = 16.72 mm, W = 15 mm, d = 1 mm, s = 1.6 mm, w_m = 2.42 mm, w_s = 1.22 mm, l_{in} = 8.2 mm, l_m = 5 mm, d_{shift} = 2 mm, l_{slot} = 12 mm, W_{slot} = 1 mm, and d_s = 8.2 mm). (© 2013 IEEE [1].)

be adjusted by varying the inset, offset, and slot position to get proper impedance matching. The excitation of the TE^{y}_{120} mode in the cavity is shown in Figure 3.5. The opposite polarity of electric field at the opposite side of the slot excites it in a similar manner as explained before.

The experimental verification of the design can be seen in Figures 3.6 and 3.7. The Rogers RT Duroid 5880 substrate (ε_r = 2.2) with a height of 0.8 mm is used for the antenna. The measured resonant frequency of the antenna is at 10.28 GHz

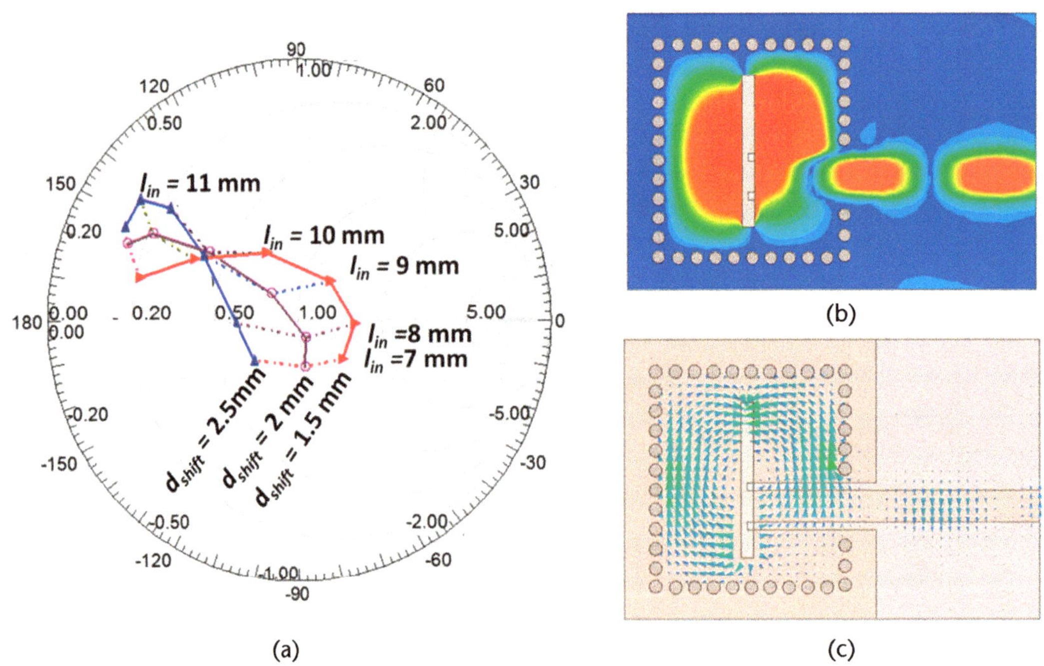

Figure 3.5 (a) Smith chart plot for variation in the shift and inset of the feed line; (b) magnitude of electric field plot; and (c) surface current plot of the slot loaded cavity. (© 2013 IEEE [1].)

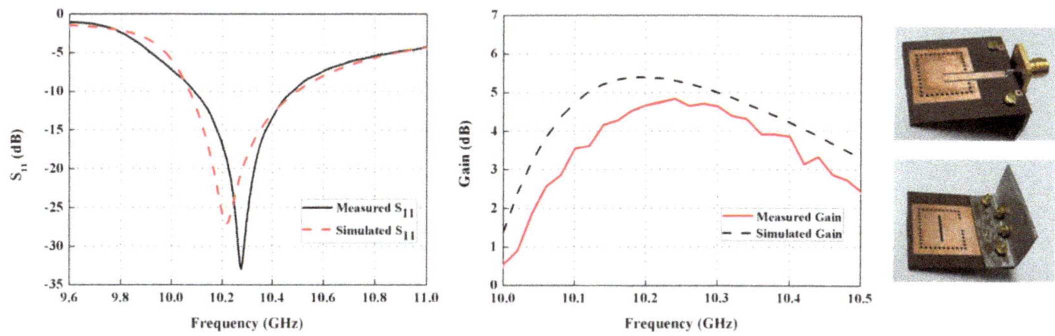

(a) (b) (c)

Figure 3.6 (a) Comparative study of simulated and measured reflection coefficient of the antenna; (b) gain profile of the antenna; and (c) fabricated SIW cavity backed slot antenna. (© 2013 IEEE [1].)

with (−10-dB S_{11}) impedance bandwidth of 400 MHz (3.9%). The measured gain of the antenna is 4.83 dBi, which is a little lower than the simulated response. The proposed antenna exhibits a broadside unidirectional radiation pattern. The measured front-to-back ratio (FTBR) of the antenna is 10 dB, which is a little lower than the simulated value of 15.6 dB. The simulated and measured cross-polarization levels of the antenna are below −30 dB and −15 dB, respectively.

The slot antenna can also be excited by the SIW circular cavity or any other type of cavity. While the operating mode in such cases may differ, the basic working principle remains same in all cases. The SIW cavity-backed antenna offers a reasonable trade-off between the gain of the antenna and the planar, low-profile configuration while we get a unidirectional radiation pattern. However, due to the high Q of the SIW cavity, such antennas offer a typically narrowband response.

3.2 Design Technique for Multiband Antennas

In the previous section, the use of dominant TE^{y}_{110} or the next higher-order mode TE^{y}_{210} mode to excite the SIW cavity-backed slot antenna is explained. However, in

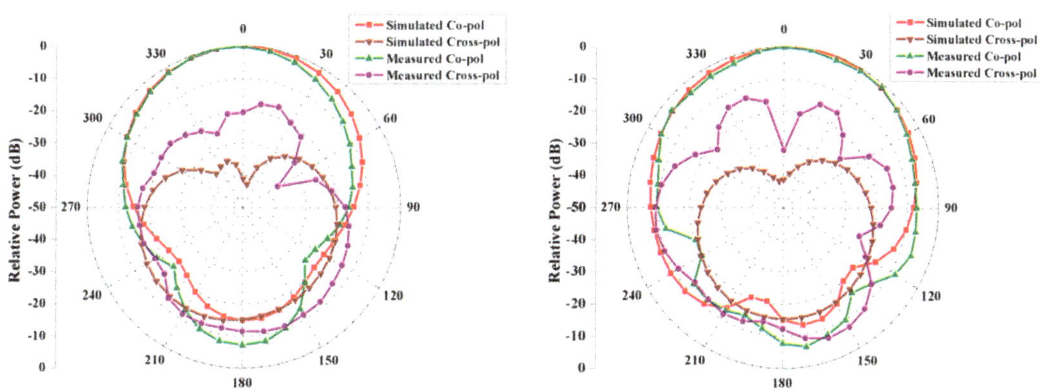

(a) (b)

Figure 3.7 Radiation pattern of the antenna: (a) E-plane; and (b) H-plane. (© 2013 IEEE [1].)

many applications, more than one resonant frequency is expected from the antenna system to support multiple subsystems simultaneously. For such applications, the SIW cavity-backed multiband slot antenna design can be employed. The multiband technique in the SIW cavity-backed slot antenna can be majorly implemented in two different approaches: (1) by adding an additional slot resonator, and (2) by exciting the same slot antenna with multiple cavity modes. Both the methods offer advantages and disadvantages in terms of design complexity and performance.

3.2.1 Multiband Response Using a Multiple Slot

In this technique, more than one slot is placed on the top or bottom metallic plate of the SIW cavity at a proper position. The addition of multiple slots may be seen as perturbation of the boundary condition of the SIW cavity, which results in the loading effect on the SIW cavity resonant mode as well as the generation of an additional hybrid mode typically not present in a conventional SIW cavity. The technique efficiently utilizes the hybrid mode to excite one of the slot antennas, whereas the conventional TE^{y}_{210} mode can be used to excite the other slot in a simultaneous fashion.

A design of the SIW cavity-backed multiple slot antenna is shown in Figure 3.8 [2]. The two slots are etched at the top metallic plate of the cavity and placed at a distance of d_s and d_{s2}, respectively, from one side wall of the cavity. A half-wavelength slot is placed at the center of the SIW cavity, which is designed to have its TE^{y}_{210} mode at 10 GHz (f_1). The parasitic slot is placed at the upper half of the cavity beside the main radiating slot with its length of approximately a half-wavelength at a higher frequency. The cavity is fed by a microstrip line similar to the previous design in Section 3.1. The inset of the feed line needs to be sufficient to excite the desired TE^{y}_{210} mode and the additional higher-order hybrid mode. The slot placed at the center of the cavity disturbs the TE^{y}_{210} mode surface current distribution with a negligible effect on the parasitic slot. The electric field and surface current distribution in this case are quite similar to the case of a single-frequency design. However, at a higher operating frequency, the field distribution is mostly concentrated on the upper half

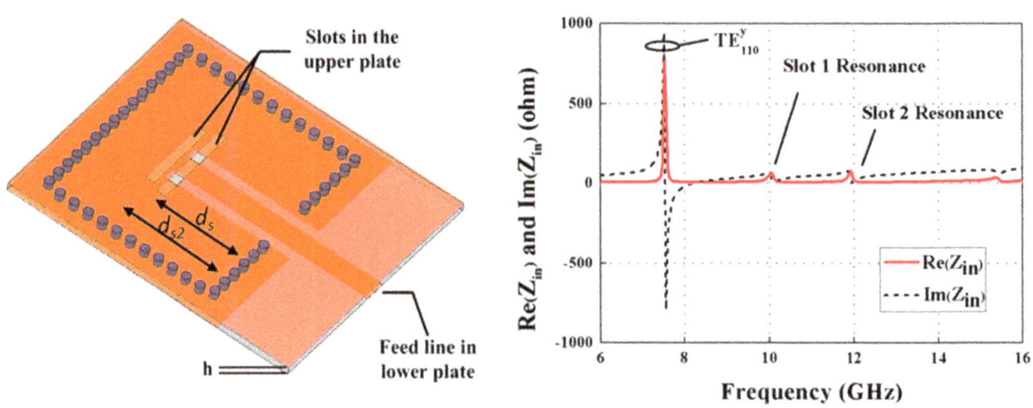

(a) (b)

Figure 3.8 (a) Dual-band cavity-backed slot antenna; (b) input impedance plot; and (c) dual-frequency design. (© 2015 IEEE [2].)

of the cavity above the parasitic slot with a negligible amount of field in the lower half as shown in Figure 3.9. The electric field in this case also is out of phase at the opposite side of the parasitic slot, which helps the slot to radiate at 11.96 GHz (f_2).

The SIW cavity is designed in the RT Duroid 5880 substrate with the dielectric constant (ε_r) of 2.2 and a thickness of 0.787 mm. The proposed design exhibits a dual-frequency response with the first and second resonances at 10.08 GHz and 11.96 GHz, with impedance bandwidths of 108 MHz (1%) and 100 MHz (0.08%), respectively. The addition of the parasitic slot does not affect the first resonance of the antenna (shown in the dotted line in Figure 3.10) and adds an additional resonance at a higher frequency. The radiation at both frequencies produces a unidirectional radiation pattern with a gain of 5.5 dBi and 5.3 dBi, respectively, and an FTBR of more than 10 dB. The lower and higher resonant frequencies are inversely proportional to the length of the slots. One important aspect of the proposed design is that the two operating frequencies of the design can be independently tuned with a negligible effect on the other resonance. This can be achieved at the design stage by tuning the resonant slot length of the corresponding operating frequency as shown in Figure 3.11. As a result, the frequency ratio (f_2/f_1) of the antenna can be tuned in a wide range of 1.18 to 1.5 by changing the lengths of the slot pair.

As expected, the high Q SIW cavity loading results in a narrowband response from the antenna at both operating frequencies. The bandwidth can be slightly improved by combining the two slot resonances by selecting them close together such that the overall bandwidth response is improved. This can be better understood

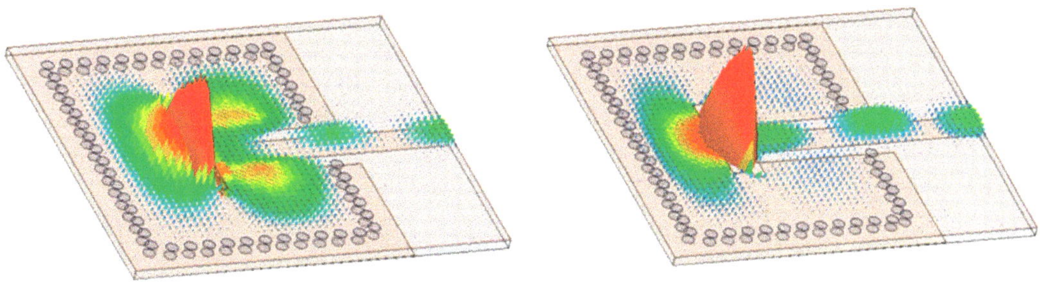

(a) (b)

Figure 3.9 Vector electric field plot at: (a) f_1 (10 GHz) and (b) f_2 (11.96 GHz). (© 2015 IEEE [2].)

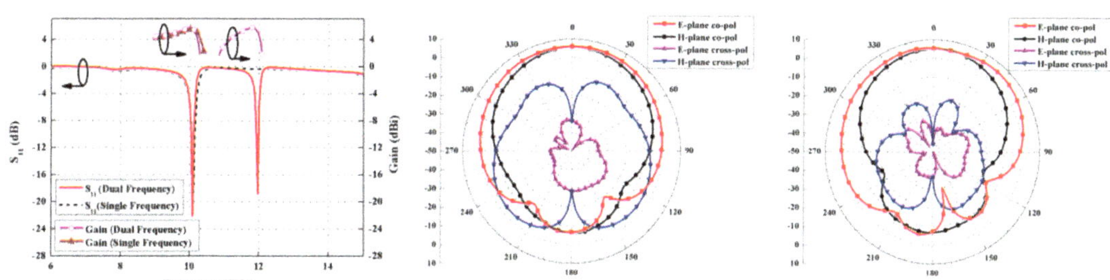

(a) (b) (c)

Figure 3.10 Performance of the antenna: (a) reflection coefficient plot; (b) radiation pattern at f_1; and (c) radiation pattern at f_2. (© 2015 IEEE [2].)

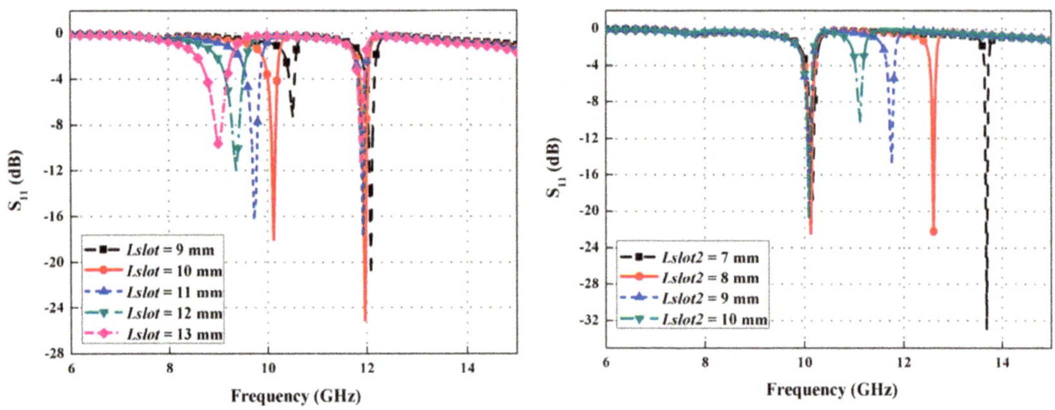

(a) (b)

Figure 3.11 Performance of the SIW cavity-backed, dual-slot antenna with a variation of slot lengths: (a) slot 1 and (b) slot 2. (© 2015 IEEE [2].)

by seeing the input impedance plot seen after de-embedding the effect of feed line. As shown in Figure 3.12, the two slot lengths are tuned to operate at close proximity to generate a stagger tuning effect. This helps to achieve a bandwidth of 380 MHz (3.6%). The radiation pattern and gain performance remain similar to the dual-frequency design.

3.2.2 Multiband Response Using a Modified Slot

In the second approach, a modified shape slot antenna is used to modify the boundary condition in the SIW cavity to generate an additional hybrid mode in addition to the conventional modes. The conventional and hybrid modes are used to excite the modified shaped slot at two different operating frequencies and produce multiband response. In this case also, the feed line is used to excite the desired modes in the cavity, which essentially excites the slot antenna. Two such modified slot shapes to generate a multiband response are described here.

In the first design example, a dumbbell-shaped slot is placed on the top metal plate of the SIW cavity [3]. The slot shape is modified by placing a circular patch

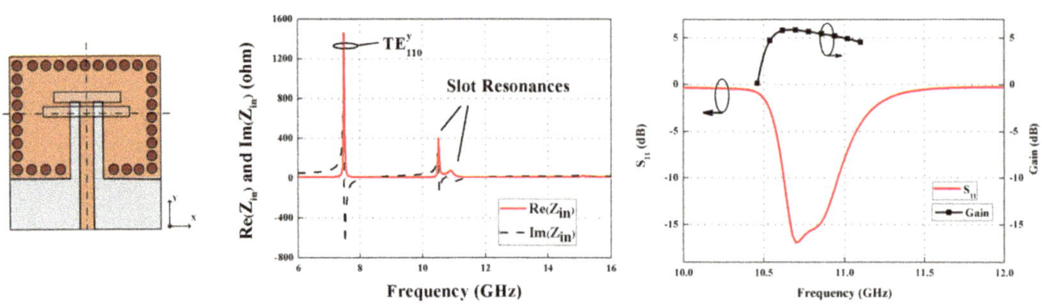

(a) (b) (c)

Figure 3.12 The SIW cavity-backed design with enhanced bandwidth: (a) input impedance plot; and (b) reflection coefficient and gain performance. (© 2015 IEEE [2].)

inside the slot along with a pair of strips connecting inner circular patch and outer conductor separated by a slot arc of length of $\varphi = 90°$, as shown in Figure 3.13. The working principle of the antenna can be explained again with the help of input impedance (Z_{11}) plot. The loading effect of dumbbell-shaped slot is visible as the higher-order TE^{y}_{210} mode shifts from 11.36 GHz to 9.28 GHz. In addition, an additional resonant mode is visible in the modified dumbbell-shaped, slot-loaded cavity, which is not present in unloaded and simple dumbbell slot-loaded cavities. A comparison between a simple dumbbell-shaped slot (generating a single-frequency radiation) and the modified dumbbell-shaped slot (dual-frequency operation) backed cavity modes can be seen in Figure 3.14. The electric field distribution of the proposed antenna at the lower resonant frequency (f_L = 9.4 GHz) is quite similar to that of the conventional slot-loaded SIW cavity (Case B). The surface current distribution at this frequency follows the outer boundary of the slot along the path *agcdhfa*, with a negligible amount of current in the inner circular patch. On the contrary, the surface current distribution at a higher resonant frequency follows the path *abcdefa*, which is essentially governed by the connecting strips in the dumbbell-shaped slot. The complex field distribution created by this alternate surface current path can be identified as a hybrid-mode resonance as it is not conventionally available in the SIW cavity. As shown in Figure 3.14, the hybrid-mode field distribution also produces out-of-phase excitation to the slot and we get higher-frequency (f_H = 13.62 GHz) radiation from the circuit.

One of the unique capabilities of the design is its tenability of individual resonance by changing particular dimension of the circuit while negligible effect on other resonance. As shown in Figure 3.15, the lower resonant frequency can be tuned by changing slot length, whereas, for the higher resonant frequency, the diameter of the inner circular patch *d_dumb1* needs to be varied. In addition, the position of slot arc section 'g' and 'h,' respectively, can also be utilized to tune the resonant frequencies. The higher-order hybrid mode gradually becomes weaker with increase in θ without affecting lower-frequency resonance. However, changing φ has a significant effect on both resonant frequencies. Overall, the design exhibits a flexibility to achieve a frequency ratio (FR) (f_H/f_L) within a range of 1.55 to 1.3.

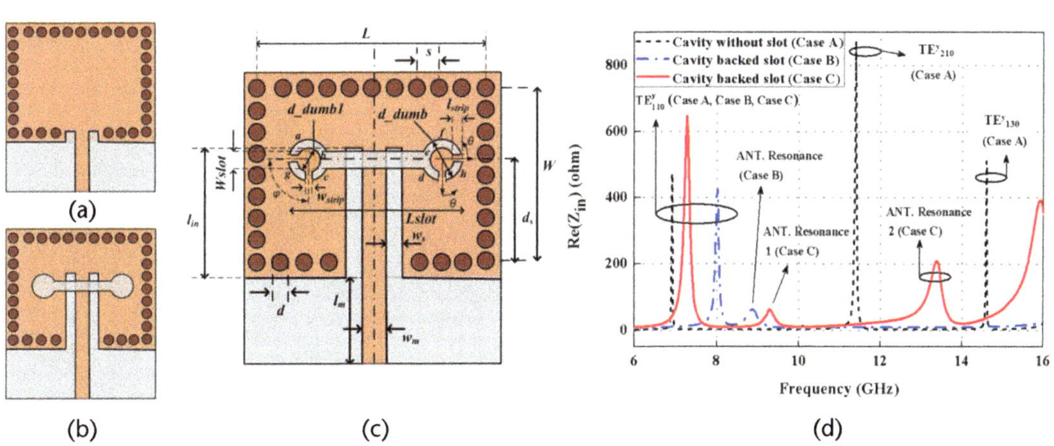

Figure 3.13 Dual-band design and its input impedance plot. (© 2015 IEEE [3].)

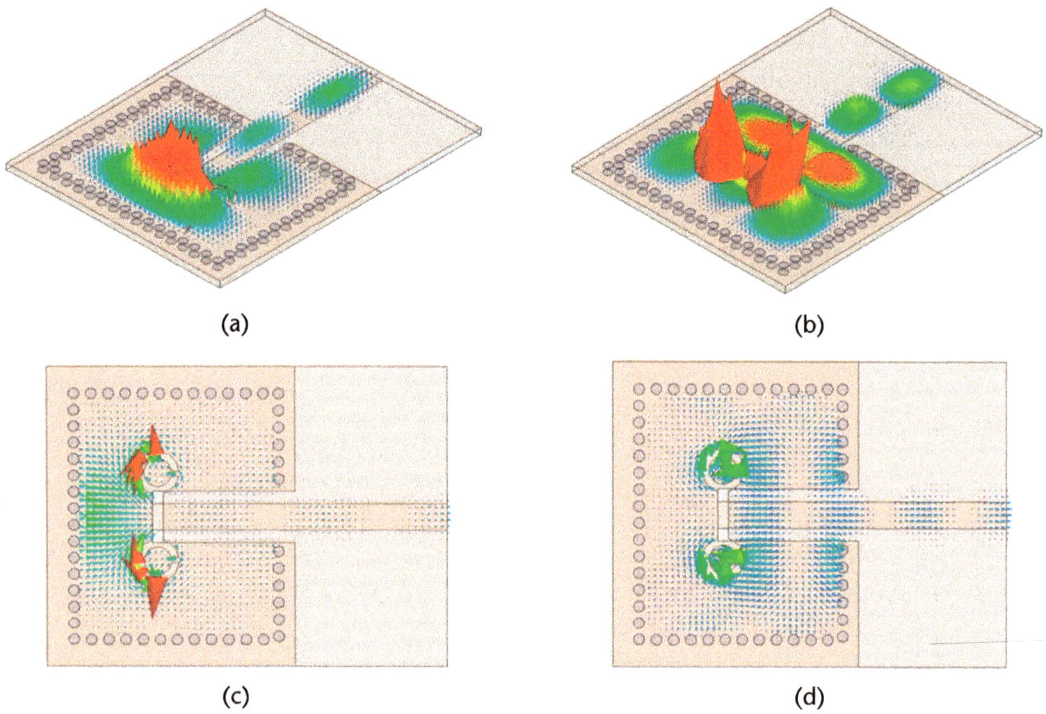

Figure 3.14 (a) Vector electric field, (b) surface current distribution plot at f_L, (c) vector electric field, and (d) surface current distribution plot at f_H. (© 2015 IEEE [3].)

The design can have further variation to achieve a third resonance from the SIW cavity mode with slight modification in the slot geometry and dimensions. The modified design is shown in Figure 3.16 where the slot is comparatively bigger than the conventional slot length of half of the operating wavelength and it is placed asymmetrically along the right half of the cavity [4]. The pair of connecting strips in the dumbbell-shaped slot is separated by a slot arc of length of $\varphi = 135°$. The 50Ω microstrip feed line inset inside the cavity excites conventional cavity modes TE^y_{110}, TE^y_{210}, and TE^y_{130} mode at 6.93 GHz, 11.4 GHz, and 14.6 GHz covering C-band

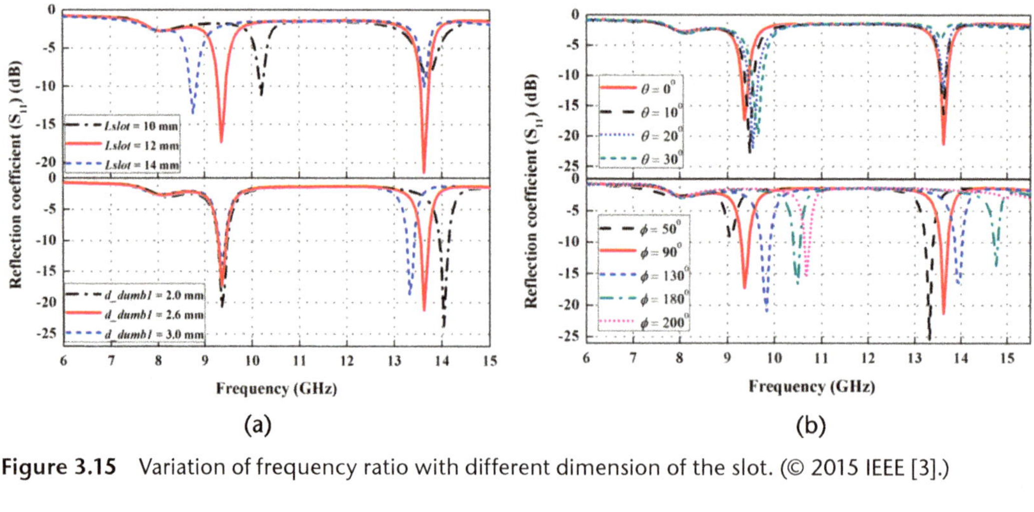

Figure 3.15 Variation of frequency ratio with different dimension of the slot. (© 2015 IEEE [3].)

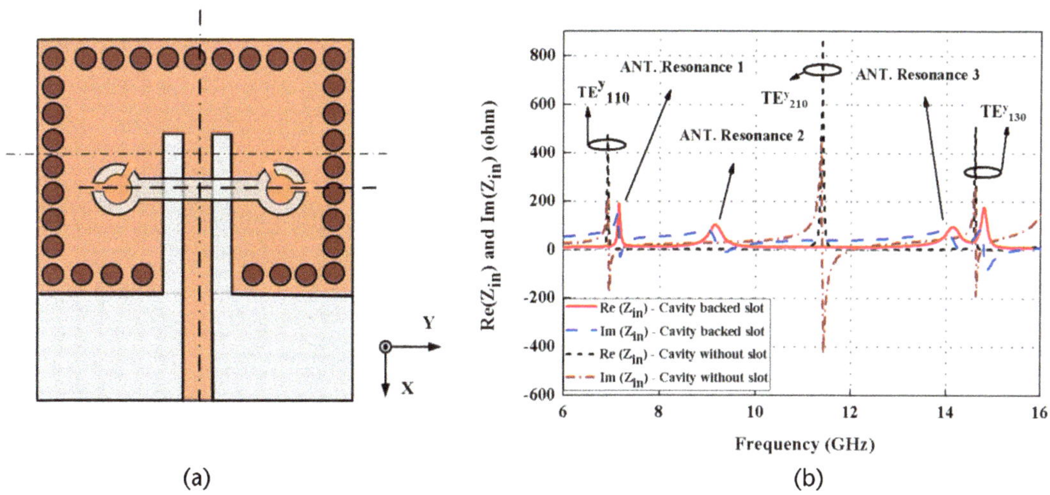

Figure 3.16 Triple-band antenna design and its input impedance plot. (© 2017 Wiley [4].)

(4 to 8 GHz), X-band (8 to 12 GHz), and Ku-band (12 to 18 GHz) without any slot loading in the cavity. The placement of the long slot perturbs the surface current distribution of the cavity to such an extent that the cavity behaves as combination of two half-cavities of equal length (L) but with unequal width as follows: $Cavity_{\text{left}}$ (width = W-d_s) and $Cavity_{\text{right}}$ (width = d_s). As a result, three hybrid cavity modes are excited at 7.16 GHz (f_1), 9.18 GHz (f_2), and 14.14 GHz (f_3). As shown in Figure 3.17, the surface current distribution at the first and third resonant modes is mainly concentrated on the left half of the cavity, and for the second mode, the surface current is mostly concentrated in the right half. The slot is excited due to the difference in magnitude and phase of the electric field distribution at the opposite side of the slot. Similar to the previous design, the frequency ratio can be tuned by varying

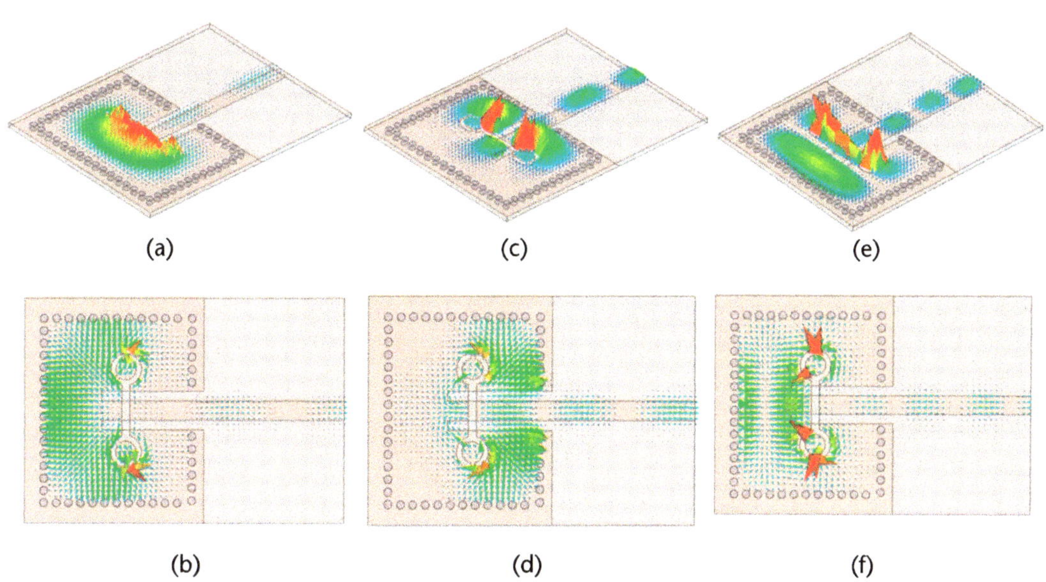

Figure 3.17 Vector electric field and surface current plot. (© 2017 Wiley [4].)

the inner circular patch diameter (*d_dumb1*) of the proposed modified dumbbell-shaped slot, within a range of $f_2/f_1 = 1.34$ to 1.37 and $f_3/f_1 = 1.84$ to 1.98. Also, by changing the position of the connecting strips denoted by θ and the length of the slot arc denoted by φ, the frequency ratio of the antenna can also be tuned ($f_2/f_1 = 1.27$ to 1.36 and $f_3/f_1 = 1.92$ to 2.01).

The working principle of the triple-band antenna (Figure 3.18) can be better explained with the help of equivalent circuit model. Each resonating mode of the cavity can be modeled by a resonant circuit with appropriate value of equivalent inductance and capacitance. Since the SIW cavity has loss (majorly conductor and dielectric loss), a resistive component also needs to be included. Hence, the three resonant modes of the proposed antenna are modeled as three parallel RLC circuits where R_{CL1a}, L_{CL1}, C_{CL1}, and R_{CL2a}, L_{CL2}, C_{CL2} correspond to hybrid-mode resonances 1 and 3 at $Cavity_{\text{left}}$ and R_{CR1a}, L_{CR1}, C_{CR1} corresponds to hybrid-mode resonance 2 at $Cavity_{\text{right}}$. The normalized input impedance (z_{in}) of the cavity can be determined from the [S] matrix as follows:

$$z_{\text{in}} = \frac{1 + S_{11}}{1 - S_{11}} \tag{3.2}$$

Considering an equivalent parallel RLC circuit,

$$\frac{1}{Z} = \frac{1}{R} + \frac{1}{j\omega L} + j\omega C = A + jB \tag{3.3}$$

where $Z = z_{\text{in}} \times 50$ (microstrip feed-line characteristic impedance = 50Ω) and $\omega = 2\pi f_{0L}$, where f_{0L} = the loaded resonant frequency of the cavity at the corresponding resonant mode:

$$A|_{f_{0L}} = \max(A)$$

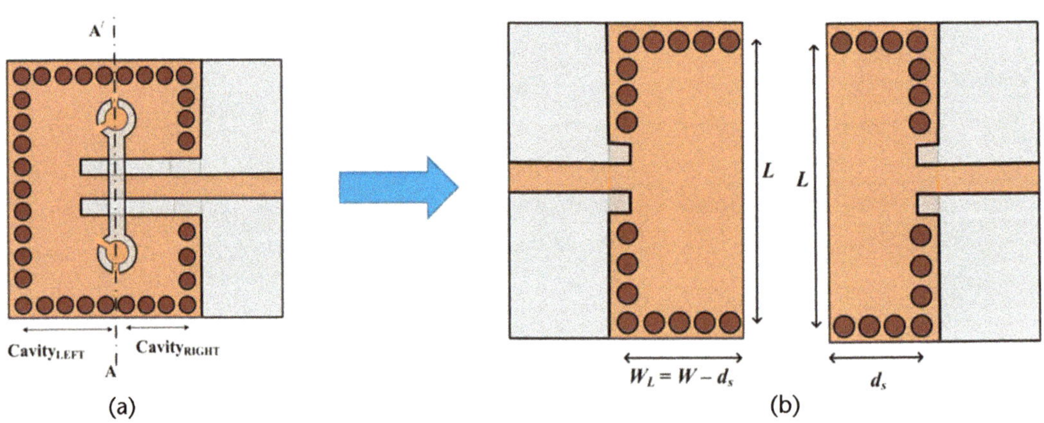

Figure 3.18 Working principle of the triple-band antenna with the help of concept of half-mode SIW cavity. (© 2017 Wiley [4].)

$$3\text{-dB bandwidth} = \Delta f = f_1 - f_2$$

where $A|_{f_1} = A|_{f_2} = 0.707 * A|_{f_{0L}}$.

Also, the Q factor of the cavity can be related with the equivalent parameter.

Say that f_{0U} = the unloaded resonant frequency of the cavity at the corresponding resonant mode and Q_U = the unloaded quality factor of the cavity. Now,

$$f_{0U} = \frac{1}{2\pi\sqrt{LC}} \tag{3.4}$$

$$Q_U \approx \frac{f_{0U}}{\Delta f} = \frac{R}{j\omega_U L} \tag{3.5}$$

where $\omega_U = 2\pi f_{0U}$.

Solving (3.3), (3.4), and (3.5), the equivalent R, L, C corresponding to each SIW cavity mode are calculated. The extracted equivalent circuit parameters are analyzed in Agilent ADS. To start with, the transformer turn ratios are taken as 1.0 corresponding to the maximum amount of coupling. However, after calculating the circuit parameters, the model is further optimized using Agilent ADS to determine the transformer turn ratio such that its response match with the performance of the proposed cavity model $Cavity_{\text{left}}$ and $Cavity_{\text{right}}$. The slot antenna, like any antenna, can be modeled as a lossy resonator with its radiation resistance as the lossy component. Hence, to model the slot antenna for the equivalent circuit of the triple-band antenna, a tank circuit with resistive component can be formulized. The input impedance of the slot antenna is modified with the inclusion of backside reflector due to loading effect [5]. Therefore, the calculation of the equivalent model of the slot should include the effect of the SIW cavity backing in the present case. For such a calculation, a simulation model as shown in Figure 3.20 can be used. The model consists of half of the dumbbell slot antenna placed in a very large ground plane (an approximate model for the infinite ground plane) and backed by a metallic plane at a distance of 0.787 mm (the height of the substrate). A perfect magnetic conductor (PMC) symmetry plane is placed in front of the half-slot antenna. To realize excitation of the proposed slot, a coaxial probe is used. To determine the unloaded resonant frequency of the antenna, a small gap is introduced between coaxial probe and slot plane to ensure weak excitation. Now (3.3) to (3.5) can be used again to determine the equivalent parameter of the slot antenna using a similar procedure.

After calculating individual cavity modes and the slot antenna equivalent model, the final model may be designed by combining each circuit together with the help of additional combining network as shown in Figure 3.21. The three RLC circuits corresponding to three cavity modes are coupled to the feed line with the help of three ideal transformers (n_1, n_2, n_3, respectively). However, all of the three RLC circuits are coupled to the slot antenna as we get triple-band resonance in the circuit. Therefore, a single ideal transformer (n_s) is used to account for the coupling between slot and cavity modes. The three RLC circuits are connected with the help of a Tee combining circuit made of an inductor and capacitor in series in each arm, which accounts for the loading effect between the three-tank circuit corresponding to the successive cavity modes. However, two distinctive changes in the final circuit may be noted. First, the two cavities $Cavity_{\text{left}}$ and $Cavity_{\text{right}}$ have open boundary at one side, whereas, in the final circuit, they are closed by the placement of face-to-face configuration. Second, the feed line is also extended into the cavity unlike the equivalent models shown in Figure 3.19, which indicates change in the coupling or the turn ratio of the ideal transformer. In addition, the loading effect of the feed line may be accounted by change in the equivalent circuit model parameters as well. Therefore, although we can start the initial model using the values deduced in initial calculation, the final model will have optimized values to account for the above effect. It is evident from the fact that the individual cavity modes have resonances at 6.35 GHz, 7.16 GHz, and 13.59 GHz, whereas the final circuit resonates at 7.2 GHz, 9.36 GHz, and 14.30 GHz, respectively, which are same as the SIW cavity-backed slot antenna resonances.

The equivalent model can also be used to predict the effect of different parameters in the performance of the antenna. Changing the slot position (d_s) from the optimized value within a small range will change the coupling between slot and three cavity modes and also between the feed line and cavity modes. Hence, the turn ratios in the equivalent circuit model will modify according to the change in d_s, which affects the reflection coefficient performance of the antenna. The comparative study of the variation of reflection coefficient of both circuit is shown in Figure 3.21. It is assumed that the equivalent R, L, C values of the corresponding equivalent circuit model of the cavity remain same considering small variation in d_s and only the transformer turn ratio values are changed in ADS circuit model throughout the analysis. Similarly, the variation of the inset feed line modifies the coupling between cavity modes and the feed line. As a result, the corresponding transformer turn ratios n_1, n_2, n_3 are changed with variation in l_{in}.

This can help us to generate design graphs as shown in Figure 3.22 for the equivalent circuit model to predict the performance of the antenna for different

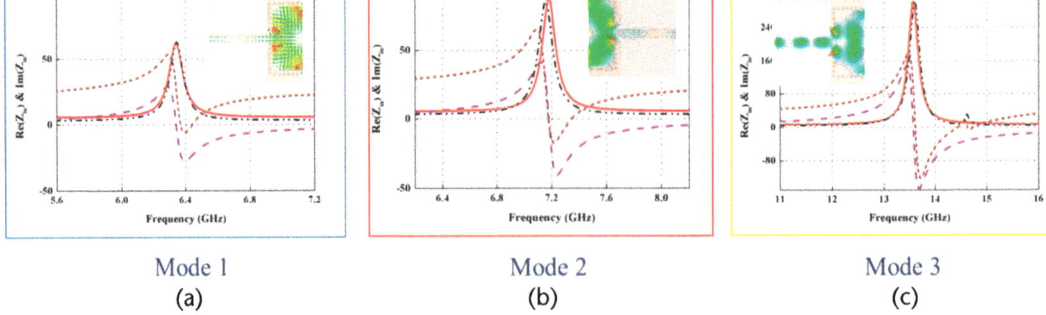

Mode 1 (a) Mode 2 (b) Mode 3 (c)

Figure 3.19 Equivalent model of cavity modes. (© 2017 Wiley [4].)

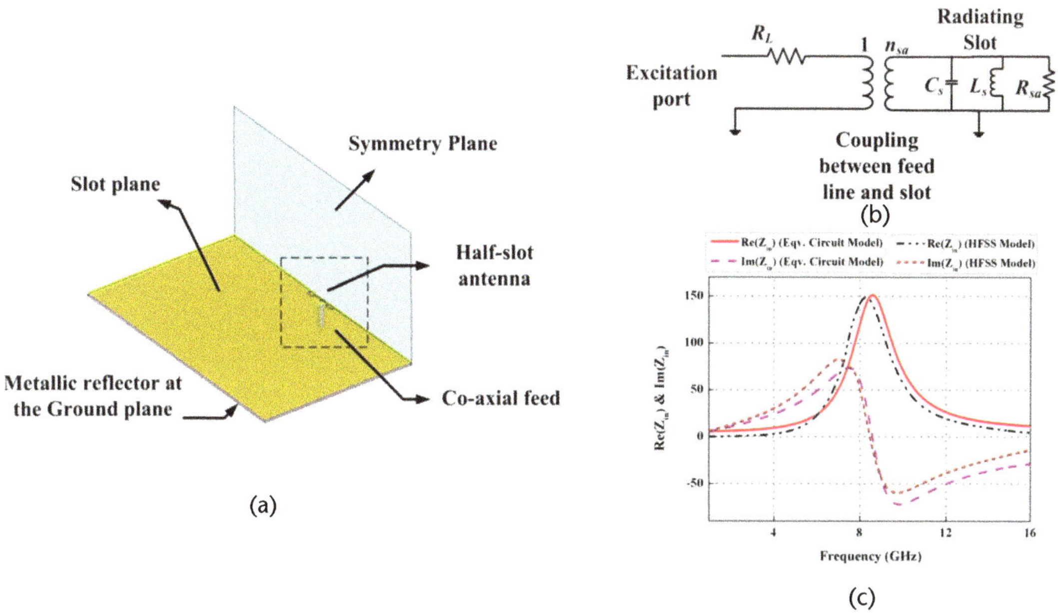

Figure 3.20 Equivalent model for slot antenna with reflector. (© 2017 Wiley [4].)

parameters and also to understand the crucial dimensions, which are sensitive to the performance of the antenna. The experimental results of both dual-band and triple-band antennas are shown in Figure 3.23. The measured results of the dual-band antenna exhibit at 9.5 GHz and 13.85 GHz with the measured impedance bandwidth of 173 MHz (1.82%) and 200 MHz (1.44%), respectively. The measured gain of the antenna at two resonant frequencies are 4.8 dBi and 3.74 dBi with unidirectional radiation pattern due to SIW cavity backing. The measured FTBR of the antenna is 10 to 12 dB at both operating frequencies. The measured results of the triple-band antenna are shown in Table 3.1. The triple-band resonances are achieved at the measured gain of the antenna at three resonances are 3.21 dBi, 4.85

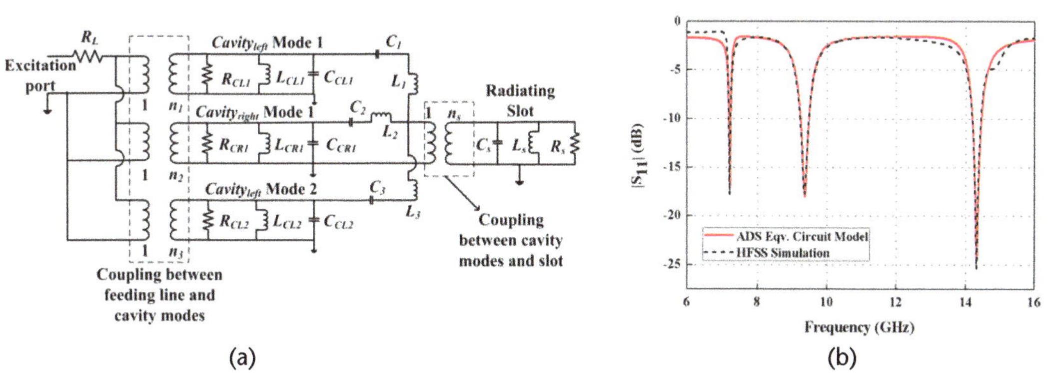

Figure 3.21 Equivalent model of triple band antenna ($n_1 = 1.79$, $n_2 = 0.897$, $n_3 = 3.96$, $n_S = 1.96$, $R_L = 4.68\Omega$, $R_{CL1} = 0.253$ kΩ, $R_{CR1} = 3.02$ kΩ, $R_{CL2} = 35.23$ kΩ, $R_S = 3.225$ kΩ, $L_{CL1} = 0.7$ nH, $L_{CR1} = 1.13$ nH, $L_{CL2} = 8.1$ nH, $L_1 = 2.45$ nH, $L_2 = 1.47$ nH, $L_3 = 1.99$ nH, $L_S = 1.72$ nH, $C_{CL1} = 0.9$ pF, $C_{CR1} = 0.435$ pF, $C_{CL2} = 0.017$ pF, $C_1 = 0.251$ pF, $C_2 = 0.249$ pF, $C_3 = 0.402$ pF, $C_S = 0.2$ pF). (© 2017 Wiley [4].)

dBi, and 4.7 dBi, respectively. The measured cross-polar component is below −20 dB at all three frequencies in the broadside direction. The measured front-to-back ratio of the antenna at three frequencies are 15 dB. The triple band antenna covers C-, X-, and Ku-band response with desired unidirectional radiation pattern, which is achieved by utilization of SIW cavity modes.

The multiband response can also be achieved using many other slot dimensions or shapes as long as it can disturb the cavity surface current distribution to generate additional hybrid mode or it can sufficiently disturb the higher-order cavity mode to excite the slot antenna. In this section, one additional example of such slot antenna is shown in Figure 3.24. Here the rectangular slot has two additional slot stub inclined at an angle 45° [6]. The central radiating slot is placed at the center of the cavity to disturb the current path of higher-order TE^{y}_{210} mode and the slot stub has minimal effect on this mode. However, at higher frequency at 15 GHz, the slot stub majorly disturbs the surface current distribution to generate an additional hybrid mode, which also helps in radiation. A dual-band response at 8.96 GHz and 15.84 GHz with simulated impedance bandwidth (−10-dB S_{11}) of 240 MHz (2.67%)

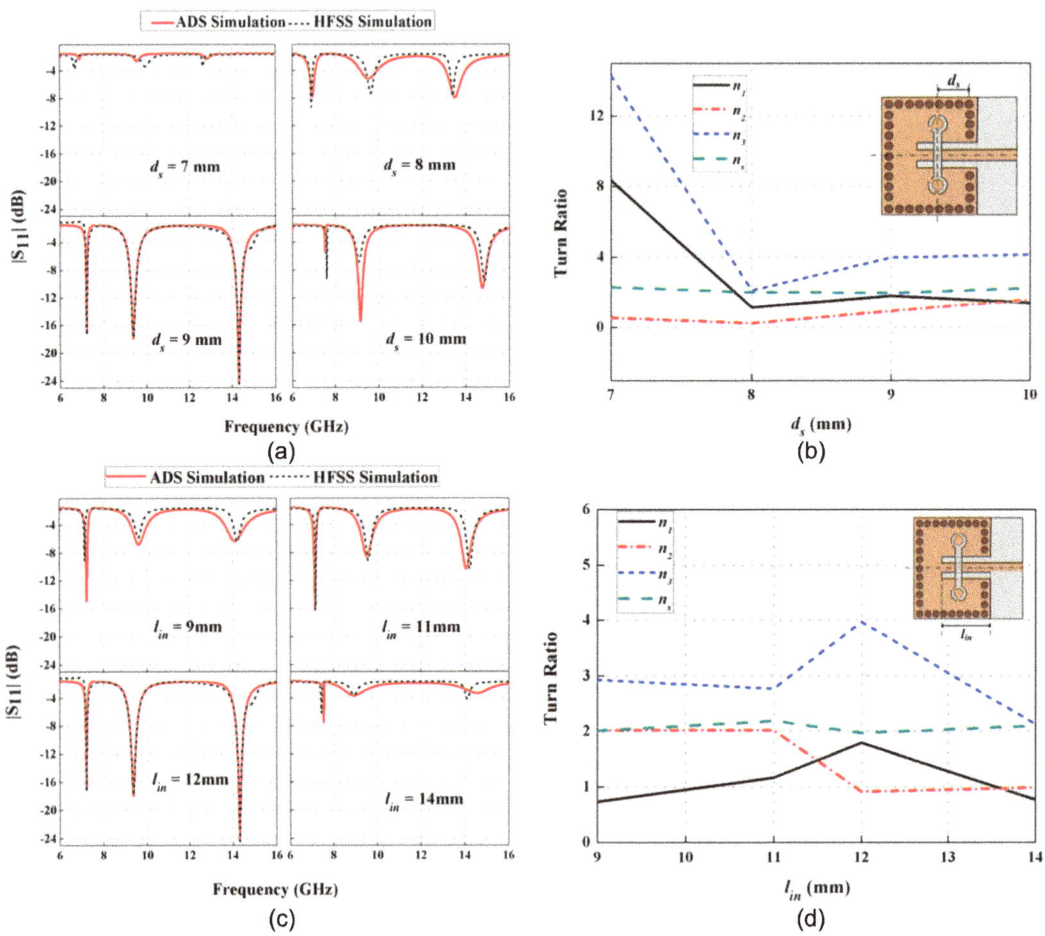

Figure 3.22 Comparison of equivalent model performance with simulation model performance. (© 2017 Wiley [4].)

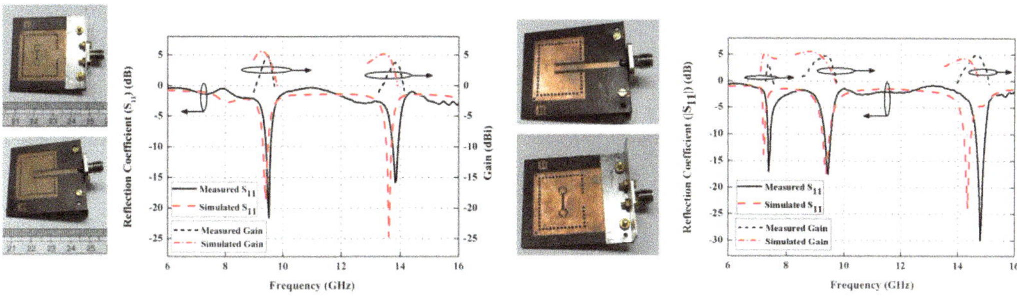

(a) (b)

Figure 3.23 Experimental result of: (a) dual-band antenna; and (b) triple-band antenna. (© 2017 Wiley [4].)

and 300 MHz (1.89%) is achieved using the same RT 5880 substrate-based SIW cavity backing. The gain of the antenna at both bands are 4 to 5 dB with FTBR in the range of 15 dB. The radiation pattern at both bands are shown in Figure 3.25. There are many more design examples to generate multiband response. The interested readers may explore all such design examples to get overall idea about multiband SIW cavity-backed slot antenna.

3.3 Design of Broadband Antenna

In the previous section, techniques to generate additional hybrid resonances in SIW cavity to achieve multiband radiation performance is discussed. The loading of different shapes and dimensions of the slot antenna at the upper plate of the SIW cavity helps in different ways to generate modes, which are not conventionally present in the cavity. In this section, techniques to achieve broadband response will be discussed. The technique majorly focuses on generation of multiple resonances in the close frequency range so that the combined response of all the resonances produces a broadband response. The technique referred to can be as similar to the well-known stagger tuning technique where multiple resonant circuits can be coupled to have a combined response. In this section, a few examples will be discussed to generate more than one resonance within a close frequency range to achieve a broadband response.

Table 3.1 Comparative Study of Performance of Equivalent Circuit Model with Simulated and Measurement Result of the Design

Properties		*Equivalent Circuit Model*	*Simulated Result*	*Experimental Result*
Resonant frequency	Lower band (f_1)	7.2 GHz	7.2 GHz	7.39 GHz
	Middle band (f_2)	9.36 GHz	9.36 GHz	9.43 GHz
	Upper band (f_3)	14.30 GHz	14.30 GHz	14.79 GHz
Bandwidth	Lower band (f_1)	56 MHz (0.77%)	60 MHz (0.83%)	100 MHz (1.35%)
	Middle band (f_2)	250 MHz (2.67%)	250 MHz (2.67%)	220 MHz (2.33%)
	Upper band (f_3)	245 MHz (1.71%)	250 MHz (1.74%)	330 MHz (2.23%)

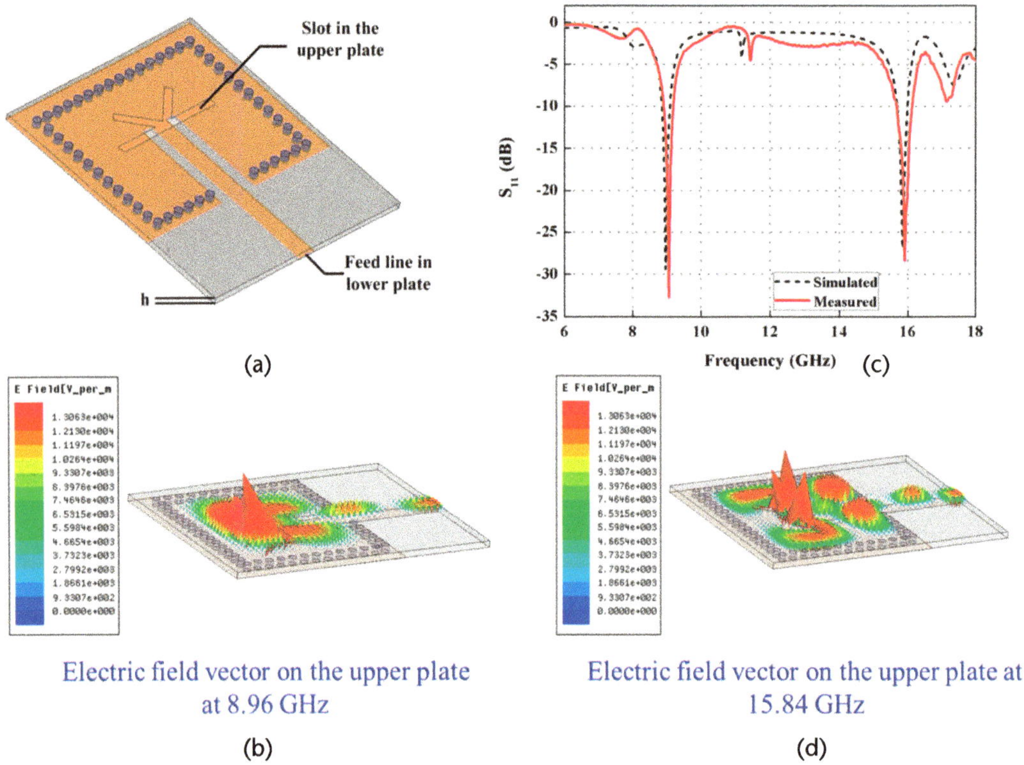

Figure 3.24 Design example 2 of dual-band SIW CBSA. (© 2014 IEEE [6].)

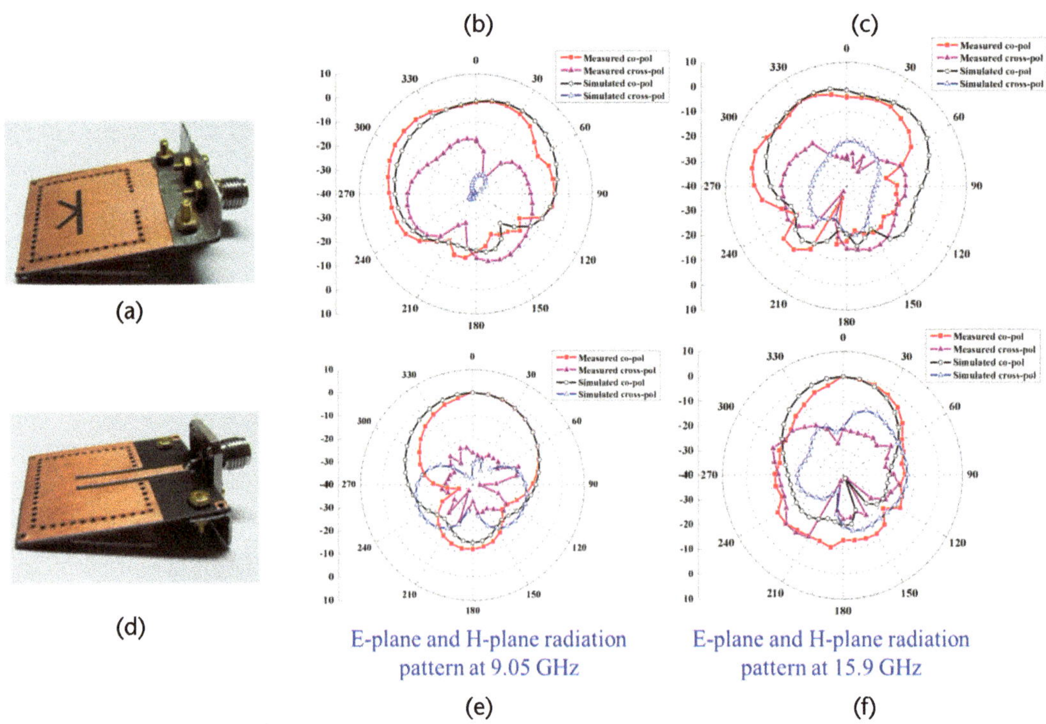

Figure 3.25 Radiation pattern of dual-band SIW CBSA Design 2. (© 2014 IEEE [6].)

The first design is shown in Figure 3.26, where a bow-tie slot is placed in the upper plate of the SIW cavity [7]. The length of the slot and the flaring angle are kept comparatively larger than the conventional half-wavelength slot antenna dimension. The larger dimension of the slot helps to have a larger loading effect in the cavity modes. The SIW cavity dimensions are chosen to have its consecutive first two resonances at 9.2 GHz (TE^y_{110}) and 14.7 GHz (TE^y_{210}). The loading effect of the slot can be seen in the input impedance plot as shown in Figure 3.27. The loaded cavity has its first two resonances close to 10 GHz. The current distribution in the lower frequency is dominant in the right side of the cavity, whereas, in the case of higher frequency, the dominant field distribution exists on the left side of the cavity. In both cases, the magnitude and phase of the electric fields are different at opposite sides of the slot, which helps it to radiate. By referring to the common equivalent circuit model in Figure 3.21, it can be stated that the two RLC circuits have been efficiently tuned to resonate at close frequency range to have broadband response. The dimensions of the bow-tie slot have significant effects on the two resonances of

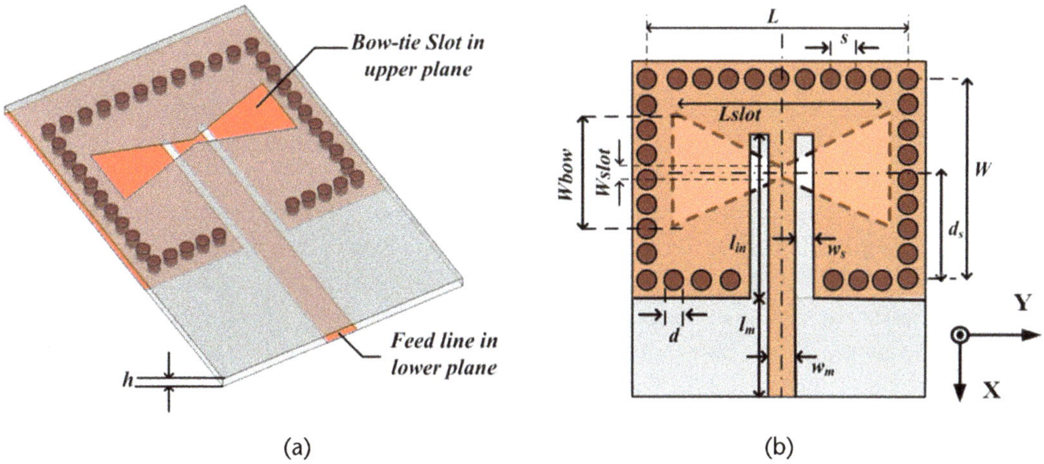

Figure 3.26 Design of SIW cavity-backed broadband slot antenna. (© 2014 IEEE [7].)

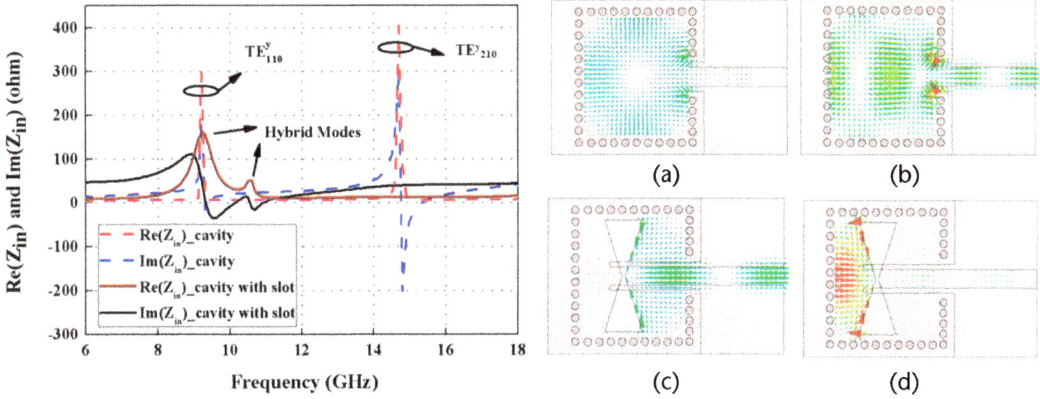

Figure 3.27 Input impedance plot and surface current distribution of SIW cavity-backed bow-tie slot antenna. (© 2014 IEEE [7].)

the antenna. The flaring of the bow-tie slot can be varied to tune the higher-frequency resonance as shown in Figure 3.28. However, as the slot length increases the lower frequency hybrid mode becomes more prominent. The measured bandwidth of the proposed antenna is 1.03 GHz (9.43%). The measured gain of the antenna over the bandwidth is between 3.5 and 5 dBi with FTBR in the range of 15 to 20 dB.

To achieve further increment in the bandwidth, we can combine a pair of SIW cavity-backed slot antenna to have more than two resonances. The idea is to have multiple resonances in close frequency range to increase the bandwidth. To have a pair of additional resonances at higher frequencies, another SIW cavity with slightly smaller dimension is chosen as shown in Figure 3.30. The resonant frequency (f_{mnp}) of the TE^{y}_{mnp} mode of the proposed cavity can be formulated as

$$f_{mnp}^{SIW} = \frac{1}{2\pi\sqrt{\mu\varepsilon}}\sqrt{\left(\frac{m\pi}{L_{eff}^{SIW}}\right)^2 + \left(\frac{n\pi}{R \times L_{eff}^{SIW}}\right)^2 + \left(\frac{p\pi}{h}\right)^2} \quad (3.6)$$

where $0.85 < R = W_{eff}^{SIW}/L_{eff}^{SIW} < 1$.

The ratio of width to length of the cavity is kept within the optimum limit, which ensures the excitation of dominant TE^{y}_{110} mode and higher-order TE^{y}_{210} mode in close proximity. However, when loaded with long ($\approx 0.65\lambda_0$) bow-tie shaped slot, the resonant frequency of TE^{y}_{210} mode decreases to interact with the dominant TE^{y}_{110} mode.

The pair of cavities can be fed by single feed line if a proper coupling mechanism can be established between these two cavities. To achieve this, the two SIW cavities are placed side by side such that they share a common side wall [8]. A H-plane coupling window is created in the common sidewall by omitting some portion of the sidewall vias. Thus, the feed line can be kept inset to Cavity 1 only and the Cavity 2 can get the desired excitation from the same feed line using the coupling window. The excitation of different cavity modes can be explained from the input impedance plot. The dominant TE^{y}_{110} and higher-order TE^{y}_{210} mode of Cavity 1 resonates at 8.5 GHz and 13.77 GHz where field is concentrated mostly in Cavity

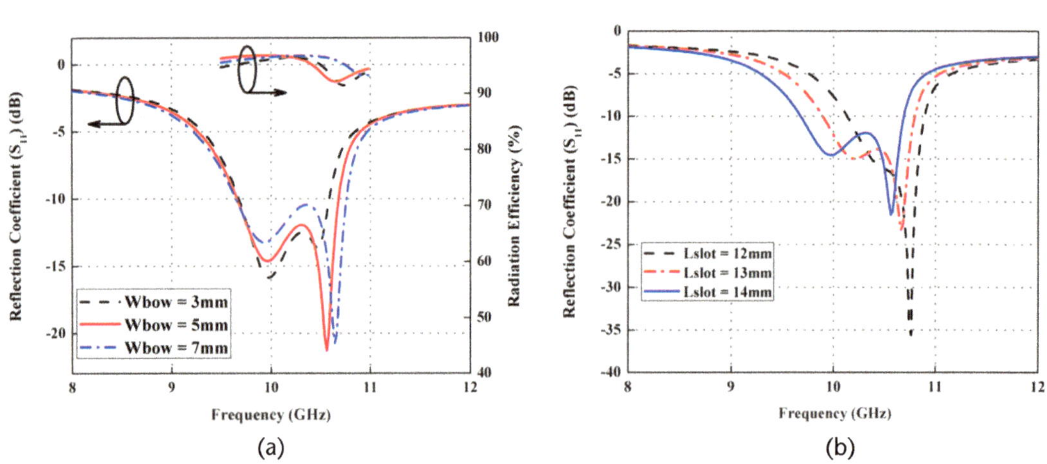

Figure 3.28 Parametric analysis of SIW cavity-backed bow-tie slot antenna. (© 2014 IEEE [7].)

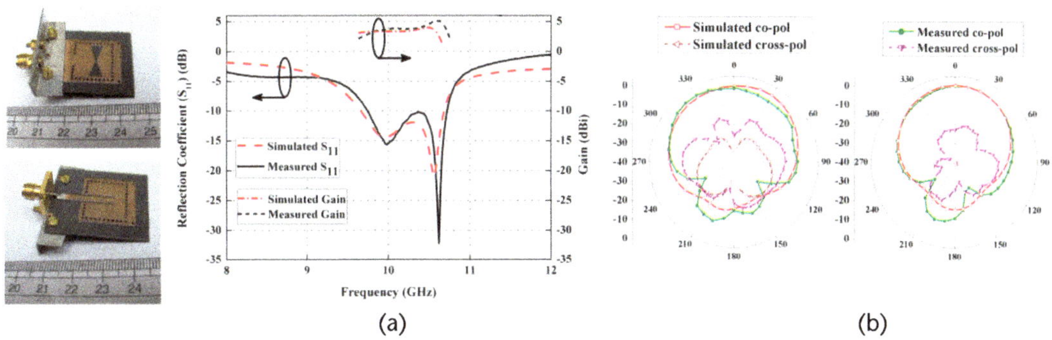

Figure 3.29 (a) Comparison of simulated and measured result. (b) Radiation pattern at 9.98 GHz. (© 2014 IEEE [7].)

1 with negligible effect of Cavity 2. Similarly, Cavity 2 also resonates almost independently at the TE^{y}_{110} and TE^{y}_{210} modes at 9.8 GHz and 16.57 GHz, respectively, with field concentrated mostly in Cavity 2 as shown in Figure 3.31. However, there is a distorted TE^{y}_{120}-like mode at 15.1 GHz, which exists in both Cavities 1 and 2.

Now a pair of bow-tie slots are placed in the corresponding cavities. Here the target is to generate similar effect as that of the previous design and generate a pair of hybrid modes in the corresponding cavities. However, additionally, it is also required that all the hybrid cavity modes should be resonating close enough so that in effect a wideband response can be achieved. By properly optimizing the cavity and slot antenna dimensions, three hybrid modes of the proposed antenna are tuned to resonate at 8.1 GHz, 9.55 GHz, and 10.41 GHz, respectively, as shown in Figure 3.31. The hybrid mode at 8.1 GHz is concentrated, mostly at right half of Cavity 1 with a negligible field in the other half. The surface current distribution of the next two hybrid modes at 9.55 GHz and 10.41 GHz concentrates in right and left halves of Cavity 2, respectively. However, placement of slot pair in the coupled cavity system makes it comparatively difficult to control the cavity modes. Although the feed line is aligned to excite only TE^{y}_{m10} mode, the large perturbation due to placement of bow-tie slot pair creates possibility of excitation of orthogonal TE^{y}_{120} mode. To prevent the unwanted mode, a shorting pin is placed

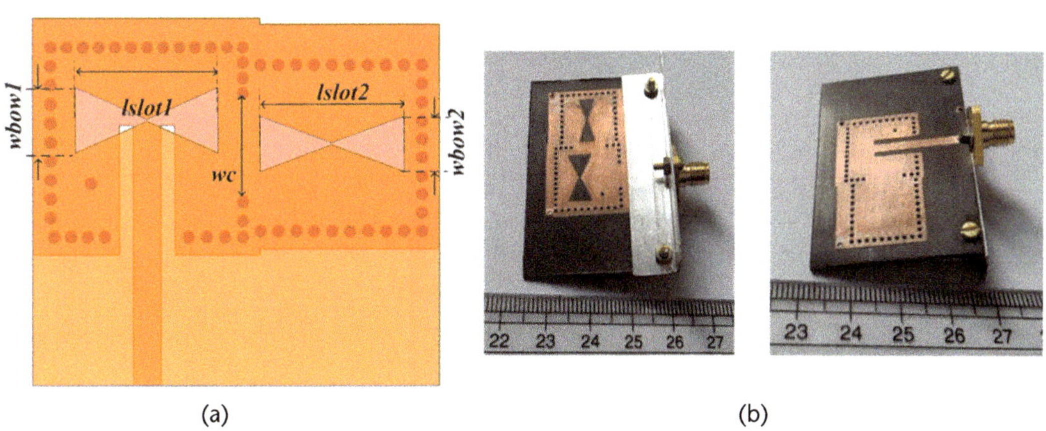

Figure 3.30 Design of wideband SIW cavity-backed slot antenna. (© 2016 IEEE [8].)

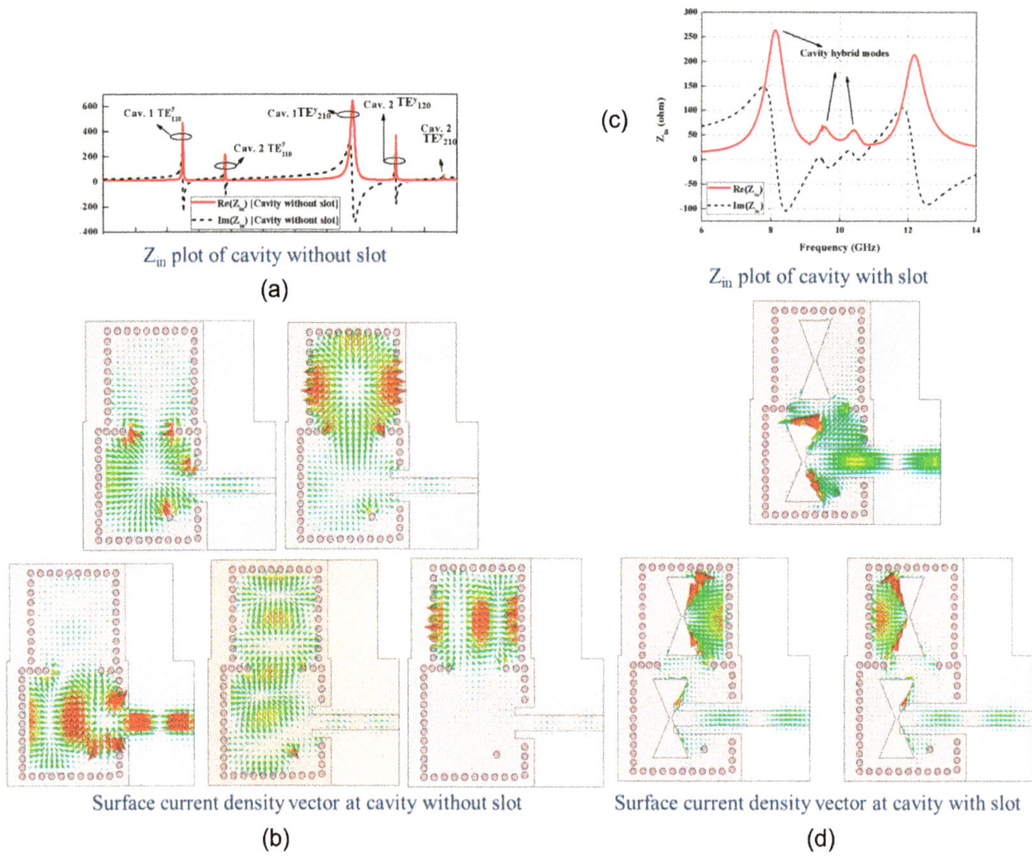

Figure 3.31 Comparison of unloaded and loaded SIW cavity-backed wideband slot antenna and surface current plot. (© 2016 IEEE [8].)

approximately at the position where the maximum of electric field of the unwanted mode is expected. However, the placement of the via will also affect the performance of the slot antenna. Hence, optimization on the via position is required with the help of simulation solver. Similar to the previous design, the slot antenna dimensions (length and flaring of both slots) affect the resonant frequencies of the hybrid modes. The additional degree of freedom in this design is the width of coupling window, which affects the impedance matching and the bandwidth of the antenna. When the coupling window is very small (wc = 7 mm), the Cavity 1 and Cavity 2 hybrid modes resonate at different frequencies and, as a result, the bandwidth of the design remains small. However, with increase of the coupling window, the loading effect between two cavities results in shifting of hybrid-mode resonant frequencies and, as a result, for wc = 9 mm, three hybrid modes resonate within the operating frequency band to give wideband performance of the antenna. However, further increase of the length of coupling window will detune the modes and the bandwidth of the proposed design will degrade.

The experimental verification of the design in shown in Figure 3.32. The antenna shows wideband response from 9.22 to 10.66 GHz with a bandwidth of 1.44 GHz

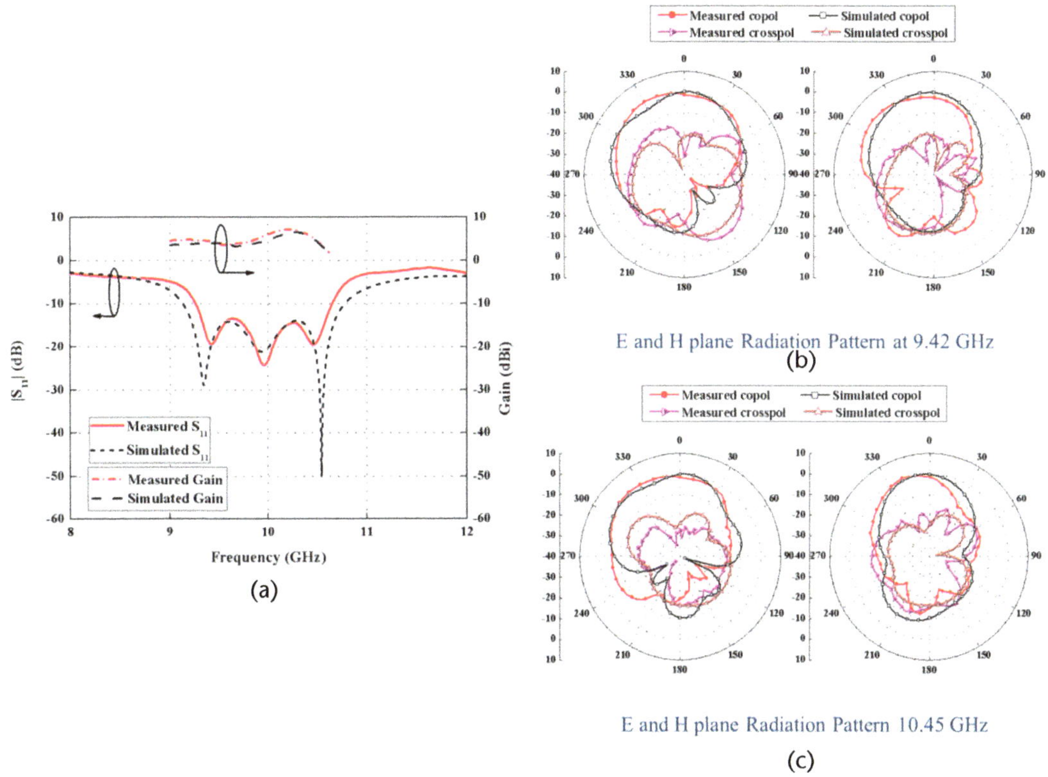

Figure 3.32 Performance of wideband SIW cavity-backed slot antenna. (© 2016 IEEE [8].)

(14.4%). The measured gain of the antenna is also very close to the simulated result with a gain above 4 dBi throughout the operating band of the antenna. The maximum measured gain of the antenna is 6.6 dBi at 10.2 GHz. The FTBR in the operating band is better than 15 dB.

Another variation of the design can be achieved by changing the coupling window width between the two cavities. A design example is shown in Figure 3.33, where the coupling window is much wider than the previous design [9]. As a result, a total of four hybrid modes could be excited in the antenna. The modes excited in the cavity pair can be explained by input impedance plot. The resonant frequencies of the consecutive modes are 7.33 GHz, 9.36 GHz, 12.33 GHz, and 15.45 GHz, respectively. The electric field distribution of the SIW cavity with and without loading of the slot is shown in Figure 3.34. The fields of the first and third modes correspond to Cavity 1 and the other two are concentrated in Cavity 2. However, the wide coupling slot in between the two cavities makes one side of the cavity almost disappeared, which distorts the field distribution to some extent. The bow-tie-shaped slot loading in the cavity generates the hybrid mode as similar to previous designs. However, contrary to the previous design, here, four hybrid modes could be excited with distinctive resonant frequencies at 7.68 GHz, 8.17 GHz, 9.22 GHz, and 10.20 GHz, respectively. The first two hybrid modes present electric field concentration majorly in the upper and lower halves of Cavity 1 and the electric field of last two

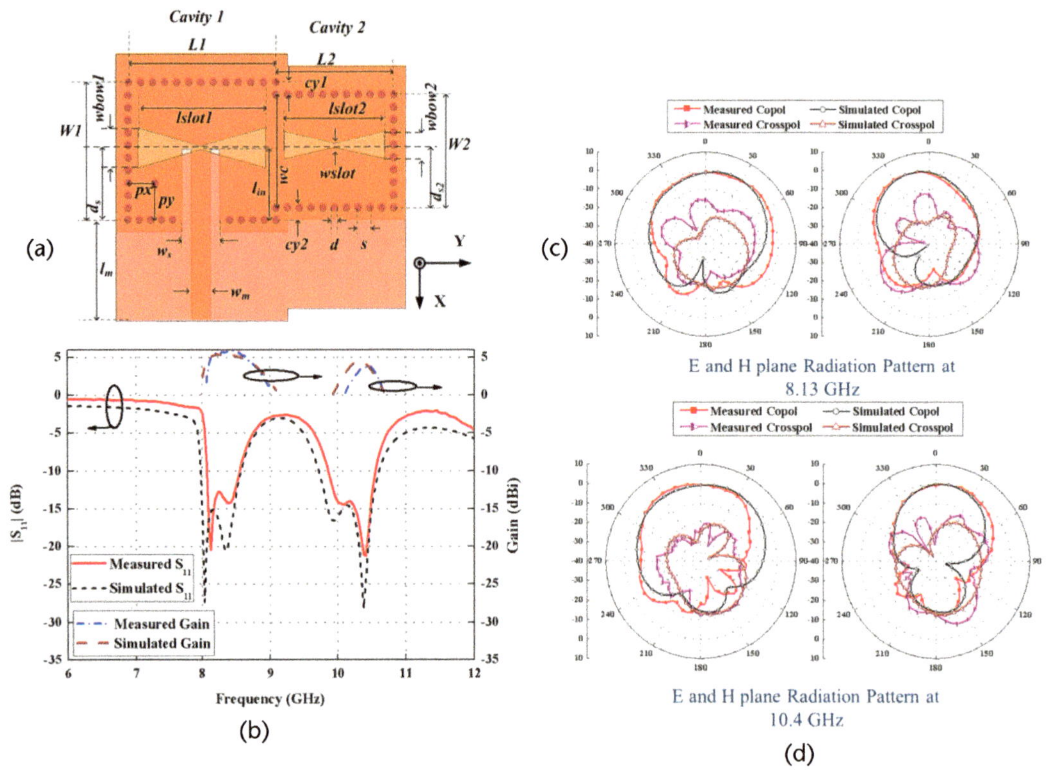

Figure 3.33 Design of dual broadband SIW cavity-backed slot antenna. (© 2016 IET [9].)

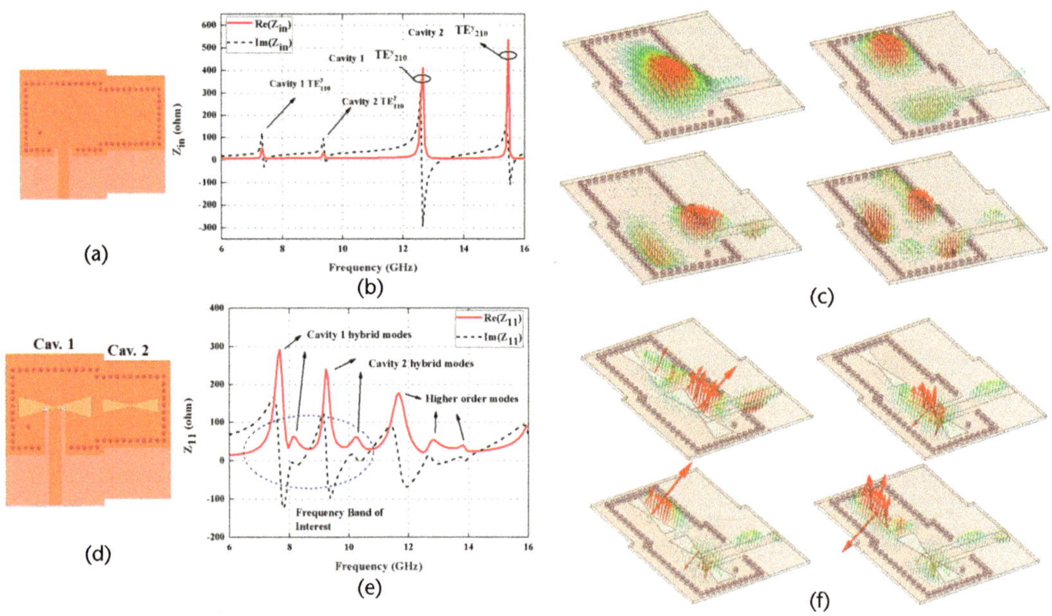

Figure 3.34 Input impedance plot of SIW cavity-backed dual-broadband slot antenna. (© 2016 IET [9].)

modes are concentrated on Cavity 2. The measured response of the antenna shows dual-band response from 8.07 to 8.58 GHz and 9.84 to 10.56 GHz with a bandwidth of 510 MHz (6.1%) and 720 MHz (7%), respectively. The measured gain of the antenna is 4.8 to 5.7 dBi and 3 to 3.9 dBi, respectively. The simulated and measured FTBR of the antenna is −15 dB and −12 dB, respectively.

A couple of design techniques to have broadband response in the SIW cavity-backed slot antenna were discussed in this section. However, many other techniques to generate additional resonance and to improve bandwidth performance of the antenna are available in the literature, which were not included here for limited scope of the book. Interested readers can explore these techniques to have holistic ideas about all available techniques of the broadband SIW antenna.

3.4 Design of the Dual-Polarized Antenna

In the previous section, the technique to achieve multiband and broadband performance using different SIW cavity modes are discussed in details. In this section, utilization of SIW cavity modes for dual polarization operation will be explored in details. Two design examples will be described here to generate dual-polarized radiation using two different techniques: (1) using orthogonal SIW cavity modes, and (2) using orthogonal orientation of the cavity-backed slot.

3.4.1 Dual-Polarization Antenna Using the Orthogonal Cavity Mode

In this technique, the consecutive orthogonal SIW cavity modes (say, TE^y_{210} and TE^y_{120}) are used to excite a pair of slot antenna which are also oriented in orthogonal orientation. The radiated electric field from each of these slots produce orthogonal polarization giving the desired dual-polarized response. As shown in Section 2.3, the resonant frequencies of these orthogonal modes can be tuned by varying the length-to-width ratio of the SIW cavity. Thus, the dual-polarized radiation can be achieved with its unidirectional radiation pattern due to cavity backing by choosing proper dimension of the cavity. Here, a design example of dual-polarized SIW cavity-backed slot antenna for polarization diversity application is discussed [10].

In case of polarization diversity application, the dual polarization is achieved at the same resonant frequency for the two antennas. Hence, degenerate orthogonal modes of the SIW cavity need to be used for this purpose. As shown in (2.16), the desired SIW cavity dimension to have degenerate orthogonal modes will require same length and width of the cavity. Hence, a square SIW cavity is chosen for this purpose. The first two higher-order modes TE^y_{210} and TE^y_{120} resonates at 10 GHz. To excite both the modes, two dedicated feeding ports followed by two microstrip feed line are inset into the cavity along the length and width, respectively, as shown in Figure 3.35. To generate radiation, a crossed slot configuration is placed at the top plate of the cavity. Since dual-polarized radiation is required at the same operating frequency, the length of both the slot is kept at half-wavelength. While one of the slot is expected to get excited by TE^y_{210} mode, the other one will be excited by TE^y_{120} mode.

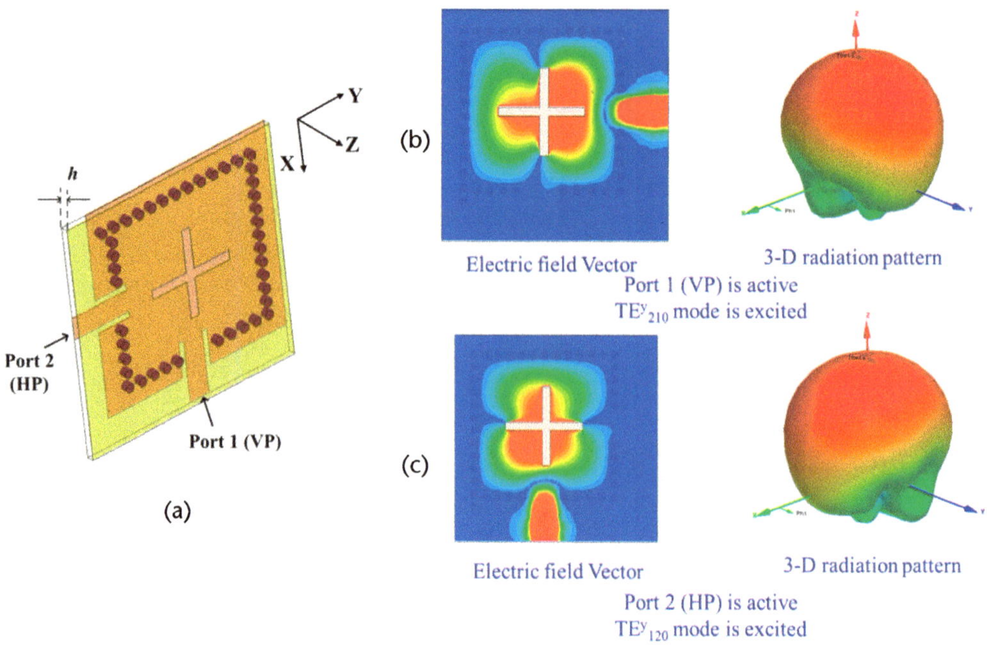

Figure 3.35 SIW cavity-backed, dual-polarized crossed slot antenna. (© 2015 IEEE [10].)

Each of the two slots can be placed at the middle plane of the cavity where the electric field minima is observed. However, the loading of crossed slot at the top plate of the SIW cavity perturbs the boundary condition, which results in a slight perturbation in field distribution as well as a shift in resonant frequency. To account for the perturbation of the field distribution, the position of the slots need to be tuned with help of EM simulator to get optimum impedance matching. Once the matching is achieved, the tuning of the slot-loaded cavity resonant frequency can be achieved by varying the length of the corresponding slot. A small inset of both the feed lines are sufficient to get the impedance matching of the antenna. The out-of-phase electric field of TE$^y_{210}$ mode at both sides of slot 1 excites it to radiate and hence the antenna produces a vertically polarized (VP) radiation pattern when it is excited through Port 1. The same phenomenon is observed for slot 2 due to the TE$^y_{120}$ mode to produce a horizontally polarized (HP) radiation pattern when fed by Port 2, with Port 1 terminated with matched load. In this manner, orthogonal orientation of radiated electric field is maintained by the crossed slot. Most of the input power is radiated through the slots with very little amount of power to be coupled in the other port, which helps to enhance the isolation between the input ports. The antenna resonates at 9.98 GHz with approximately 2% bandwidth and a gain of 4.2 dBi. The details of the simulated and measured results can be seen in Table 3.2. Due to SIW cavity backing, the radiation pattern shows unidirectional phenomenon while maintaining polarization diversity with dedicated feeding ports for both polarizations.

The design is shown in Figure 3.38. Here, four total SIW cavities are arranged in two pairs where one pair of cavities are arranged in 45° orientations and the other pair in −45° orientation [11]. Just like previous dual broadband design in Section 3.3, the coupling window between the two cavities in a pair are much wider

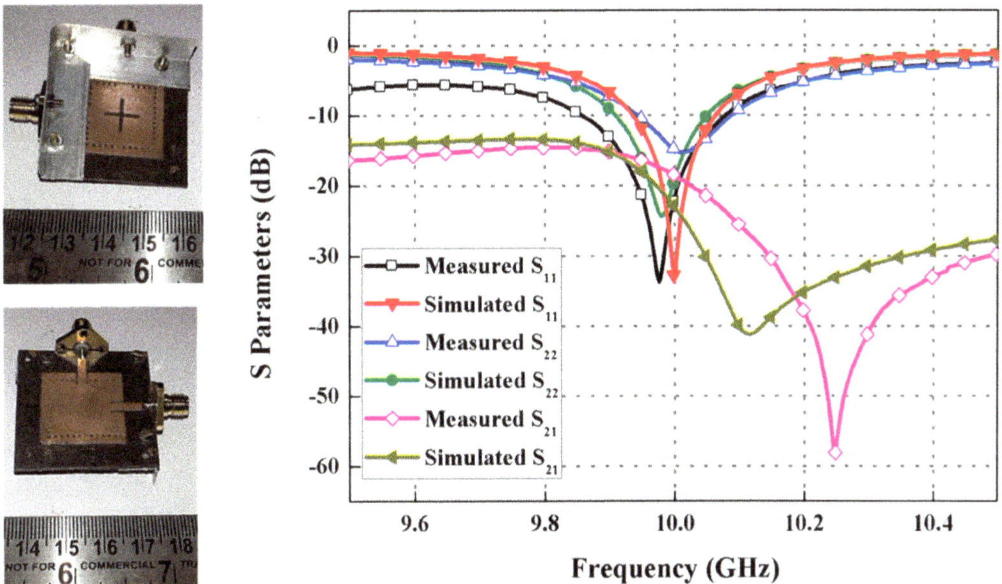

Figure 3.36 Comparison of simulated and measured result of dual-polarized SIW cavity-backed crossed slot antenna. (© 2015 IEEE [10].)

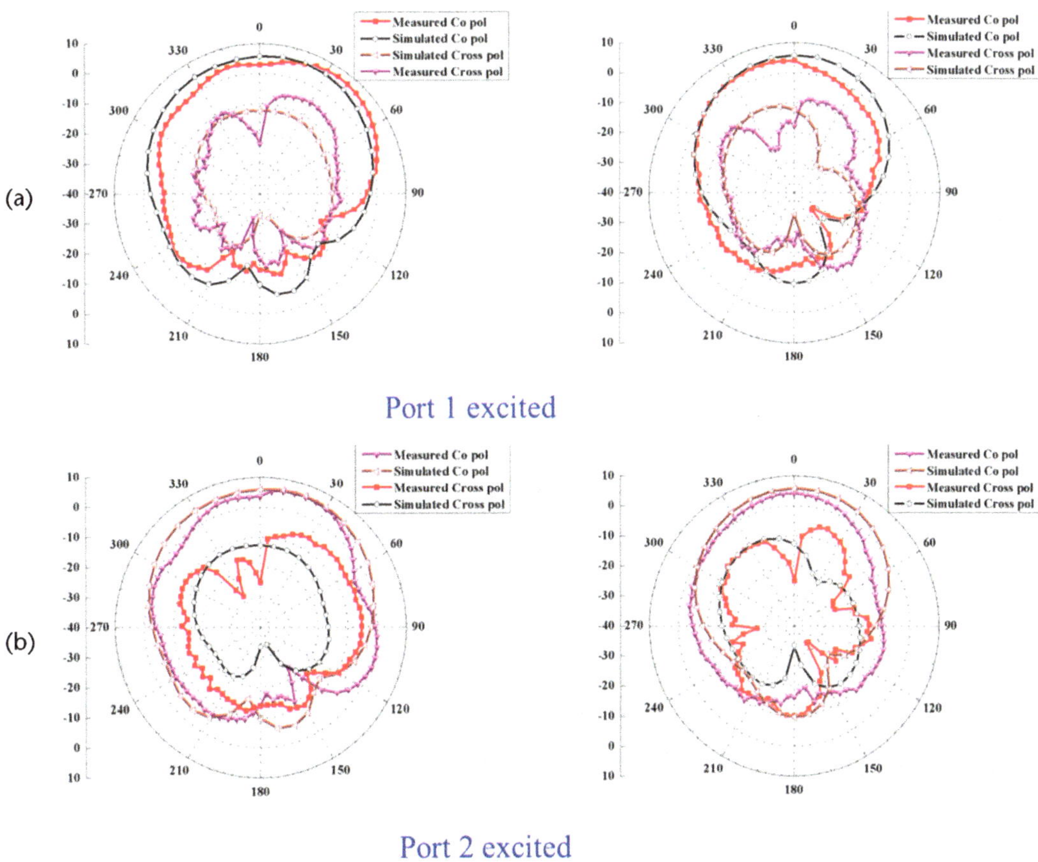

Figure 3.37 Radiation pattern for Port 1 and 2 excitations of the SIW cavity-backed crossed slot antenna. (© 2015 IEEE [10].)

Table 3.2 Comparison Between Simulated and Measured Result

		Port 1	*Port 2*
Frequency	Simulated	9.98 GHz	9.98 GHz
	Measured	9.97 GHz	10 GHz
Bandwidth	Simulated	160 MHz (1.6%)	160 MHz (1.6%)
	Measured	170 MHz (1.7%)	190 MHz (1.9%)
Isolation	Simulated	−20 dB	
	Measured	−18 dB	
Polarization		Vertical	Horizontal
Gain	Simulated	4.3 dBi	4.2 dBi
	Measured	4 dBi	3.8 dBi

to ensure excitation of all possible hybrid modes to achieve optimum bandwidth. The additional part of the design is inclusion of SIW power divider network in the feed section. The feeding port of the antenna is connected to a small microstrip line section, which is used to excite the SIW section by a similar transition as shown in Section 2.2. The power divider is unique in a sense that its output ports are orientated in ±45°. Therefore, the impedance matching at the three-port junction needs to be designed accordingly. The power dividing network is special in a sense since it is meant to direct the EM wave towards Cavity 1 and Cavity 2 only and not the other pair for lower band. The vice versa effect is expected for the upper band. Thus, the feed network also facilitates a frequency diplexing operation and therefore the impedance matching at the input port needs to take account of the design. Three vias are placed at the optimum position in the three-port junction. These vias are orientated such a way that the center and left sides via form a coupling window dedicated to lower band and the center and the right side via form a coupling window dedicated to the upper band. Essentially, these vias ensure directing the EM wave at the lower band towards Cavities 1 and 2 and vice versa for the upper band. Therefore, their location is very important and needs to be optimized with the help of EM simulation software.

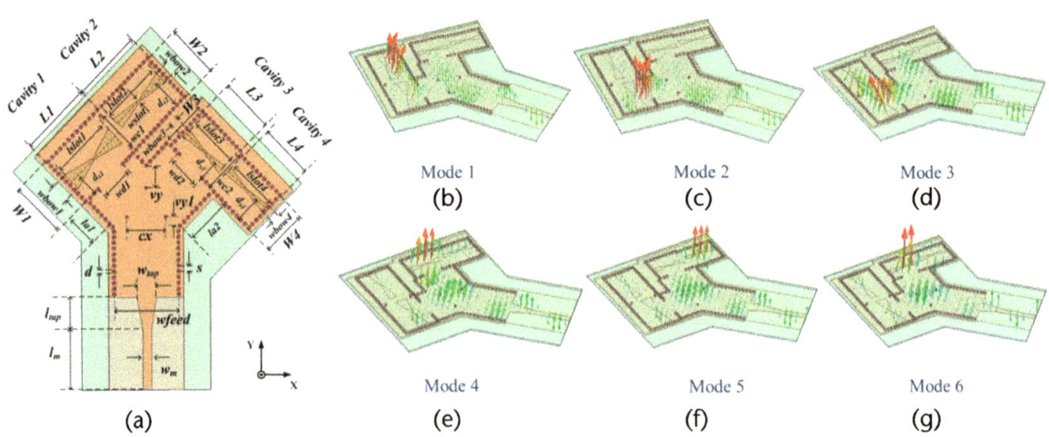

Figure 3.38 Dual-polarized SIW cavity-backed slot antenna design example 2 and its modes. (© 2016 IET [9].)

Two pairs of bow-tie-shaped slots are placed at the top metallic plate of both cavity pair as per similar design procedure explained in (3.6). As a result, a pair of three hybrid modes are excited. It is to be noted that to suppress unwanted orthogonal mode, mode suppressing vias are used in both the cavity pair as well. The electric field vector of first three modes at 7.66 GHz, 7.8 GHz, and 8.07 GHz are mostly concentrated in the Cavity 2 lower half, Cavity 1 lower half, and Cavity 1 upper half, respectively. The next three higher-frequency hybrid mode electric field vector at 9.17 GHz, 9.58 GHz, and 10.07 GHz are mainly concentrated in Cavity 3 upper half, Cavity 4 lower half, and Cavity 3 lower half, respectively. Typically, the coupling window-based excitation exhibits a band-limited response. As a result, in this design, it appears difficult to achieve four hybrid modes in contrast to designs explained in Section 3.3.

The difference of the magnitude of the electric field in the opposite side of the slot will ensure radiation at the hybrid-mode resonant frequencies. In addition, the

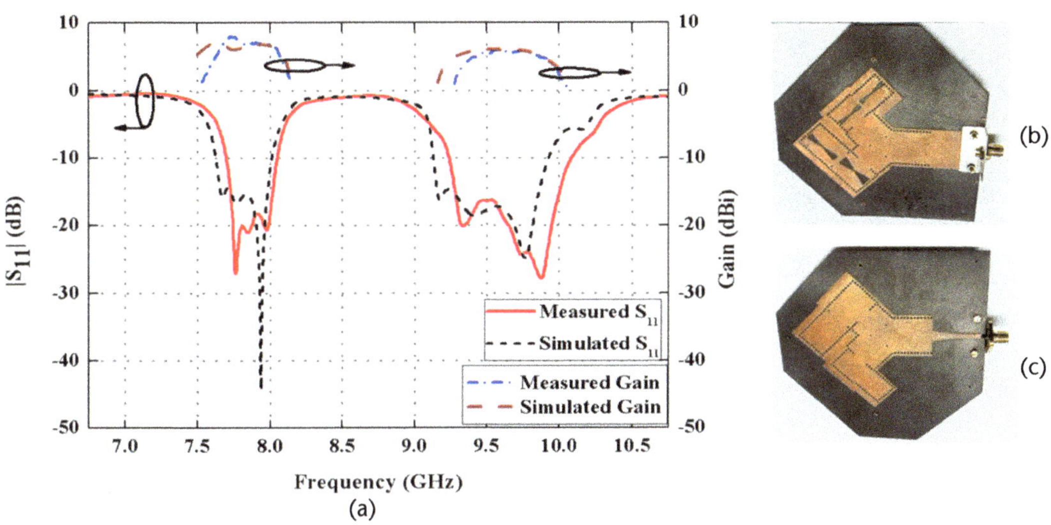

Figure 3.39 Performance of the antenna. (© 2016 IET [9].)

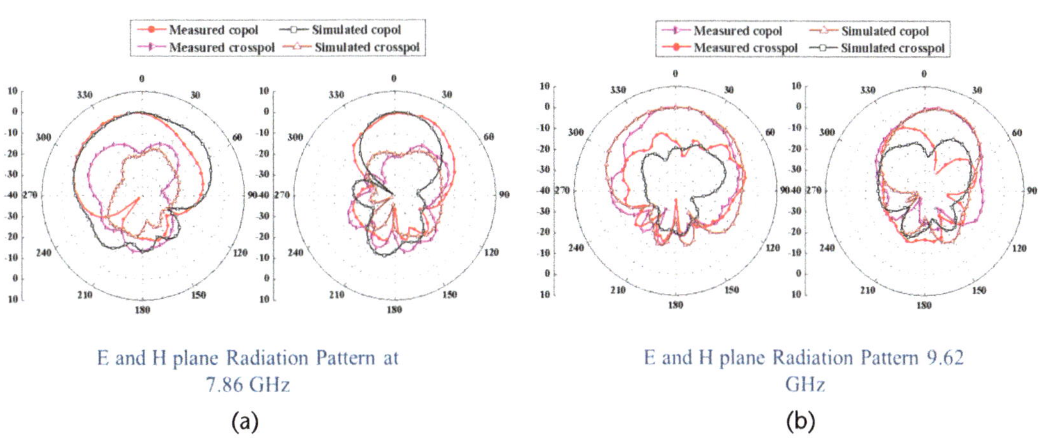

Figure 3.40 Radiation pattern of dual-polarized dual-broadband antenna. (© 2016 IET [9].)

orthogonal orientation of the cavity pair will ensure dual-polarized radiation at ±45° polarization. Thus, a dual-broadband dual-polarized radiation is achieved in this design. The antenna exhibits 4.5% and 7.9% bandwidth in lower and upper bands with 4.7 to 6.7-dBi gain over both the bands. The radiation pattern remains unidirectional due to SIW cavity backing. Thus, the design produces some unique results by exploiting the inherent properties of SIW technology.

The designs presented in this chapter are majorly oriented towards some basic approaches to achieve broadband and multiband unidirectional radiation using SIW cavity modes. However, several other techniques such as using the multilayer cavity, HMSIW, QMSIW, and the combination of SIW and other types of resonator (e.g., microstrip patch) have been explored by researchers as well. All such techniques essentially try to have multiresonance response by implementing efficient feeding mechanism. The interested readers are encouraged to study the other available techniques to have broader picture about SIW cavity-backed slot antenna with enhanced performance.

References

[1] Mukherjee, S., A. Biswas, and K. V. Srivastava, "Bandwidth Enhancement of Substrate Integrated Waveguide Cavity Backed Slot Antenna by Offset Feeding Technique," *IEEE Applied Electromagnetics Conference (AEMC)*, Bhubaneswar, India, 2013.

[2] Mukherjee, S., A. Biswas, and K. V. Srivastava, "Substrate Integrated Waveguide Cavity Backed Slot Antenna with Parasitic Slots for Dual-Frequency and Broadband Application," *9th European Conference on Antennas and Propagation (EuCAP)*, Lisbon, Portugal, 2015.

[3] Mukherjee, S., A. Biswas, and K. V. Srivastava, "Substrate Integrated Waveguide Cavity-Backed Dumbbell-Shaped Slot Antenna for Dual-Frequency Applications," *IEEE Antennas and Wireless Propagation Letters*, Vol. 14, 2015, pp. 1314–1317.

[4] Mukherjee, S., and A. Biswas, "Computer Aided Equivalent Circuit Model of SIW Cavity Backed Triple Band slot Antenna," *Int. J. RF Microw. Comput. Aided Eng.*, Vol. 27, 2017.

[5] Long, S., "The Impedance of a Cavity Backed Slot Antenna," *Proc IEEE Antennas and Propagation Society Int. Symp.*, 1973.

[6] Mukherjee, S., A. Biswas, and K. V. Srivastava, "Substrate Integrated Waveguide Cavity Backed Slot Antenna for Dual-Frequency Application," *44th European Microwave Conference*, Rome, Italy, 2014.

[7] Mukherjee, S., A. Biswas, and K. V. Srivastava, "Broadband Substrate Integrated Waveguide Cavity-Backed Bow-Tie Slot Antenna," *IEEE Antennas and Wireless Propagation Letters*, Vol. 13, 2014, pp. 1152–1155.

[8] Mukherjee, S., and A. Biswas, "Design of SIW Cavity Backed Slot Antenna for Wideband Application," *Asia-Pacific Microwave Conference (APMC)*, New Delhi, India, 2016.

[9] Mukherjee, S., and A. Biswas, "Design of Dual Band and Dual-Polarised Dual Band SIW Cavity Backed Bow-Tie Slot Antennas," *IET Microwaves, Antennas & Propagation*, Vol. 10, No. 9, 2016, pp. 1002–1009.

[10] Mukherjee, S., and A. Biswas, "Substrate Integrated Waveguide (SIW) Cavity Backed Slot Antenna for Polarization Diversity Application," *IEEE Applied Electromagnetics Conference (AEMC)*, Guwahati, India, 2015.

[11] Mukherjee, S., and A. Biswas, "Design of Dual Band and Dual-Polarised Dual Band SIW Cavity Backed Bow-Tie Slot Antennas," *IET Microwaves, Antennas & Propagation*, Vol. 10, No. 9, 2016, pp. 1002–1009.

CHAPTER 4

Advanced SIW Antennas

In Chapter 3, novel antenna designs utilizing the SIW cavity were discussed. Several design techniques to achieve multiband and broadband response with a unidirectional radiation pattern were presented. In this chapter, the design technique to realize antennas with an improved performance will be discussed. Specifically, the design technique to enhance the gain of the antenna, the integrated multiplexer antenna approach, will be discussed in detail. Also, SIW implementation of conventional waveguide slot array antenna will be discussed in detail.

4.1 SIW High-Gain Antenna

In Chapter 2, the design of SIW rectangular cavity is explained with (2.16), and the SIW cavity-backed slot antenna is discussed in Section 3.1. The typical gain of such an antenna ranges from 4 to 5 dBi. In this section, the design technique to enhance the gain of the antenna will be explored in detail. In Sections 3.2 and 3.3, hybrid modes are excited in the SIW cavity, which was utilized to generate multiband and broadband responses. In this section, another efficient use of the SIW cavity hybrid mode to enhance the gain of the antenna will be presented. The design technique can be explained with the help of a few design examples.

4.1.1 SIW Elliptical Cavity-Backed, High-Gain Slot Antenna

To demonstrate the idea, the first design example is taken as an SIW elliptical cavity-backed slot antenna. The beauty of the structure is that we can choose different eccentricities of the cavity and accordingly unconventional behavior can be extracted from such a cavity design [1]. The design implements an SIW cavity in the planar substrate by placing rows of metallic vias elliptically as shown in Figure 4.1, which, along with the top and bottom copper cladding layers, completes the metallic enclosure of the cavity. The dimension of the proposed elliptical cavity is chosen to keep the TM_{c110} mode resonant frequency within the X-band (8 to 12 GHz), which can be derived from (4.1) [2].

$$f = \frac{c}{\sqrt{\varepsilon_r}}\frac{1}{\lambda_c} \tag{4.1}$$

$$\lambda_c = \frac{\pi L e}{2\sqrt{q}} \tag{4.2}$$

where c = speed of light, λ_c = cutoff wavelength of the corresponding resonant mode, L = major axis of the SIW cavity, and e = eccentricity of the cavity.

Due to the small height of the SIW cavity, there is no variation along this direction of the cavity. However, due to the presence of rows of metallic vias instead of a continuous metallic wall, there is a small change in the effective length and eccentricity of the proposed cavity. The parameter q for the TM_{cmn} mode is the nth parametric zero of the modified Mathieu function (even) of the first kind and order m [3].

The excitation of the elliptical SIW cavity by the 50Ω microstrip line is shown in Figure 4.1. As shown in Figure 4.1(a), a small inset of the 50Ω microstrip line excites the small elliptical SIW cavity to resonate at the TM_{c110} mode. The corresponding field distribution of the cavity is shown in Figure 4.1(c). The normalized field distribution along AA_1 in the small elliptical cavity is shown in Figure 4.2 as a dotted line. As shown in Figures 4.1(c) and 4.2, the electric field distribution in the small cavity is evenly distributed along AA_1, which takes a sinusoidal form.

However, for the long elliptical cavity, a nonsinusoidal hybrid mode is excited with a modified electric field distribution as shown in Figure 4.1(d). In this case, the dimensions L and W are chosen such that the eccentricity of the cavity becomes

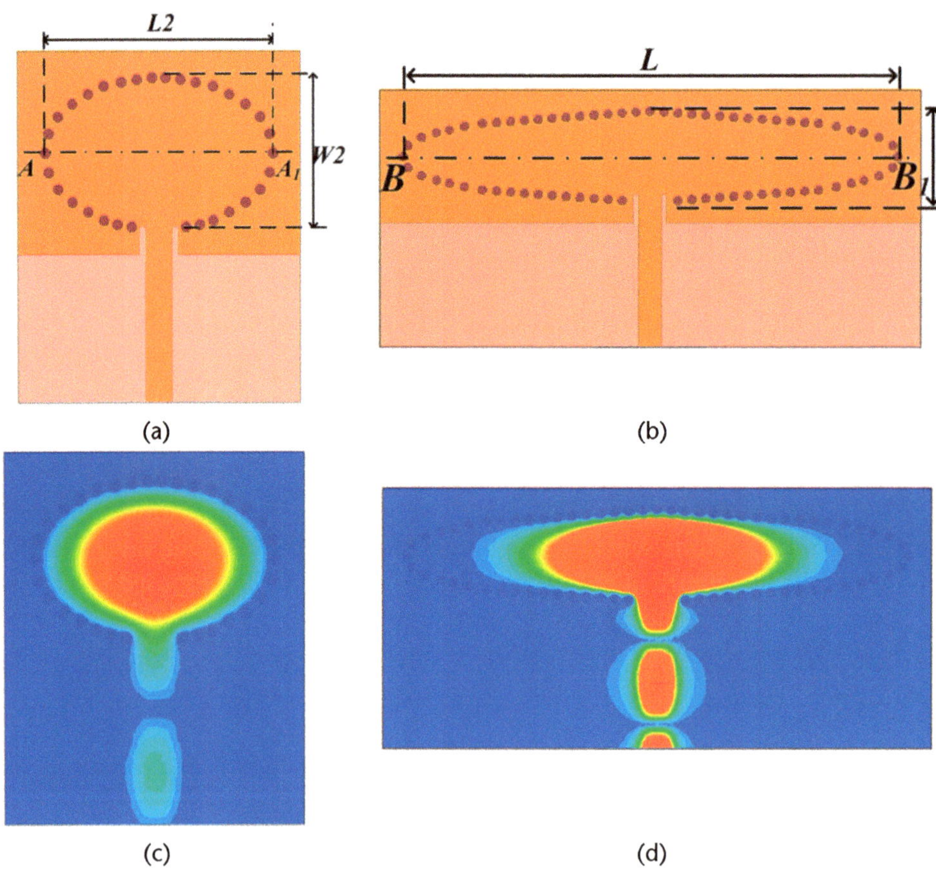

Figure 4.1 Analytical model of an elliptical SIW cavity: (a) small cavity (e = 0.77); (b) long cavity (e = 0.98); and magnitude of electric field at top metallic plate of (c) a small cavity or (d) a long cavity.

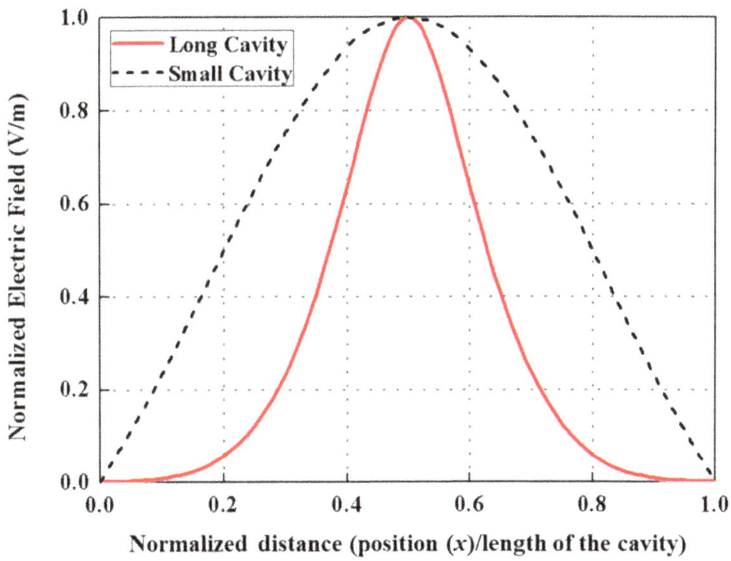

Figure 4.2 Normalized electric field across the center line (AA_1 and BB_1, respectively) of small and long elliptical cavity.

large. The electric field is mostly concentrated along the center of the cavity with a negligible field at the edges. As a result, the normalized field distribution along BB_1 extracted from the Ansoft HFSS takes a form as shown in Figure 4.2 as a solid line. Thus, by changing the eccentricity of the elliptical SIW cavity, the electric field distribution of the proposed cavity can be modified accordingly.

The field distribution in the small and long SIW elliptical cavities can be used to excite planar slot antenna by placing them at the top metallic plate. As shown in Figure 4.3(a, b), a narrow rectangular slot is etched at the top metallic plate of the cavity. The inset of the feed line to the antenna is optimized to get the proper

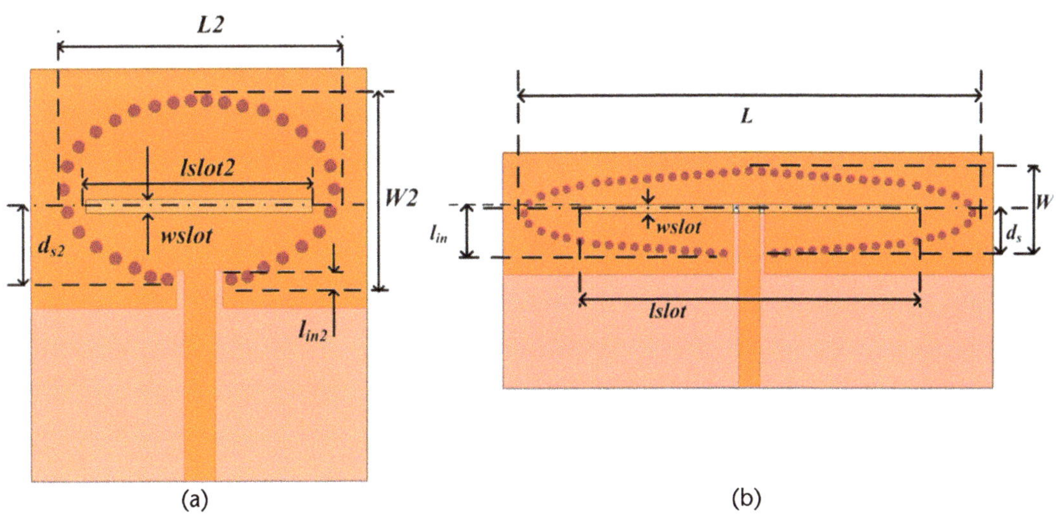

Figure 4.3 Top view of the SIW elliptical cavity-backed slot antenna: (a) small cavity; and (b) long cavity.

impedance matching of the antenna in a similar way as explained in Section 3.1. The placement of the slot disturbs the surface current distribution at the top metallic plate of the cavity, which helps to excite the slot antenna. In the case of a small SIW elliptical cavity with low eccentricity, the length of the slot is kept in the order of half of the guided wavelength, as shown in Figure 4.3(a). A small inset l_{in2} of the feed line is used to get impedance matching of the antenna, which excites the TM_{c110} mode in the cavity, as shown in Figure 4.4(a, b). As shown in Figure 4.4(b), the surface current at the top metallic plate of the cavity is disturbed by the placement of the slot. The corresponding electric field across the slot takes a sinusoidal distribution with an almost negligible field at the opposite side of the slot. The difference in magnitude of the electric field at the opposite sides of the slot helps to excite it and the slot antenna starts to radiate into free space.

The magnitude of the electric field and surface current density in both small and long SIW elliptical cavity-backed slot antennas is shown in Figure 4.4. However, in the case of the SIW elliptical cavity with a larger eccentricity, the major axis of the proposed cavity is extended such that the eccentricity of the cavity becomes high, as shown in Figure 4.3(b). A narrow rectangular slot of length twice that of the guided wavelength is placed at the top metallic plate of the cavity. To get impedance matching, the feed line is extended into the cavity up to a length l_{in}, which helps to excite a hybrid mode in the cavity. As a result, the electric field distribution in the cavity takes a modified form. The electric field in the proposed design is concentrated near the center of the slot with an almost negligible field at the edges

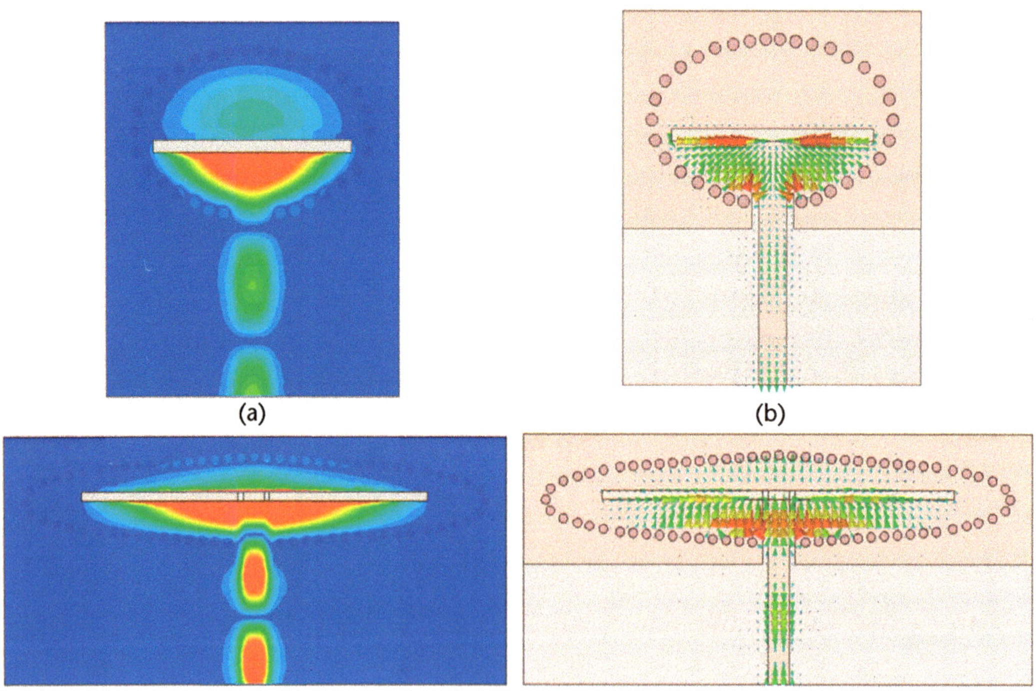

(a) (b) (c) (d)

Figure 4.4 (a) Magnitude of electric field; (b) surface current density at the top metallic plate of small elliptical cavity-backed slot antenna; (c) magnitude of the electric field; and (d) surface current density at the top metallic plate of a long elliptical cavity-backed slot antenna.

as shown in Figure 4.4(c, d). The placement of the slot disturbs the current distribution at the top metallic plate of the cavity in a similar way as in the case of the small elliptical cavity. The difference in magnitude of electric field at the opposite side of the long slot as shown in Figure 4.4(c) helps it to radiate. However, due to the hybrid-mode excitation field, the far-field radiation pattern of the antenna is also modified with narrower beamwidth and improved directivity, which leads to higher gain of the antenna.

The response of the proposed antenna is shown in Figure 4.5. The proposed antenna using a long elliptical SIW cavity resonates at 11.2 GHz with a bandwidth of 1 GHz (9%), as shown in Figure 4.5 as a solid line. The response of the small elliptical cavity-backed slot antenna is also shown in Figure 4.5 as a dotted line, which also resonates at 11.18 GHz with a bandwidth of 341 MHz (3.05%). The radiation pattern of the two antennas is shown in Figure 4.6. Both antennas exhibit unidirectional radiation pattern due to SIW cavity backing behind the slot. A comparative study of the performance of the proposed long elliptical cavity-backed slot antenna and that of the small cavity-backed slot antenna is shown in Table 4.1. The proposed slot antenna exhibits a radiation pattern with narrower beamwidth and higher gain. The proposed design technique achieves a high gain of 7 dBi while maintaining a planar form and a simple feeding technique, which makes it attractive for an X-band application.

4.1.2 SIW Rectangular Cavity-Backed, High-Gain Slot Antenna

The concept of the generation of a hybrid mode and utilizing it for a high-gain slot antenna can also be applied in rectangular cavity as well. However, contrary to the conventional SIW rectangular cavity used in Chapter 3, the aspect ratio (length-to-width ratio) needs to be modified to a higher number to generate a similar effect. A design example along with theoretical analysis of the same is described here for a better understanding of the working principle [4].

The top view of the narrow SIW cavity is shown in Figure 4.7. Following (2.16) in Chapter 2, we can choose multiple combinations of the length-to-width ratio of the SIW cavity to keep its dominant mode at a particular desired frequency. Here,

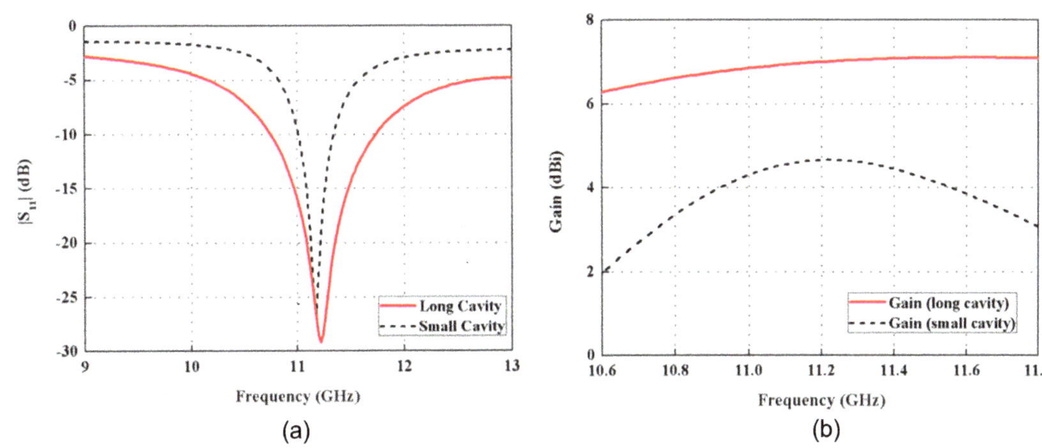

(a) (b)

Figure 4.5 Performance of the small and long elliptical cavity-backed slot antennas.

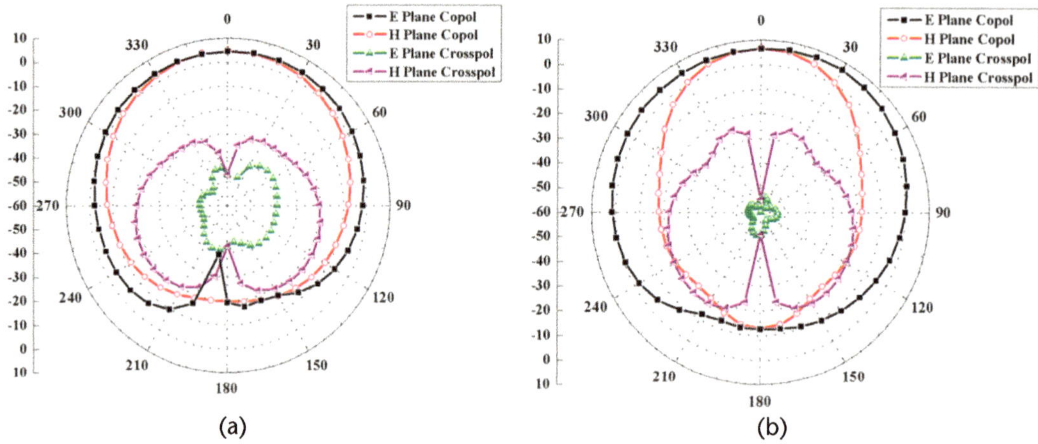

Figure 4.6 Radiation pattern of: (a) small cavity-backed slot antenna; and (b) long cavity-backed slot antenna.

Table 4.1 Comparative Study of the Performance of the Antennas

Properties		*Long elliptical cavity backed slot antenna (e = 0.98)*	*Small elliptical cavity-backed slot antenna (e = 0.77)*
Resonant frequency		11.2 GHz	11.18 GHz
Bandwidth		1 GHz (9%)	341 MHz (3.05%)
Gain		7 dBi	4.5 dBi
HPBW	E-plane	113°	118°
	H-plane	38°	70°
FTBR		16 dB	24 dB
Cross-polar level		34 dB	35 dB

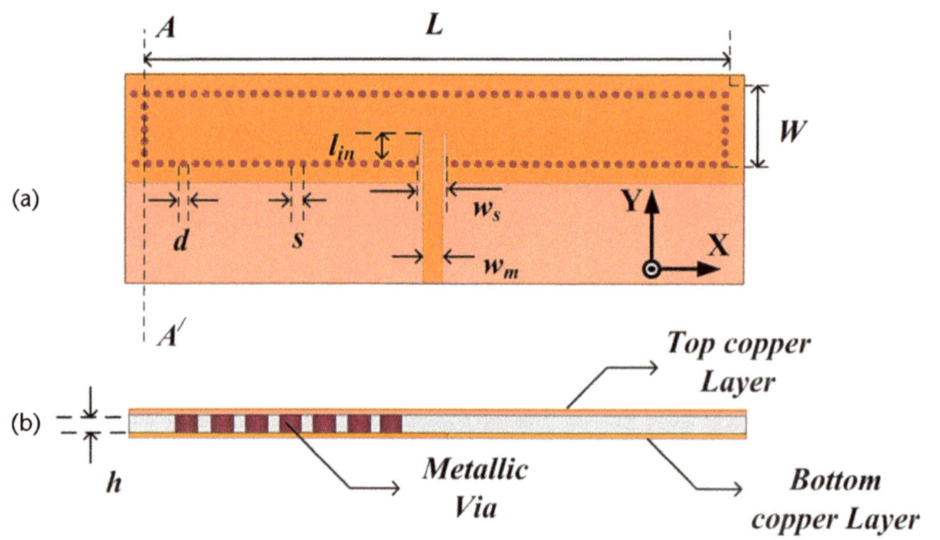

Figure 4.7 (a) Top view and (b) side view of the proposed SIW narrow rectangular cavity along AA¢. (Substrate used: RT Duroid 5880 with dielectric constant (ε_r) of 2.2.) (© 2018 IEEE [4].)

in this design example, the length is much higher than the width, keeping the ratio much higher than 1 while the dominant-mode resonant frequency remains at X-band (8 to 12 GHz). The cavity is fed by a small section of 50Ω microstrip line at the center of the cavity, which is suitable for the excitation of TE^{y}_{m10} modes of the cavity where m is an odd integer.

4.1.2.1 Theoretical Study of the Hybrid Mode in the SIW Cavity with a High Aspect Ratio

When the SIW cavity with a high aspect ratio is excited with the microstrip feed line at the center, the generated field distribution exhibits a unique phenomenon. Unlike the conventional SIW rectangular cavity, the field distribution is not sinusoidal along the extended length of the cavity. This is a similar effect, as seen in the elliptical SIW cavity case in Figure 4.2. A comparative study by plotting the normalized field distribution of the conventional SIW cavity and the design with an extended length is shown in Figure 4.8. The electric field in the proposed narrow rectangular cavity is concentrated at the middle of the cavity with a negligible field at both ends as shown in Figure 4.8 as a solid line. If we minutely inspect the field distribution graph, the best fit to such distribution would be the summation of multiple sinusoidal functions. Therefore, it can be stated that, for the conventional cavity mode TE^{y}_{m10}, the normalized electric field distribution can be written as

$$E_{zn} = \sin\left(\frac{m\pi x}{L}\right) * \sin\left(\frac{\pi y}{W}\right) \tag{4.3}$$

However, for the proposed narrow SIW cavity, the field distribution is modified by the interaction between higher-order modes and the resultant hybrid-mode field distribution can be written as

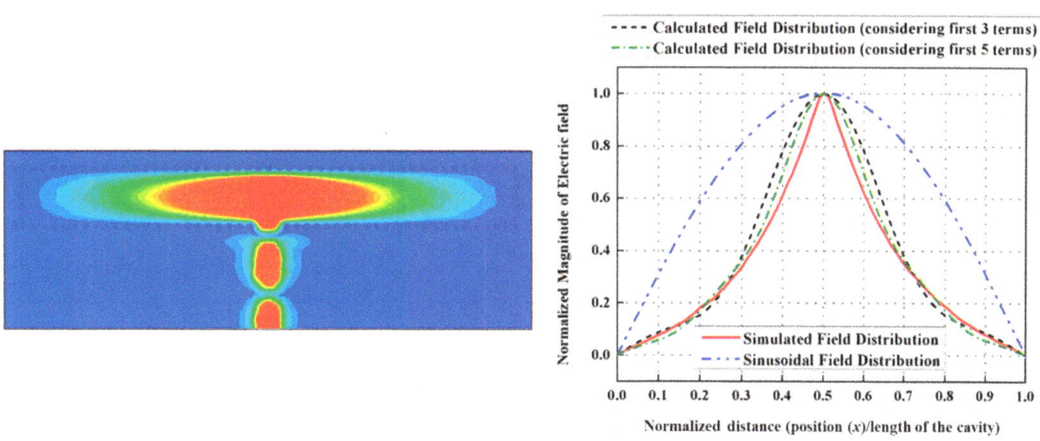

(a) (b)

Figure 4.8 (a) Magnitude of the electric field at the top metallic plate of the cavity; and (b) comparison between the calculated normalized electric field distribution from (4.3) and the simulated normalized electric field distribution from Ansoft HFSS along with the reference sinusoidal field distribution. (© 2018 IEEE [4].)

$$E_{zn}\Big|_{\text{mod}} = f(x) * \sin\left(\frac{\pi y}{W}\right) \tag{4.4}$$

$$\begin{aligned} f(x) = \sin\left(\frac{\pi x}{L}\right) - 0.375 * \sin\left(\frac{3\pi x}{L}\right) + 0.125 * \sin\left(\frac{5\pi x}{L}\right) \\ -0.0625 * \sin\left(\frac{7\pi x}{L}\right) + 0.03125 * \sin\left(\frac{9\pi x}{L}\right) \end{aligned} \tag{4.5}$$

where $f(x)$ in (4.5) represents resultant electric field distribution along the x-axis due to the overlapping of TE^y_{110}, TE^y_{310}, TE^y_{510}, TE^y_{710}, and TE^y_{910} modes in the SIW cavity. Essentially, if we refer to Figure 2.22 in Chapter 2, as the length-to-width ratio of the cavity is increased, all $TE^y_{m10\ (m\,=\,\text{odd integer})}$ modes resonate in close proximity with a similar orientation of an electric field in the cavity. As a result, these field distributions interact with each other and generate a resultant hybrid-mode field distribution similar to that expressed by $f(x)$ in (4.5). A plot of $f(x)$ is seen to be overlapping with the original hybrid-mode field distribution, as shown in Figure 4.8. (The naming of the mode is done based on the axis defined in the particular design. One should carefully follow the axis convention for each design while looking at the mode names.)

4.1.2.2 Prediction of the Far-Field Radiation Pattern in the SIW Cavity with a High Aspect Ratio

The working principle of the generation of the hybrid mode in the SIW cavity with a large aspect ratio can be utilized to get radiation by using a slot, usually longer than the conventional half-wavelength slot. Typically, when one attempts to excite a long slot longer than a half-wavelength, multiple minima points are expected (due to excitation of higher-order sinusoidal distribution) as compared to the conventional half-wavelength slot where the two minima are at the both end of the slot. As a result, the long slot is expected to generate a radiation pattern with the multiple grating lobe. However, in the present case, a hybrid-mode extending over a large length may be utilized to excite such a long slot with only two minima at two ends.

The excitation of a long slot with the help of the new hybrid mode will be explained here. As shown in Figure 4.9(a), the electric field across the slot E_{slot} can be calculated from the potential difference between $(d_s - w_{\text{slot}}/2)$ and $(d_s + w_{\text{slot}}/2)$.

$$\phi\Big|_{(d_s - w\text{slot}/2)} = E_{zn}\Big|_{\text{mod}} * h = f(x) * \sin\left(\frac{\pi\left(d_s - \frac{w\text{slot}}{2}\right)}{W}\right) * h \tag{4.6}$$

$$\phi\Big|_{(d_s + w\text{slot}/2)} = E_{zn}\Big|_{\text{mod}} * h = f(x) * \sin\left(\frac{\pi\left(d_s + \frac{w\text{slot}}{2}\right)}{W}\right) * h \tag{4.7}$$

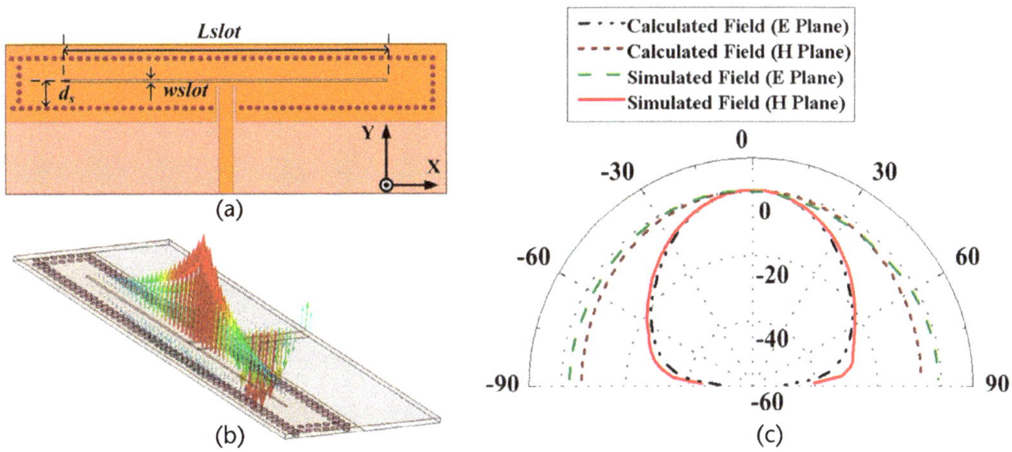

Figure 4.9 (a) Proposed narrow SIW cavity-backed slot antenna; (b) electric field vector at the top metallic plate of the proposed antenna; (c) comparison between the predicted far-field radiation pattern from (4.17) and (4.18) and the simulated radiation pattern extracted from Ansoft HFSS. (L_{slot} = 60 mm, w_{slot} = 0.4 mm, d_s = 5.5 mm.) (© 2018 IEEE [4].)

$$E_{\text{slot}} = \frac{d\phi}{dw} = \hat{y}\frac{1}{w\text{slot}} * h * f(x) * \left\{ \sin\left(\frac{\pi\left(d_s - \frac{w\text{slot}}{2}\right)}{W}\right) - \sin\left(\frac{\pi\left(d_s + \frac{w\text{slot}}{2}\right)}{W}\right) \right\} \tag{4.8}$$

Hence the equivalent magnetic current density across the slot is [5]:

$$M_s = -2\hat{n} \times E_{\text{slot}} \tag{4.9}$$

Now the far-field radiation pattern from the slot can be calculated from (4.9) following the procedure described in [5]. The final calculated fields can be expressed as (4.10) and (4.11).[1]

$$E_\phi = l\text{slot} * w\text{slot} * ki * \frac{e^{-ikr}}{4\pi r} * \left(-2 * \cos(\theta) * \cos(\phi) * w\text{slot}\right) * I \tag{4.10}$$

$$E_\theta = l\text{slot} * w\text{slot} * ki * \frac{e^{-ikr}}{4\pi r} * \left(-2 * \sin(\phi) * w\text{slot}\right) * I \tag{4.11}$$

$$I = (I1) - 0.375 * (I2) + 0.125 * (I3) \tag{4.12}$$

1. Only $m \geq 3$ is considered in this calculation.

$$I1 = -\frac{l\text{slot}^2}{4} * \frac{1}{X^2 - \left(\frac{\pi}{2}\right)^2} * \frac{\pi}{l\text{slot}} * \left(1 + e^{i2X}\right) \quad (4.13)$$

$$I2 = -\frac{l\text{slot}^2}{4} * \frac{1}{X^2 - \left(\frac{3\pi}{2}\right)^2} * \frac{3\pi}{l\text{slot}} * \left(1 + e^{i2X}\right) \quad (4.14)$$

$$I3 = -\frac{l\text{slot}^2}{4} * \frac{1}{X^2 - \left(\frac{5\pi}{2}\right)^2} * \frac{5\pi}{l\text{slot}} * \left(1 + e^{i2X}\right) \quad (4.15)$$

$$X = \frac{k * l\text{slot}}{2} * \sin(\theta) * \cos(\phi) \quad (4.16)$$

where only $m \geq 3$ is considered in this calculation.

However, the proposed slot antenna is backed by the SIW cavity. Hence, to account for the effect of the back-side reflector, the expression in (4.10) and (4.11) is modified following the procedure described in [5, Chapter 4.7], which is shown below:

$$E_{\phi r} = f_r(\theta) * E_\phi \quad (4.17)$$

$$E_{\theta r} = f_r(\theta) * E_\theta \quad (4.18)$$

$$\begin{aligned} f_r(\theta) &= 2 * \cos\left(kh * \cos(\theta)\right);\ |\theta| < \frac{\pi}{2} \\ &= 0;\ |\theta| > \frac{\pi}{2} \end{aligned} \quad (4.19)$$

A comparative plot of the calculated and simulated far-field radiation pattern proves the working principle explained above. It can be noted that the radiation pattern exhibits narrower half-power beamwidth (HPBW), which improves the directivity of the antenna. The HPBW in both the calculated and simulated H-plane patterns is 32° and in the E-plane, the calculated and simulated HPBWs is 96° and 102°, respectively. This is due to the fact that the longer slot makes the effective aperture area larger by making a larger length in the Y direction and, accordingly, the directivity in the H-plane is increased. The effectiveness of the present design is that it can generate high directive radiation beam with no side lobe or grating lobe due to the utilization of the hybrid mode in the cavity.

4.1.2.3 SIW High-Gain Rectangular Slot Antenna

Based on the working principle stated above, a practical realization of the SIW cavity-backed, high-gain slot antenna is shown in Figure 4.9. The long slot is etched in the top plate of the SIW cavity with a high length-to-width ratio. The position of the slot (d_s) is required to be optimized to ensure excitation of the hybrid mode and, at the same time, the slot should disturb the mode field distribution to get excitation. Also, the inset feed line also plays a role for excitation of the desired mode and essentially the impedance matching of the antenna. The performance of the antenna is shown in Figure 4.11 where an additional resonance is seen at the higher frequency at 10.48 GHz. This happens due to the high aspect ratio of the SIW cavity where higher-order modes can get excited by perturbation done by the slot antenna. To prevent the excitation of the unwanted mode, the shape of the slot is changed, as shown in Figure 4.10. The proposed antenna resonates at 9.5 GHz with a bandwidth of 700 MHz along with an improved out-of-band performance. The measured response of the antenna matches well with the simulated response along with a measured resonant frequency at 9.6 GHz with a bandwidth of 670

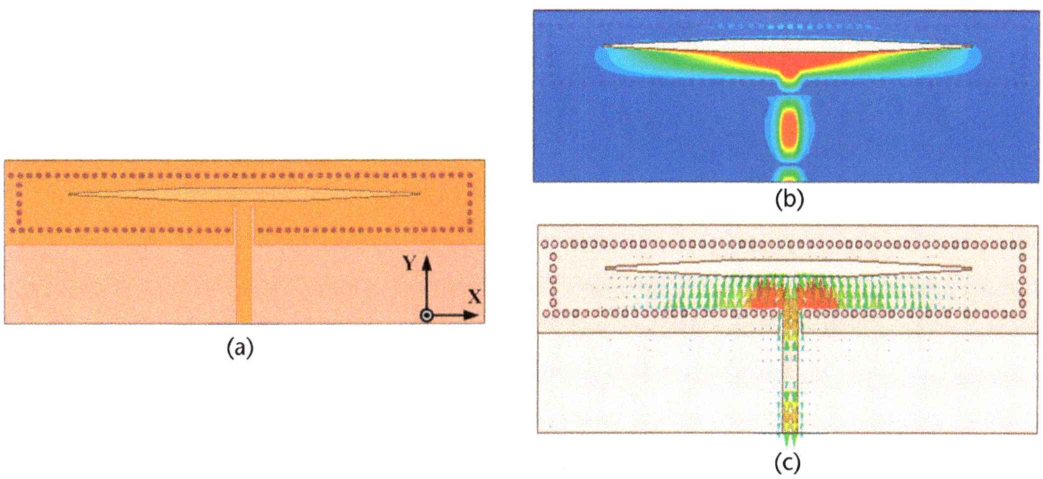

Figure 4.10 (a) Top view of the proposed SIW cavity-backed, high-gain slot antenna; (b) surface current density plot at the top metallic plate of the cavity; and (c) magnitude of the electric field at the top metallic plate of the cavity. (© 2018 IEEE [4].)

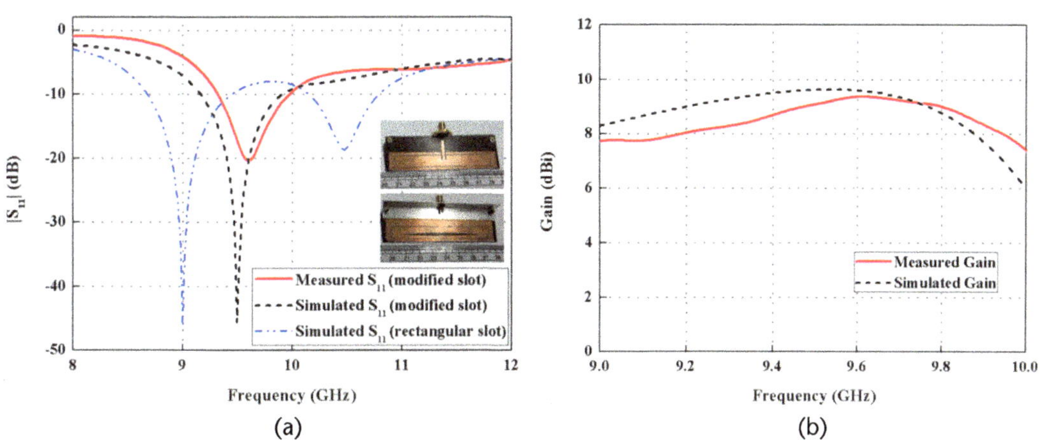

Figure 4.11 Comparative study of the simulated and measured responses of the proposed antenna. (© 2018 IEEE [4].)

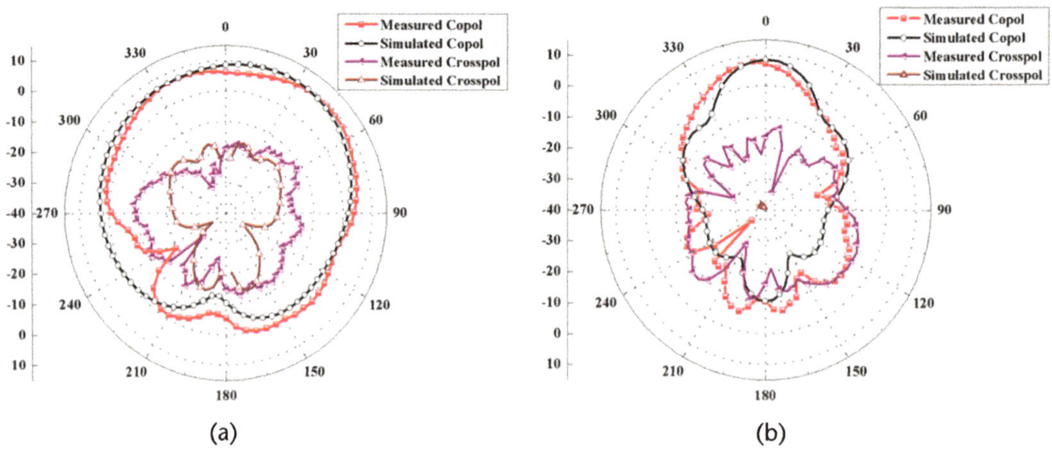

Figure 4.12 Comparison between the simulated and measured radiation patterns: (a) E-plane; and (b) H-plane. (© 2018 IEEE [4].)

MHz. The radiation pattern in the H-plane produces a narrower main beam with a simulated HPBW of 28° only. The measured H-plane pattern also follows the simulated response with measured HPBW of 30°. The simulated and measured E-plane HPBWs are 92° and 95°, respectively. This is as expected in calculations in (4.17) and (4.18). The measured gain of the antenna is 9.3 dBi. The proposed design improves the gain by 4.2 dBi as compared to the conventional SIW cavity-backed slot antenna as described in Section 3.1 while maintaining its planar form as well as a compact size (volume ~ 0.019 λ_0^3). The design provides a good trade-off between the gain of the antenna and the complexity of feed network while maintaining a planar profile.

4.1.3 HMSIW High-Gain Antenna

In the previous section, extending an effective aperture area to achieve high-gain radiation without generating a grating lobe is discussed. The advantage of using a hybrid-mode generation-based simple feed network to achieve high gain is described. However, the extended length of the antenna caters to a larger footprint to achieve a better gain performance.

The larger footprint can be tackled if the miniaturization technique is used. In Chapter 1, Section 1.4 described the half-mode SIW (HMSIW) technique to miniaturize the SIW components. The technique can easily be adopted in the SIW cavity as well. In this case, a full SIW cavity can be bisected along the middle plane to generate a half-mode cavity, as shown in Figure 4.13.

The cut plane can be aligned along the principal axis (X or Y) or can be diagonally aligned as well. The difference between these three cases is that the field distribution generated in the HMSIW cavity will not be similar. The open end of the cavity resembles a perfect magnetic conducting (PMC) boundary wall. As a result, the cavity modes for which a PMC symmetry wall can be maintained will be excited in the HMSIW cavity. All other modes will not exist, as the boundary condition will not support them. For example, if we refer to Figure 2.20 in Chapter

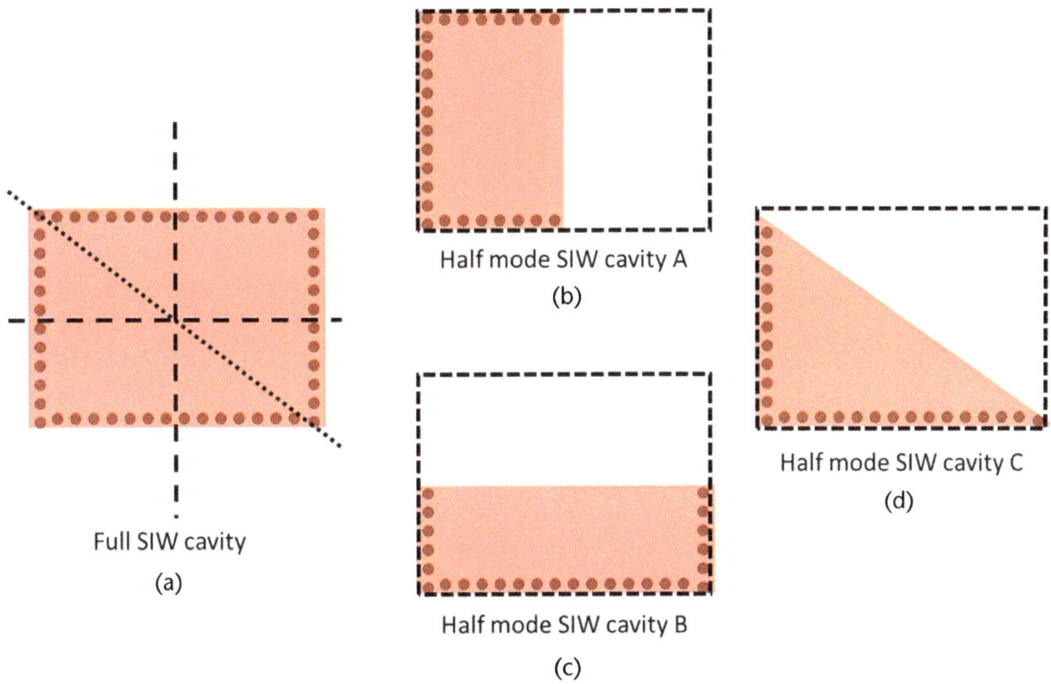

Figure 4.13 Different configuration of the HMSIW cavity from the full SIW cavity.

2, the HMSIW cavity in Case A will support TE^X_{110}, TE^X_{210}, TE^X_{310}, and TE^X_{130} but will not support TE^X_{120} and TE^X_{220}. The concept of hybrid mode-based high-gain antenna can also be applied in the HMSIW cavity as well.

The generation of the hybrid mode in the SIW cavity is majorly dependent on a high length-to-width ratio as described in the previous section. We can start with the same design in Section 4.1.2.1 and try to miniaturize the antenna by using the half-mode concept. Here, a PMC symmetry wall can be drawn along the middle of the cavity, as shown in Figure 4.14. Now, if we bisect the cavity with same length and half of the width, we get a half-mode SIW cavity with a high aspect ratio, which can be fed in a similar way by a small inset of microstrip line. This HMSIW cavity has an open aperture at the open end, which can generate radiation. The field distribution of the HMSIW cavity with a high aspect ratio is shown in Figure 4.15, which confirms the generation of the hybrid mode in the cavity and the radiation of the same through the open end of the cavity. However, the resonant frequency of the cavity is shifted due to change in the effective width of the cavity (refer to (1.3)).

While directly employing half-mode concept miniaturizes of the design, further miniaturization may also be possible while maintaining a trade-off between an antenna footprint and the gain of the antenna. The electric field of the hybrid mode generated in the HMSIW cavity is majorly concentrated in the middle of the cavity with very small magnitude of field at the both ends along the length of the cavity. Hence, it can be deduced that the major contribution of the narrow beam high-gain radiation pattern is due to concentration of the field at the center and, therefore, even if the field distribution at the end is omitted by a design modification, a similar high-gain pattern may be expected [6]. This reasoning gives a new

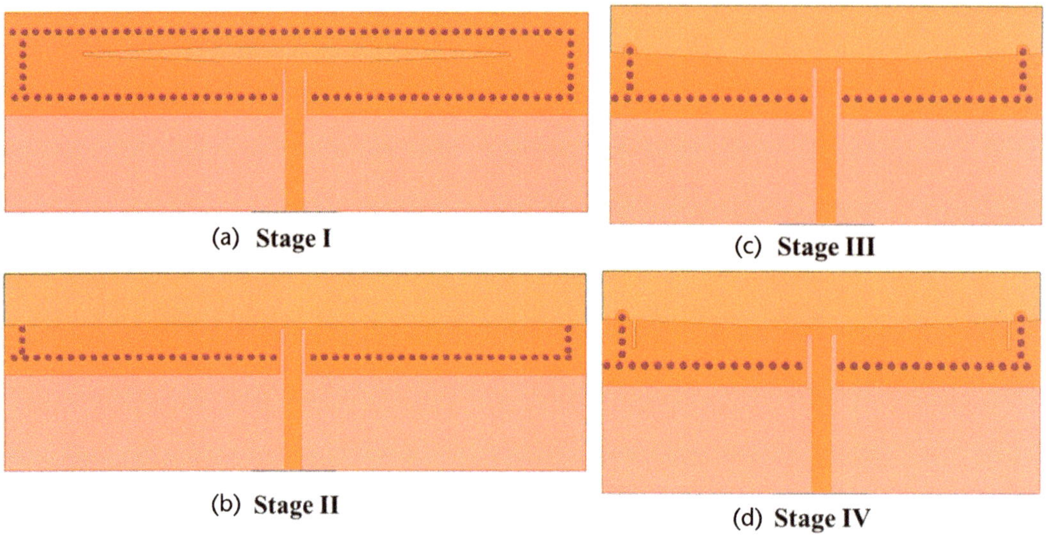

Figure 4.14 Step-by-step design procedure from full-mode, high-gain SIW cavity-backed slot antenna to HMSIW compact high-gain antenna (Stage I: full-mode SIW cavity-backed slot antenna; Stage II: half-mode SIW cavity; Stage III: compact HMSIW cavity; Stage IV: compact HMSIW cavity with slot stub). (© 2019 Wiley [6].)

direction to miniaturize the antenna along its length by truncating the both ends since the field distribution is negligible at these places.

To realize the concept, a HMSIW cavity of shorter length is proposed where the length of the cavity in Stage III is made $1.62\lambda_0$ length long only while the cavity length of the Stage-II cavity is $2.47\lambda_0$ long. Now, following the conventional excitation procedure by the inset microstrip feed, the generated field distribution becomes sinusoidal and not similar to Stage II. This can be attributed to the fact that the length-to-width ratio (aspect ratio) is not enough to bring all the higher-order

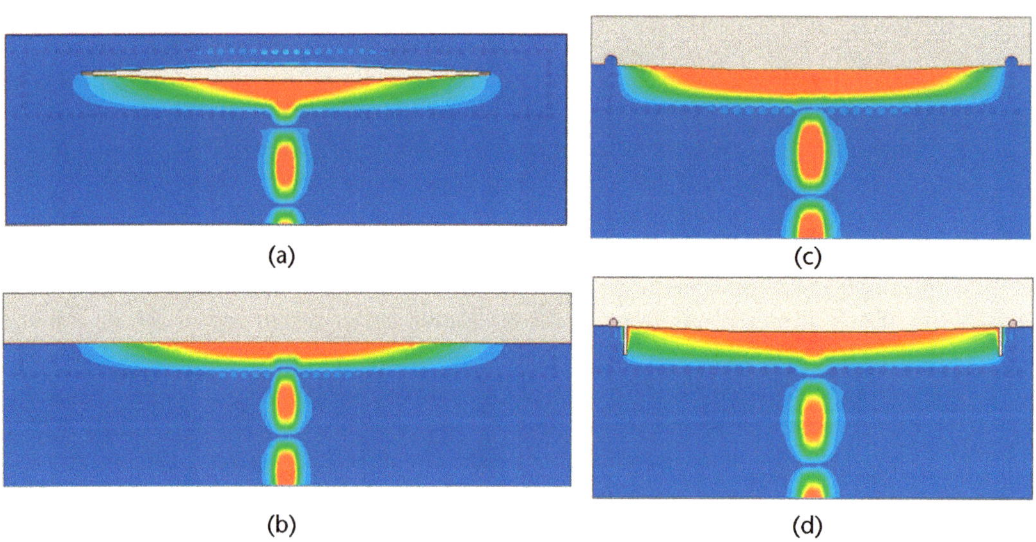

Figure 4.15 Magnitude of the electric field at the top plate of the cavity for: (a) Stage I; (b) Stage II; (c) Stage III; and (d) Stage IV. (© 2019 Wiley [6].)

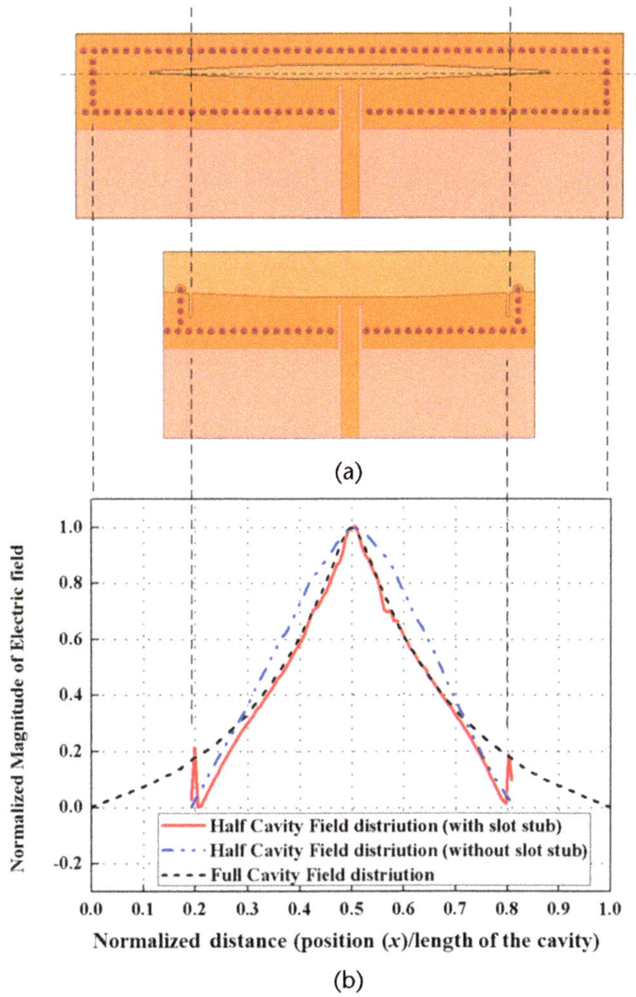

Figure 4.16 (a) Realization of the proposed HMSIW high-gain antenna from the full-mode SIW cavity-backed slot antenna; and (b) comparison of the normalized electric field of the HMSIW cavity with that of the full-mode SIW cavity. (© 2019 Wiley [6].)

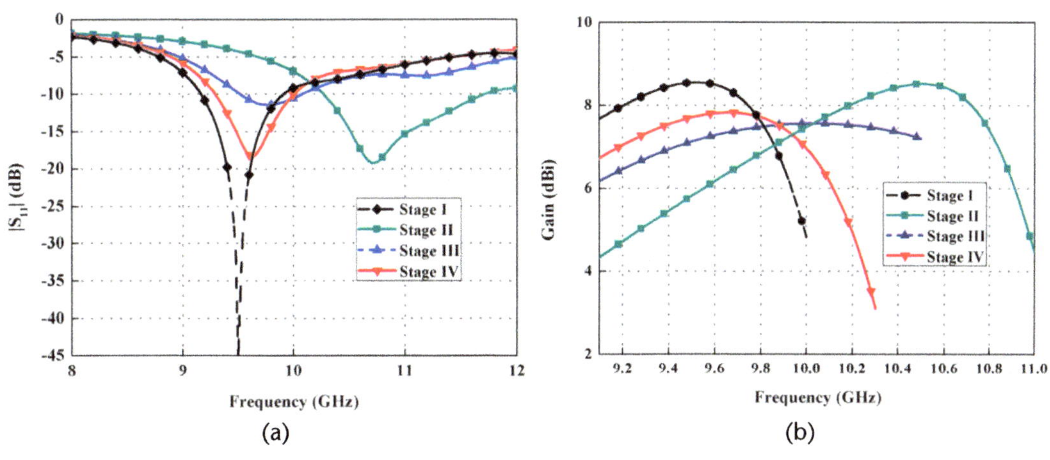

Figure 4.17 Reflection coefficient and gain response of the different design stages of the antenna. (© 2019 Wiley [6].)

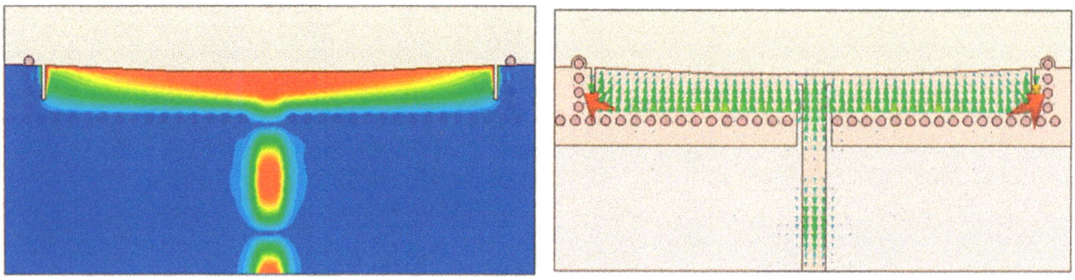

Figure 4.18 (a) Magnitude of electric field; and (b) vector surface current at the top plate of the cavity. (© 2019 Wiley [6].)

modes close together to generate the desired hybrid mode as described in (4.5). This can also be explained in the way that since we have forcefully introduced a PEC boundary at $\pm 0.81\lambda_0$ making a zero field at this position, the original hybrid-mode distribution boundary condition is not mimicked here and, as a result, a sinusoidal distribution is generated. The gain of the antenna is also reduced.

Therefore, the next modification of the design should lead to mimic the similar boundary condition in the range $\pm 0.81\lambda_0$ along the length of the cavity, which is expected to produce the desired hybrid-mode distribution. To implement it, a pair of slot stubs is placed at both ends of the cavity (Stage IV), which essentially modifies the boundary condition at these positions from PEC to PMC boundary making the electric field at these points nonzero. The field distribution of different stages is shown in Figure 4.15, which shows the final HMSIW can generate almost similar hybrid mode with much smaller length leading to a compact design. To account the effect of frequency shifting, the width of the cavity is slightly extended to bring back the resonant frequency as the design in Section 4.1.2.3. Also, the open edge can be made slanted due to a similar design as in Section 4.1.2.3 to ensure good out-of-band performance.

The fabricated prototype of the proposed design along with its response is shown in Figure 4.19. The measured resonant frequency of the antenna is at 9.94 GHz

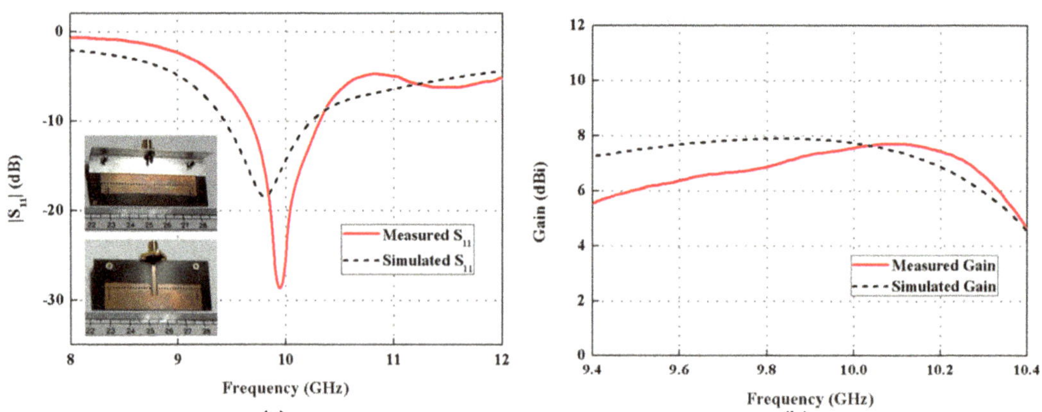

Figure 4.19 Comparative study of the simulated and measured responses of the antenna. (© 2019 Wiley [6].)

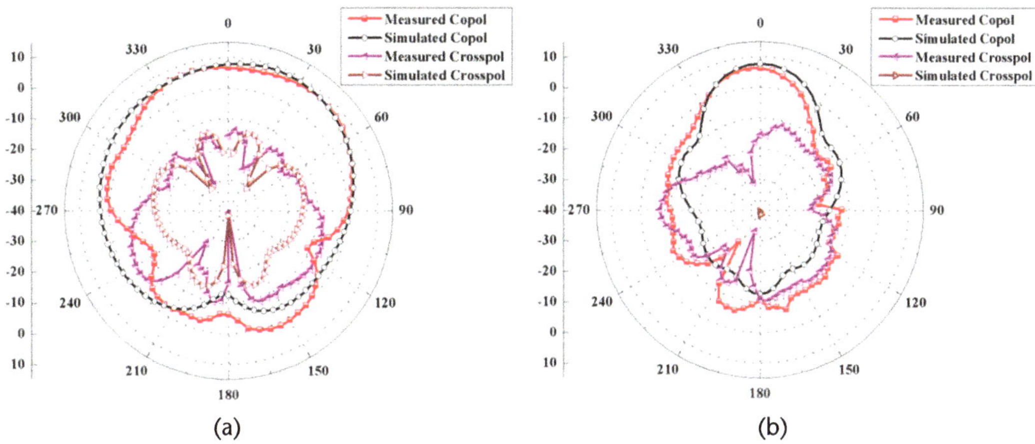

Figure 4.20 Comparison between simulated and measured radiation patterns of the HMSIW antenna: (a) E-plane; and (b) H-plane. (© 2019 Wiley [6].)

with a measured bandwidth at 650 MHz. Due to the presence of a hybrid-mode field distribution in the proposed HMSIW cavity, the H-plane radiation pattern produces narrower main beam with measured HPBW 38°. The measured gain of the antenna is 7.4 dBi.

4.2 SIW Self-Multiplexing Antenna

The RF multiplexer finds a wide variety of application in microwave and mm-wave transceiver architecture for establishing connection between multiple RF paths with a dedicated port or vice versa. One of the major uses of such a multiplexer is frequency division multiplexing where the two different RF channels operating at distinctive frequencies are used for enhanced channel capacity, signal transmission efficiency, and system performance. A typical configuration of a multichannel RF transceiver system is shown in Figure 4.21 where the antenna receives a wideband signal, which then passes through the multiplexer to distribute the received signal into the corresponding RF chain.

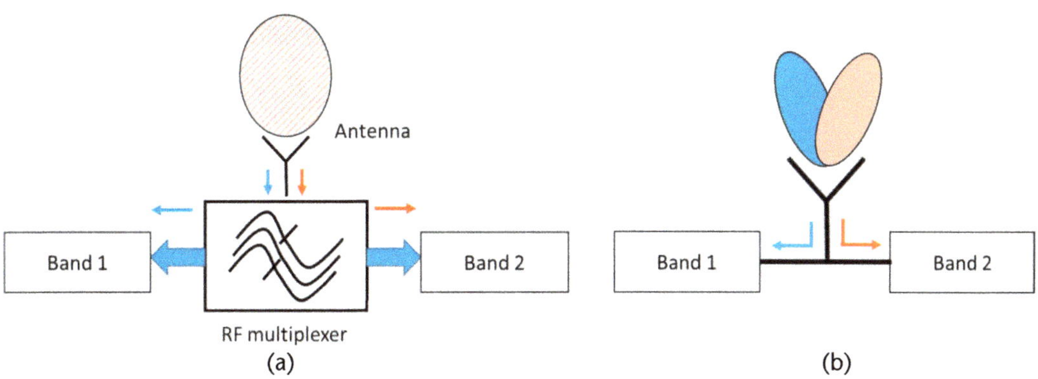

Figure 4.21 (a) Conventional architechture of antenna-diplexer configuration in the transciever system; and (b) conceptual representation of the self-diplexing two-port antenna.

While such a system is very much suitable for robust, large-scale applications, for small-scale, handheld device applications, the miniaturization of such architecture may be more suitable. Such modification will introduce a trade-off between the cost and performance of the circuit, which opens up a new direction of a suitable application scenario. In this section, two design examples of SIW-based, miniaturized multiplexing technique are discussed.

4.2.1 SIW Self-Diplexing Antenna

The term "self-diplexing antenna" refers to a special kind of circuit where the combination of antenna and feed network serves the diplexing capability by diverging the frequency-dependent channel information into a dedicated path. Such a technique resolves the requirement of the diplexer behind the antenna, and a multiport antenna network is directly connected to dedicated RF channel. Such systems may find their application in wearable electronics, retrodirective antennas, or on chip-active antennas [7, 8] where the reduction of a circuit footprint is a major issue and a trade-off between performance and the cost of device is exercised by using such a technique.

In Chapter 3, SIW cavity-backed slot antenna is discussed. It has been shown that multiple cavity modes can be utilized efficiently to extract desired performance of the antenna. Here, in this section, a technique to efficiently utilize SIW cavity mode to implement self-diplexing antenna will be described. As the name suggests, unlike the earlier cases, the antenna here will have two feeding ports dedicated for lower and higher frequencies. The idea is that the lower-frequency signal should be fed by the dedicated port and radiated by the antenna without being transmitted to the other port and vice versa. Similarly, the received signal at the lower band should be collected by the dedicated port only. Thus, the mechanism should support the diplexing mechanism internally by the antenna itself to maintain the self-diplexing design configuration. The design is shown in Figure 4.22. A bow-tie-shaped slot with enlarged length is placed on the top wall of a SIW cavity at a distance d_s from the lower side wall of the cavity [9]. The cavity dimensions are chosen to keep its dominant mode in X-band. The modification in the design from its counterpart is that it is fed by two microstrip lines at the opposite side of the slot antenna. The feed lines are placed in an offset from the center line of the cavity. The elongated slot perturbs the SIW cavity in such a way that it behaves as a pair of half-mode cavities joined face to face to form the design. This leads to excitation of a pair of resonating modes with two different resonating frequencies with perturbed field distribution as shown in Figure 4.22. The surface current at 9 GHz corresponding to the excitation port 1 is mainly concentrated in the lower part of the cavity with a negligible field in the upper part. Similarly, the excitation by input port 2 helps the surface current at 11.2 GHz to concentrate only in the upper half of the cavity. The perturbed field distribution exhibits unequal magnitude at the opposite side of the slot and, as a result, these two field distributions excite the slot to radiate. Since most of the energy is radiated through the slot, a negligible amount of the field enters the other port and, thus, the isolation between two feeding ports are maintained, keeping a simple planar profile.

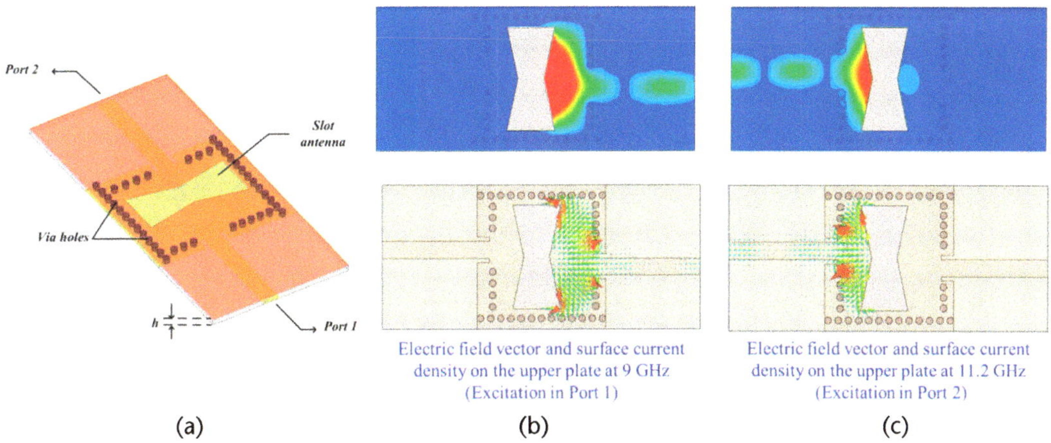

Figure 4.22 Design of the SIW self-diplexing cavity-backed slot antenna and its field response. (© 2016 IEEE [9].)

The perturbed field distribution can be analyzed with the help of a half-mode cavity field distribution, as discussed in Section 4.1.3. To demonstrate the idea, a pair of HMSIW cavities with its dimension equal to the two halves of the original design is analyzed. As shown in Figure 4.23, the two HMSIW cavities have equal length (L) and unequal width ($W - d_s$ for HMSIW Cavity 1 and d_s for HMSIW Cavity 2). The unequal dimension of the HMSIW cavities refers to the placement of the bow-tie slot in the original SIW cavity, which was slightly shifted from the center of the cavity to ensure better impedance matching. Now the analysis of the pair of HMSIW cavities exhibits their dominant-mode excitation at two distinctive resonant frequencies. This can be computed from (2.16) considering the unequal width of the cavity pair. It is interesting to note that the resonant frequencies of the dominant TE^{HM}_{110} mode of the individual half-mode cavities (9.03 GHz and 11.3 GHz, respectively) match the resonant frequencies of the original cavity loaded with an elongated bow-tie slot. In addition, the field distribution of the HMSIW cavity also matches to that of the original one. Therefore, it may be concluded that the long bow-tie slot perturbs the SIW cavity such that it behaves as a combination of two HMSIW cavities of two different resonant frequencies and radiating at those frequencies from the slot aperture. Thus, a single design is implemented, delivering the antenna radiation mechanism along with diplexing characteristics, which refers to the self-diplexing antenna. The frequency ratio of the antenna varies within a range of 0.77 to 1.37 by changing the slot position d_s while maintaining good isolation between two input ports. The frequency ratio of the proposed antenna can also be tuned by changing the slot-flaring W_{bow}, which adds additional flexibility to the design. The isolation between two input ports depends on the slot width w_{slot} as shown in Figure 4.25. With increase of the slot width w_{slot}, an isolation of better than −25 dB can be achieved.

The simulated and measured resonant frequencies of the antenna are at 9 GHz and 8.98 GHz, respectively, when excited by Port 1 with Port 2 match-terminated and similarly at 11.2 GHz and 11.22 GHz, respectively, when excited by Port 2, only

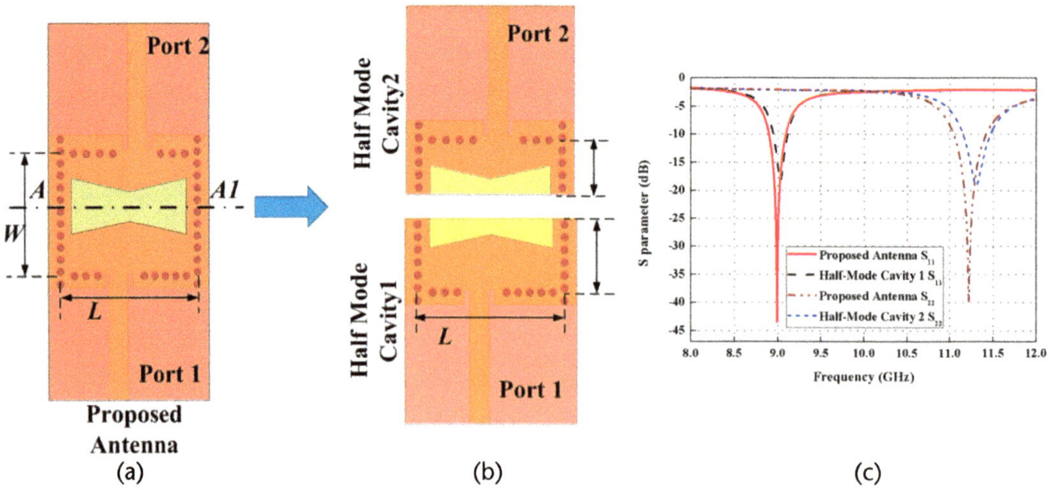

Figure 4.23 (a) Analytical model of the proposed design and its equivalence using two half-mode cavities; and (b) comparative study of the response of the proposed design and half-mode cavities. (© 2016 IEEE [9].)

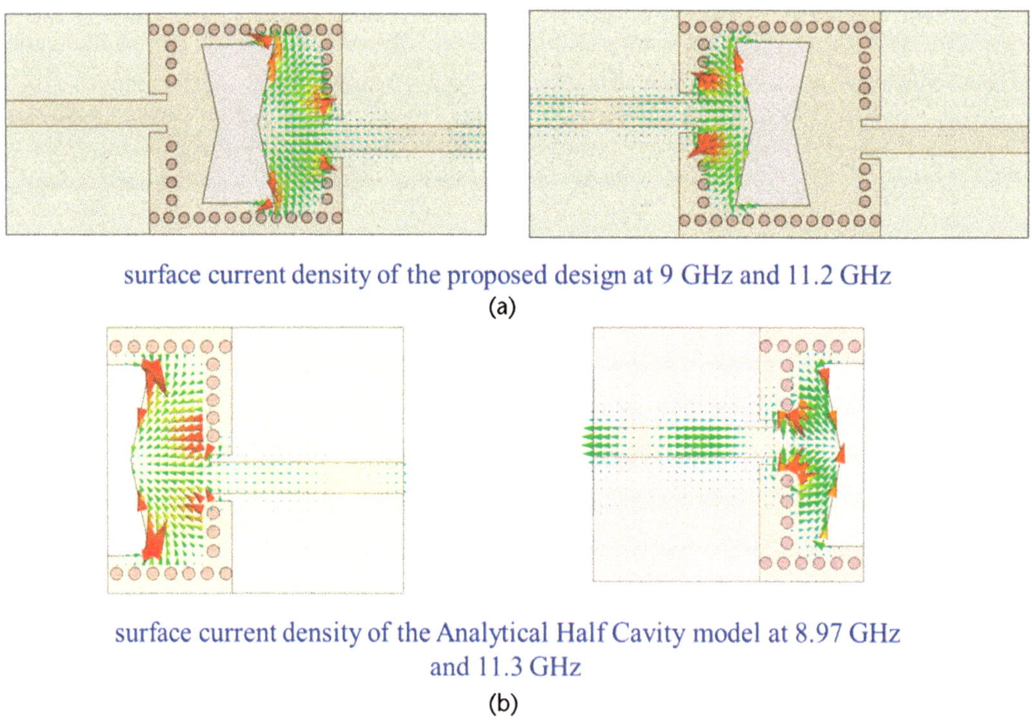

Figure 4.24 Comparative study of surface current distribution of the half-mode Cavity 1 at 8.97 GHz and half-mode Cavity 2 at 11.3 GHz. (© 2016 IEEE [9].)

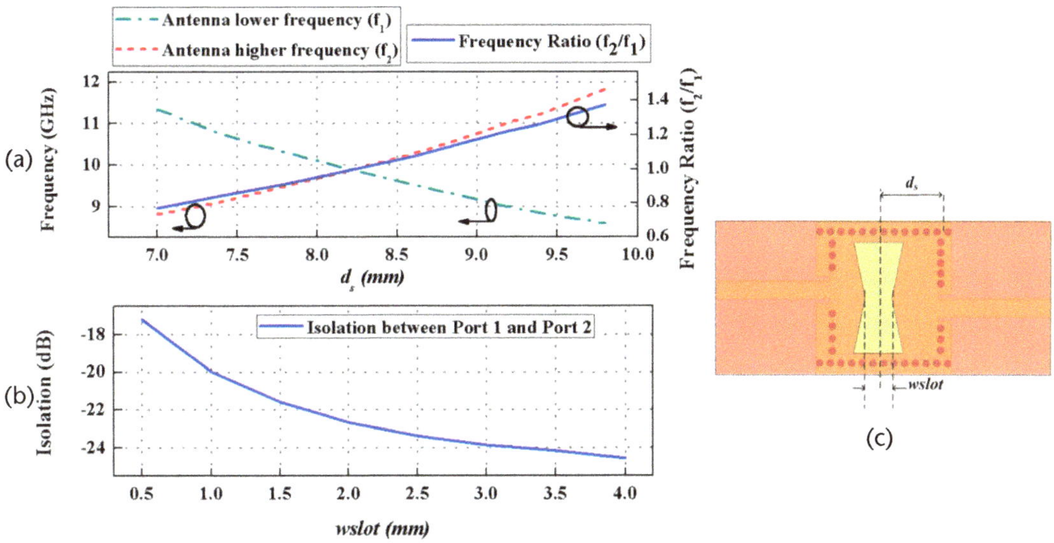

Figure 4.25 Variation of resonant frequency and frequency ratio of the antenna with the change in the slot position (d_s) and isolation with change in slot width (w_{slot}). (© 2016 IEEE [9].)

with Port 1 match-terminated. The measured isolation between two input ports of the antenna is below −24 dB and −22 dB at lower and upper resonant frequencies, respectively. The measured gain at lower and higher resonant frequencies of the antenna is 4.3 dBi and 4.2 dBi, respectively. However, it can be seen from Figure 4.26 that the gain of the antenna at a higher frequency is negligible when excited by Port 1. Similarly, when excited by Port 2, the gain at the lower frequency is much

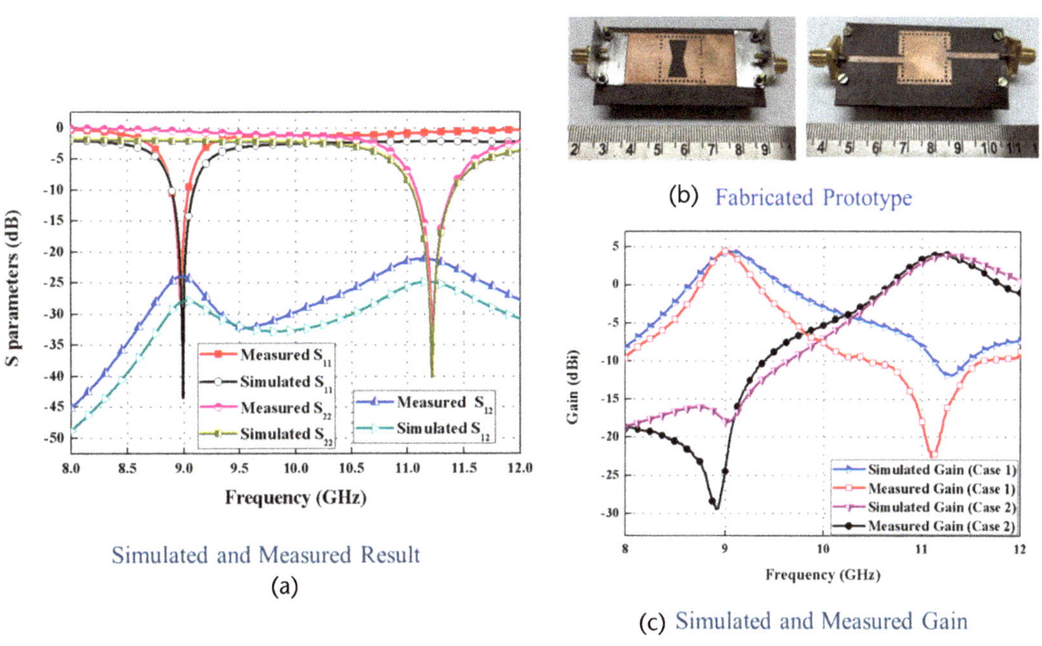

Figure 4.26 Response of the self-diplexing antenna. (© 2016 IEEE [9].)

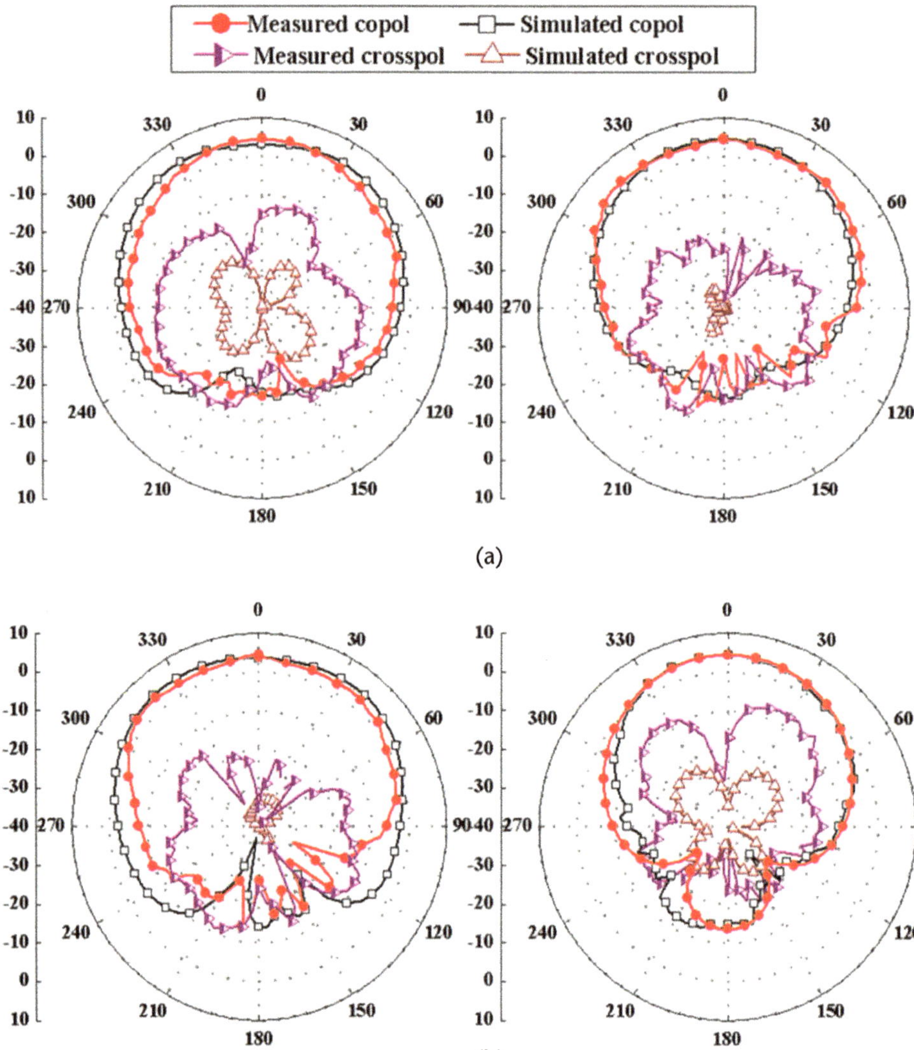

(a)

(b)

Figure 4.27 Radiation pattern plot of self-diplexing antenna. (© 2016 IEEE [9].)

less, which confirms the self-diplexing phenomenon of the proposed antenna. The radiation pattern at the two frequencies exhibit high-gain unidirectional radiation due to the SIW cavity backing. The pattern maintains low cross-polar level and front-to-back ratio (FTBR) in the range of 19 to 21 dB.

4.2.2 SIW Self-Triplexing Antenna

The concept of multiplexing the transmit/receive signal before the antenna can be extended to the triplexing architecture. Just like its diplexer counterpart, here the antenna is fed by three distinct ports dedicated to three different bands. The design is shown in Figure 4.28. The basic working principle remains the same where a large slot is placed in the top plate of the cavity to perturb its field distribution and

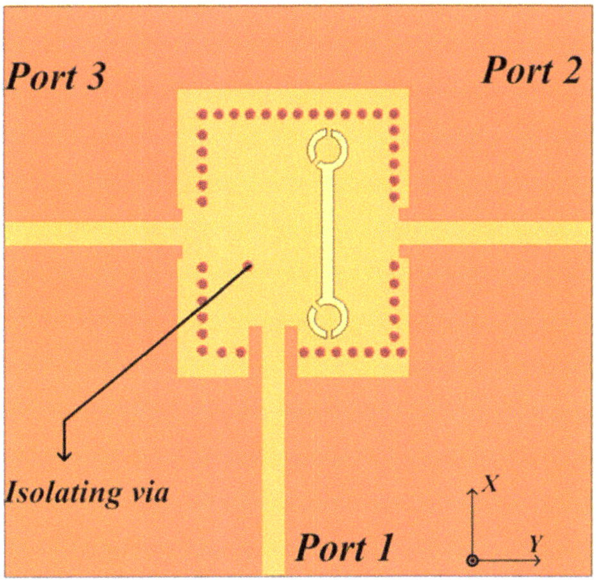

Figure 4.28 Design of self-triplexing SIW cavity-backed slot antenna. (© 2021 EuCAP [10].)

create two half-mode cavities [10]. The additional third band is realized by effectively utilizing higher-order mode of one of the half-mode cavities.

Similar to Section 3.2.2, a dumbbell-shaped slot with two connecting strips is chosen here. However, the placement of the slot is different from the previous design. Unlike aligning the slot along the center of the cavity, it is shifted along one of the side wall so that a pair of asymmetric HMSIW cavity is realized. Also, the separation between connecting strips in the dumbbell-shaped slot is managed by a slot arc of $\varphi = 135°$. The three microstrip feed lines are extended inside the cavity with different inset to ensure maximum isolation between the ports. One of the major challenges in the design is to maintain isolation between Ports 1 and 3 as they both enter into the same HMSIW Cavity 1. This is accomplished by placing a shorting via at an appropriate position in the cavity. The idea is to perturb the field distribution in the cavity such that the modes excited by Port 3 is confined in the right half of the cavity so that negligible field enters Port 1 and, in case of excitation by Port 1, the negligible field enters in Port 3. By this method, a good isolation of 15 dB can be achieved between these two ports.

For isolation between Ports 2 and 1 as well as Ports 2 and 3, the working principle similar to the self-diplexing antenna is employed. The long slot perturbs the field to get confined in the left or right half of the cavity leading it to behave as a pair of HMSIW cavities. Due to different resonant frequencies of the corresponding modes of these two cavities, the fields do not get coupled to the other side and, as a result, good isolation is observed between ports residing on opposite side of the slot. Three SIW cavity modes with modified current distribution are excited in the proposed design at 6.77 GHz (f_1), 8.57 GHz (f_2), and 13.37 GHz (f_3). The surface current distribution of these three cavity modes are shown in Figure 4.29. This helps the proposed antenna to exhibit a triple-band response at C-band (4 to

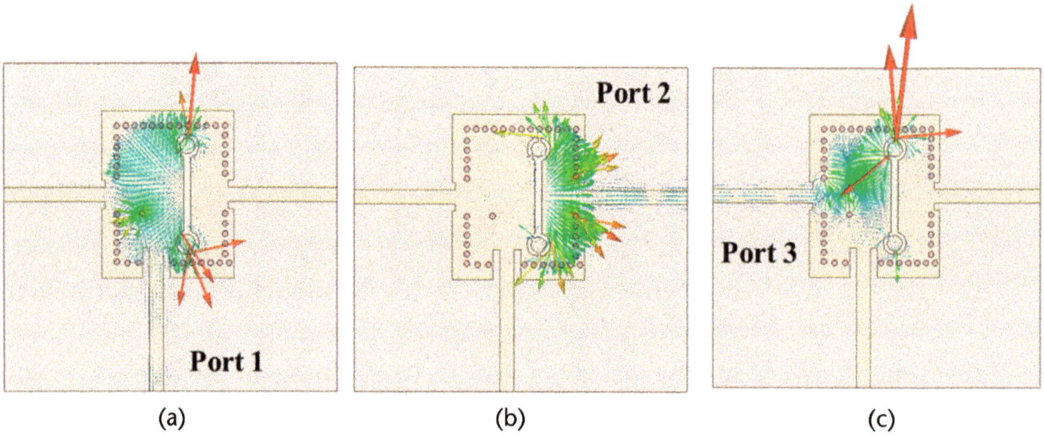

Figure 4.29 Surface current distribution in the cavity with different port excitation. (© 2021 EuCAP [10].)

8 GHz), X-band (8 to 12 GHz), and Ku-band (12 to 18 GHz) when excited by Port 1, Port 2, and Port 3, respectively, with a high isolation between these ports.

The resonant frequency of the HMSIW cavity can be calculated from (4.20). However, the correction factor due to the firing field in the open end of the HMSIW cavity similar to (1.3) in Chapter 1 will not be applicable in this case, as the pair of HMSIW cavities is placed face to face, making a closed environment. Therefore, the effective dimensions for the HMSIW cavities will be as follows:

$$f_{mn0}^{\mathrm{HMSIW}} = \frac{c}{2\sqrt{\varepsilon_r}}\sqrt{\left(\frac{m}{L_{\mathrm{eff}}^{\mathrm{HMSIW}}}\right)^2 + \left(\frac{n}{W_{\mathrm{eff}}^{\mathrm{HMSIW}}}\right)^2} \tag{4.20}$$

$$L_{\mathrm{eff}}^{\mathrm{HM}} = L - \frac{d^2}{0.95s} \tag{4.21}$$

$$W_{\mathrm{eff}}^{\mathrm{HM}} = 2 \times d_s \qquad \left(\text{for Cavity}_{\text{right}}\right) \tag{4.22}$$

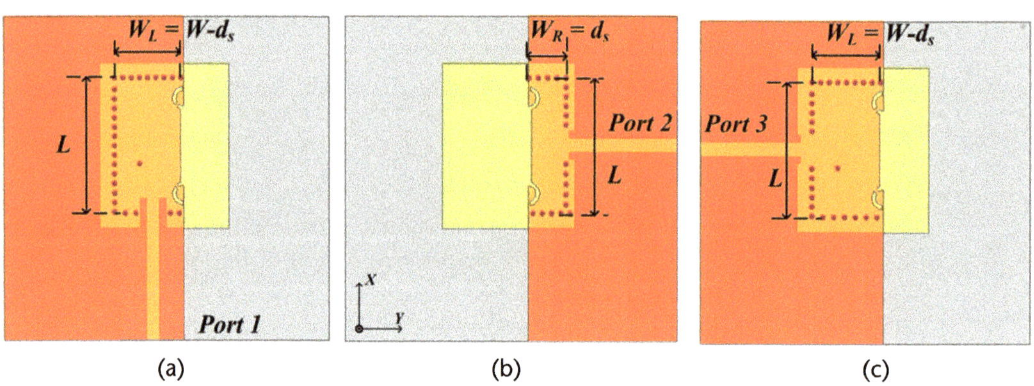

Figure 4.30 Analytical model of the HMSIW cavities. (© 2021 EuCAP [10].)

$$W_{\text{eff}}^{\text{HM}} = 2 \times \left(W - d_s\right) \quad \left(\text{for Cavity}_{\text{left}}\right) \tag{4.23}$$

$$W_{\text{eff}}^{\text{HMSIW}} = W_{\text{eff}}^{\text{HM}} - \frac{d^2}{0.95s} \tag{4.24}$$

The dimensional correction due to the fringing field is not considered in (4.21) and (4.24). The inset and offset of the feeding port play a crucial role to excite a particular mode while not exciting the higher-order or lower-order modes. Port 2 is placed at the center of the $Cavity_{\text{right}}$ with its inset adjusted to excite only the dominant TE^{yHM}_{110} mode. However, Port 1 is shifted slightly towards the open end of the $Cavity_{\text{left}}$ and its inset is adjusted to excite the dominant mode only. The challenging part is to adjust the inset and position of Port 3 as it is required to excite only the TE^{yHM}_{130} mode at $Cavity_{\text{left}}$ and not the dominant mode. As shown from (4.22) and (4.23), due to the unequal width of $Cavity_{\text{right}}$ ($W_{\text{eff}} = 2 \times d_s$) and $Cavity_{\text{left}}$ ($W_{\text{eff}} = 2 \times (W - d_s)$), the corresponding resonant frequencies at the TE^{yHM}_{110} and TE^{yHM}_{130} modes are different. The resonant frequencies and surface current distribution of the HMSIW cavities for the dominant and higher-order modes match well to the response of the original antenna. A comparative study of the proposed antenna resonance and that of the HMSIW cavity resonances are mentioned in Table 4.2. It is to be noted that the calculation of the HMSIW cavity-resonant mode does not account for the effect of slot antenna loading as well as loading effect of the other HMSIW cavity at the open end due to the face-to-face placement of the two cavities. Hence, a small variation between the original antenna and its analytical counterparts is observed.

The antenna shows measured resonant frequency at 6.76 GHz, respectively, when excited by Port 1 with all other ports match-terminated. Similarly, the excitation through Ports 2 and 3 with all other ports match-terminated results in measured antenna resonances at 8.62 GHz and 13.33 GHz, respectively. The measured isolation between Ports 1 and 2 is below 24 dB. Also, the measured isolation between Ports 2 and 3 is below 16 dB and between Ports 1 and 2 is below 15.3 dB. The

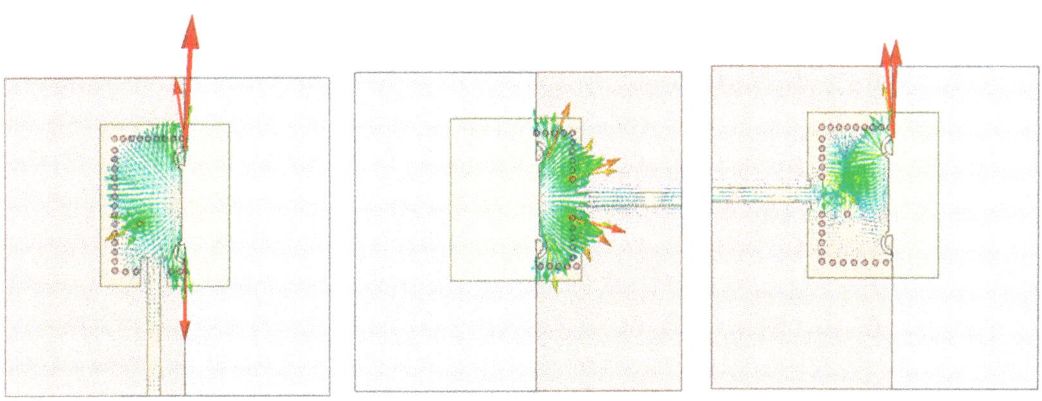

(a) (b) (c)

Figure 4.31 Surface current distribution in the analytical model of HMSIW cavities. (© 2021 EuCAP [10].)

Table 4.2 Comparison of the Resonant Frequency of the Proposed Antenna and the Half-Mode SIW Cavities

Antenna Center Frequency (GHz)	*Half-Mode SIW Cavity* *$Cavity_{left}$ ($W_{eff} = 2 \times (W - d_s)$)* From (4.1)	*From HFSS*	*$Cavity_{right}$ ($W_{eff} = 2 \times d_s$)* From (4.1)	*From HFSS*
(f_1) 6.77	6.13 GHz	6.66 GHz	—	—
(f_2) 8.57	—	—	8.73 GHz	8.54 GHz
(f_3) 13.37	13.7 GHz	13.49 GHz	—	—

measured gains of the antenna at three resonances are 3.78 dBi, 4.46 dBi, and 5.6 dBi, respectively.

As shown in Figure 4.33, the gain of the antenna when excited by Port 1 has peak value at only 6.76 GHz with negligible gain at higher frequency. Similar behavior can be observed for other two excitation ports also which ensures that only the desired band radiation is achieved with the corresponding excitation port and thus the self triplexing phenomenon is achieved in the proposed design. Unidirectional radiation pattern is observed in three resonant frequencies due to SIW cavity backing with good co-to-cross polarization level.

Two basic design strategies to implement self-multiplexing antenna designs have been explained in this section. The performance of such antennas can be enhanced in terms of adding a higher number of feeding ports, adding polarization diversity, or introducing circular polarization. There are many works reported in literature in this direction. Interested readers may read these works in details from the available literature.

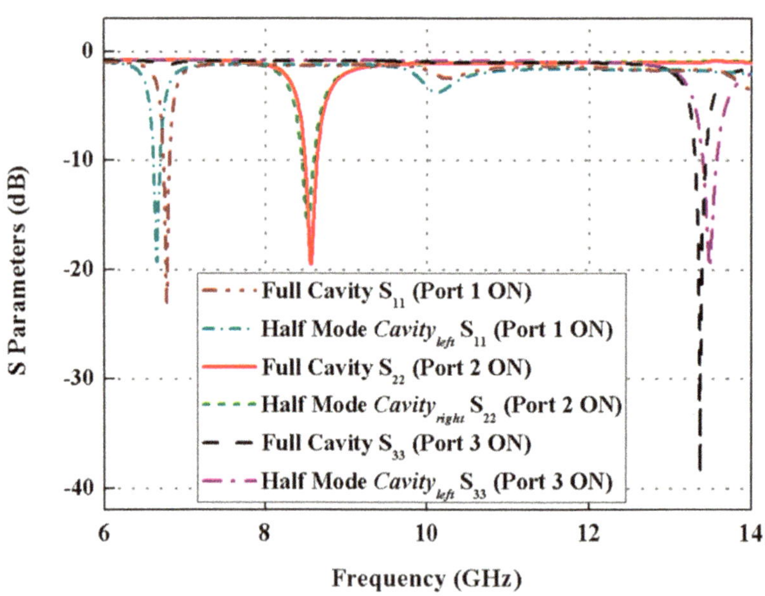

Figure 4.32 Comparative study of analytical model of HMSIW cavity and full SIW cavity design. (© 2021 EuCAP [10].)

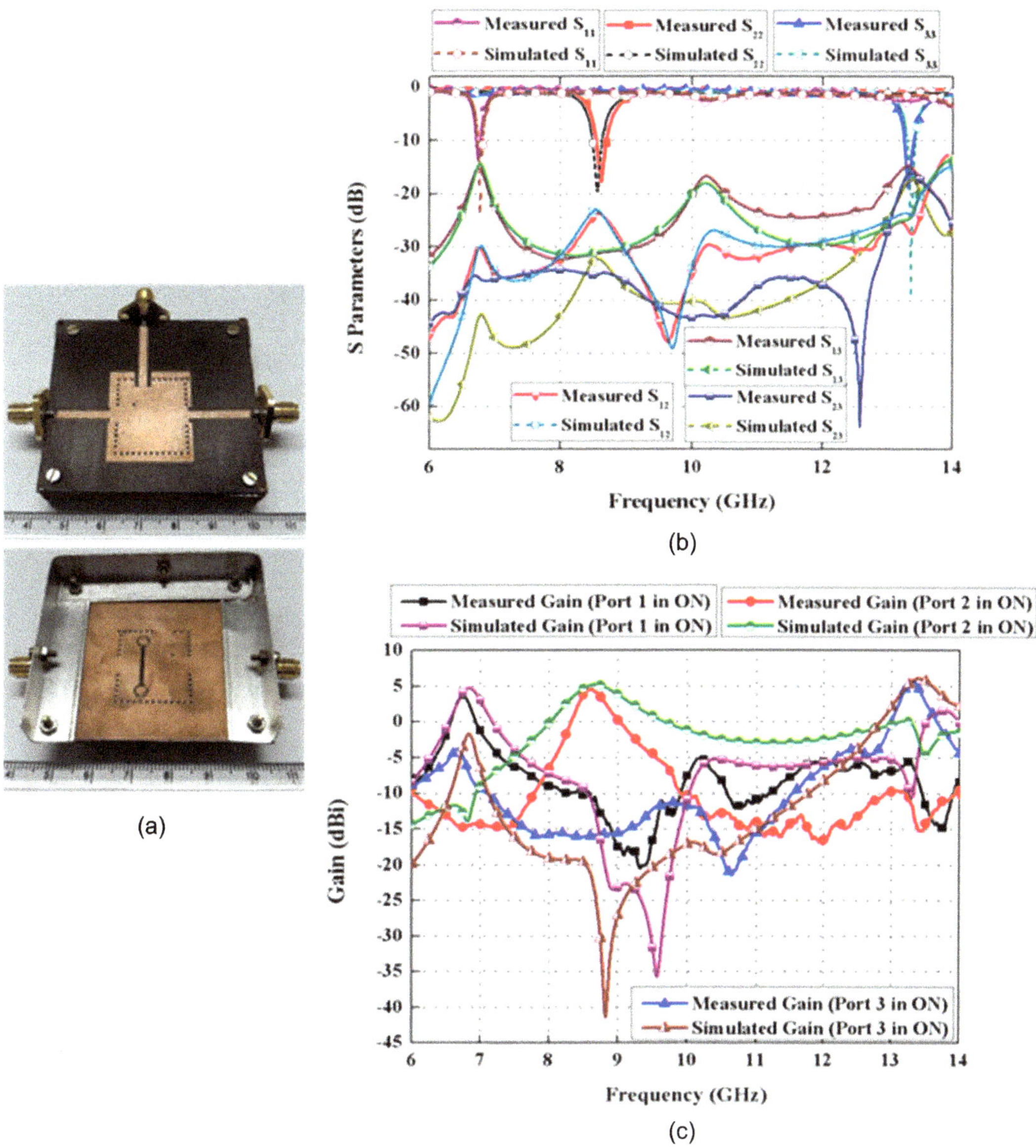

Figure 4.33 (a) Fabricated prototype. (b) Response of SIW cavity-backed self-triplexing antenna. (c) Gain profile of the antenna. (© 2021 EuCAP [10].)

4.3 SIW Slot Array Antenna

The slot antenna realized by utilizing SIW cavity modes produces unidirectional radiation pattern with comparatively moderate gain. Several design examples with multiband, broadband, or wideband response of such antennas have been discussed in previous sections. A design of a high-gain slot antenna with the help of an SIW cavity hybrid mode was discussed in Section 4.1. However, all these designs can achieve a gain in the order of 4.5 to 9 dBi. For practical applications operating at X-band and above, typically much higher gain (10 to 15 dBi) is required. Therefore,

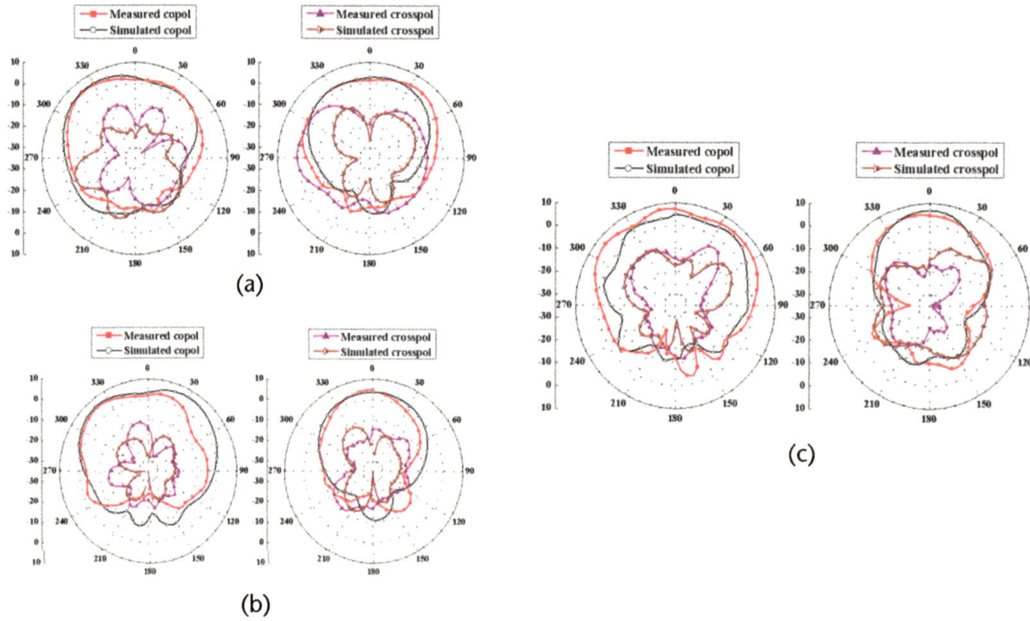

Figure 4.34 Radiation pattern of the antenna at three resonant frequency. (© 2021 EuCAP [10].)

array antenna technique must be applied to realize such system. The SIW slot array antenna can be realized in two ways: (1) utilizing the SIW cavity-backed slot as basic element and designing the feed network accordingly, and (2) implementing the SIW counterpart of the conventional waveguide slot array antenna. Both design strategies are popular for realizing high-gain antenna systems.

4.3.1 SIW Cavity-Backed Slot Array Antenna

To realize the SIW cavity-backed slot array antenna, a design of the SIW cavity-backed slot antenna previously explained in Section 3.2.1 is chosen. A dual-band SIW cavity-backed slot antenna utilizing a pair of slots on the top plate of the SIW cavity is used here. The advantage of using SIW technology to realize the array antenna is that the complete circuit including the antenna and the feed network can be realized using the same technology removing a lossy transition while implementing a completely shielded feeding network with negligible mutual coupling between elements. To design the feed network, the SIW-based power divider network needs to be realized.

4.3.1.1 The SIW Power Divider

The SIW power divider follows a similar working principle of the conventional waveguide-based power divider. Typically, a Tee or Y-type three-port junction is designed as a basic building block for such a power divider [11]. The typical configuration of both the junctions is shown in Figure 4.35. In the Tee junction, a via is placed at the junction center line, which acts as an inductive matching element to

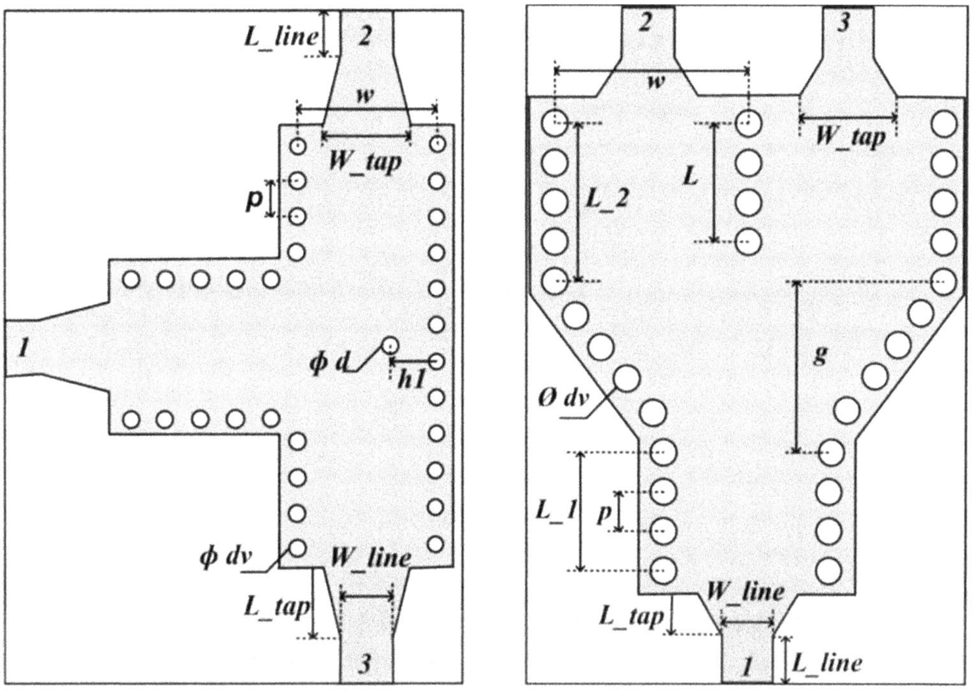

(a) (b)

Figure 4.35 SIW Tee and Y-type power divider. (© 2013 IEEE [11].)

compensate for the junction capacitance. In addition, sometimes additional vias are placed at the other end of the junctions for better matching performance. The position and diameter of the vias play a significant role to provide a better impedance-matching performance. The via combinations act as a coupling window between input and output ports, which determine the coupling of the dominant TE_{10}-mode wave towards the output ports. Since a pair of coupling windows is placed side by side, the input EM wave is coupled to two output ports with a certain power division ratio decided by the via placement and dimensions. It is evident that shifting the central via location towards one end of the junction will create two coupling windows of unequal width. Accordingly, the coupling to the output ports will be unequal in this case. Thus, by changing the via locations, either equal or unequal power division is achieved in the SIW Tee power divider. Here a study of the effect of the via diameter over the bandwidth performance of the Tee power divider is shown in Figure 4.36.

The SIW Y-type power divider also follows a similar working principle as its waveguide counterpart. The EM wave with the TE_{10} mode from the input port is fed to a transition section with a larger width. Such a change in the waveguide dimension helps to excite the higher-order TE_{20} mode in this section with its peak value of half of the input EM wave. The idea is to create a PEC wall in between the wide waveguide section at the output so that the EM wave is divided due to the change in the boundary condition and is fed into two output ports with an individual TE_{10}-mode wave of a magnitude having half of the input wave. Thus, the power division is achieved in this three-port junction. It is evident that the transition section and

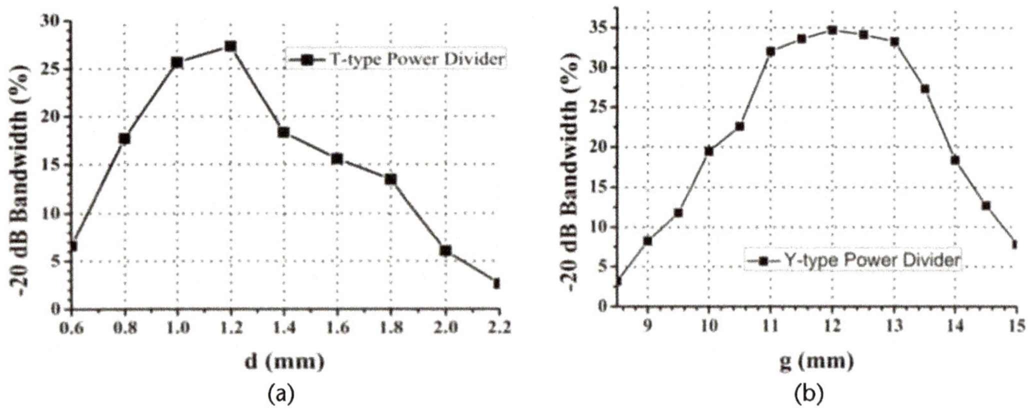

Figure 4.36 Effect of dimension on the performance of SIW power divider. (© 2013 IEEE [11].)

the placement of the central PEC wall play a crucial role for the EM wave to excite a higher-order mode and then excite the individual TE_{10} mode in each output port. A study of the effect of the placement of the central wall on the performance of such a power divider is shown in Figure 4.36.

A typical SIW antenna array feed network typically includes both type of junctions. A typical cascaded 1-to-4 power divider is shown in Figure 4.37 where, along with SIW Tee and Y power divider, an additional SIW bend is used to place the output ports side by side so that the network can feed antenna arrays. The performance of the complete system depends on the individual section as well as on some additional minor optimization during the integration of the different

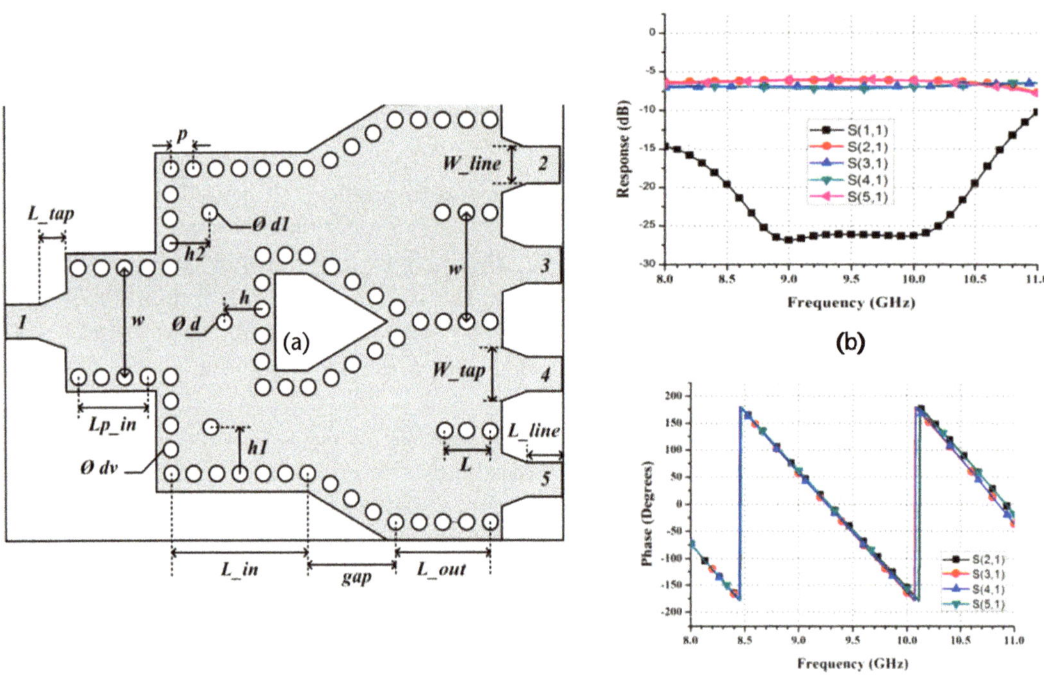

Figure 4.37 Design of the SIW 1-to-4 power divider and its response. (© 2013 IEEE [11].)

subsystem. A good antenna array feed network requires a power divider section with minimum amplitude and phase imbalance in the output ports arising from the junction reactances.

4.3.1.2 SIW Cavity-Backed Slot Array

The SIW cavity-backed slot antennas have been discussed in detail in Chapter 3. In this section, one of the designs is picked as a basic antenna element for realizing a slot array antenna. Here, we have taken the design of the SIW cavity-backed dual-slot antenna for dual-frequency application as basic antenna element. A 1×4 linear array is realized by placing the SIW cavities side by side with the slot pair on the top metallic plate of each cavity as shown in Figure 4.38 [12]. Since the complete design of the antenna array and feed network is realized using SIW technology; therefore, the feeding of the SIW cavity in this design is done through a section SIW instead of microstrip line feed as it was done in Section 3.2.1. A coupling window is realized in the junction of the SIW section and the cavity by forming a pair of H-plane steps.

The window helps to couple the desired TE^{y}_{110} mode in the SIW cavity from the dominant TE_{10}-mode field in the SIW feed section. The width of the window needs to be adjusted in order to ensure maximum coupling and accordingly to achieve good impedance matching performance of the antenna. The feed network consists of an SIW 1-to-4 power divider whose output ports are directly connected to the antenna leaving no need to have additional transition between the two sections. However, as shown in Figure 4.38, additional reflections cancelling H-plane steps are used in different sections of the power divider to improve its bandwidth

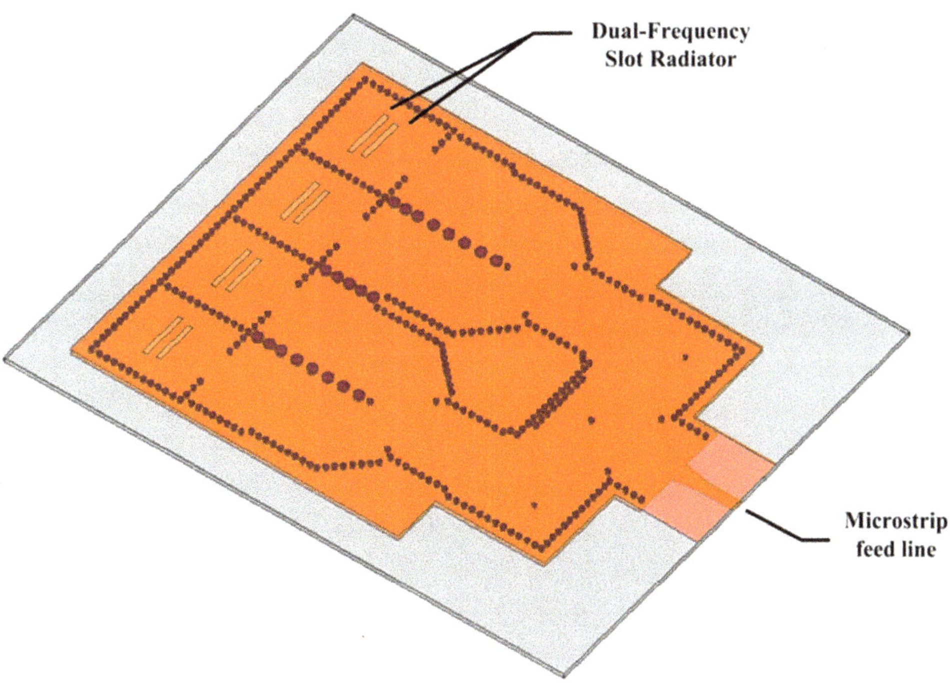

Figure 4.38 The SIW cavity-backed slot antenna array. (© 2015 IEEE [12].)

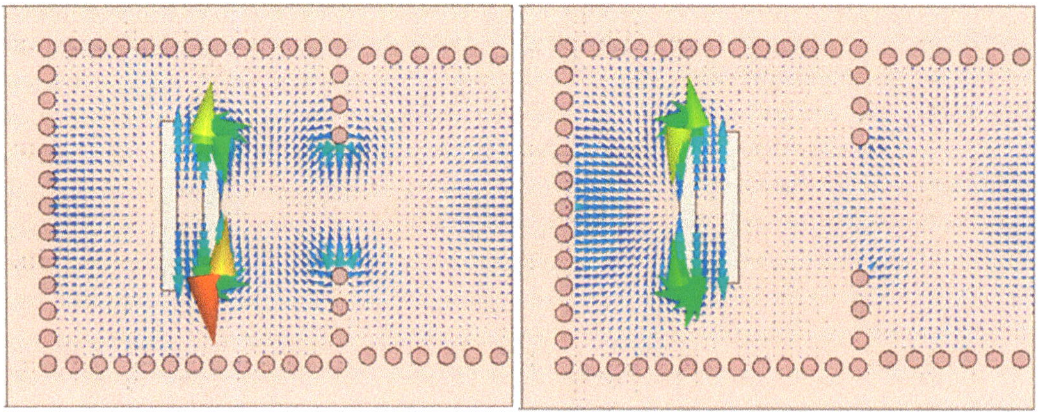

(a) (b)

Figure 4.39 Surface current distribution at two resonant frequencies of the antenna array element. (© 2015 IEEE [12].)

performance. The final design is fed by a small section of microstrip line at the input port of the power divider.

The slot pair in the single cavity resonates at two distinctive frequencies of 9.88 GHz and 10.76 GHz with a gain of 6 dBi and 5.9 dBi, respectively. The power divider exhibits −15-dB S_{11} bandwidth of 1.6 GHz with output power level of −6.33 dB. The phase imbalance between the output ports are within 6°. The advantage of using SIW-based slot array antenna is that the self-shielded structure of the SIW technology helps to eliminate unwanted mutual coupling between the elements and as a result, the final array design almost mimics its basic element response. The proposed antenna exhibits a gain of 11.3 dBi and 11.8 dBi, respectively, at two resonances with a sidelobe level below 12.5 dB at both resonances. The cross-polar level of the array antenna is below −23 dB at both frequencies.

4.3.1.3 SIW Section-Based Slot Array Antenna

In the previous sections, the SIW-based antenna array feed network is used along with an antenna array made of basic cavity-backed slot antenna element to realize

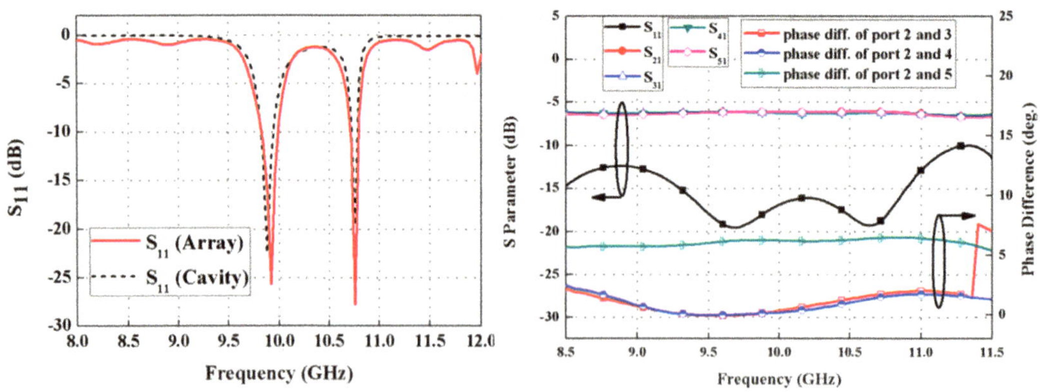

(a) (b)

Figure 4.40 Response of the antenna array and feed network. (© 2015 IEEE [12].)

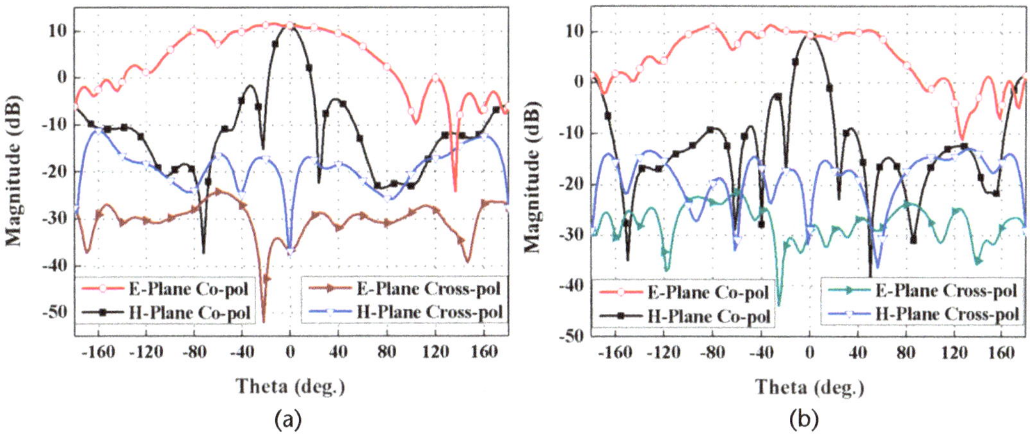

Figure 4.41 Radiation pattern of the antenna array at: (a) 9.88 GHz and (b) 10.76 GHz. (© 2015 IEEE [12].)

the slot array. The same configuration can be used to implement array antenna using any other basic element, which may be fed by a similar SIW-based feeding network.

Another approach to design an array antenna that utilizes fundamental properties of waveguide structure is the slot array on the top wall of the waveguide. In this case, the waveguide itself acts as feed network for all the antenna elements without the need of any three-port junctions as compared to the previous design. The same can be realized in a planar substrate using SIW technology. The SIW is typically excited by the dominant TE_{10} mode, which generates a surface current component in longitudinal and transverse directions at the top plate of the waveguide. By properly placing the slot antenna element, the surface current path can be disturbed, which, in effect, helps to excite the slot antenna element to radiate. A design example of exciting a slot antenna by the dominant TE_{10} mode of the SIW is shown in Figure 4.42.

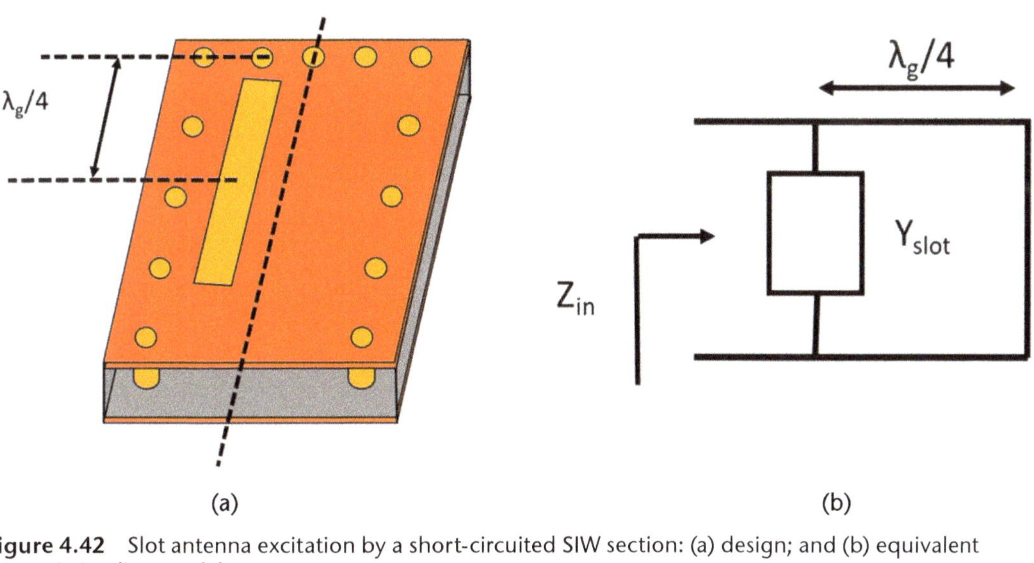

Figure 4.42 Slot antenna excitation by a short-circuited SIW section: (a) design; and (b) equivalent transmission line model.

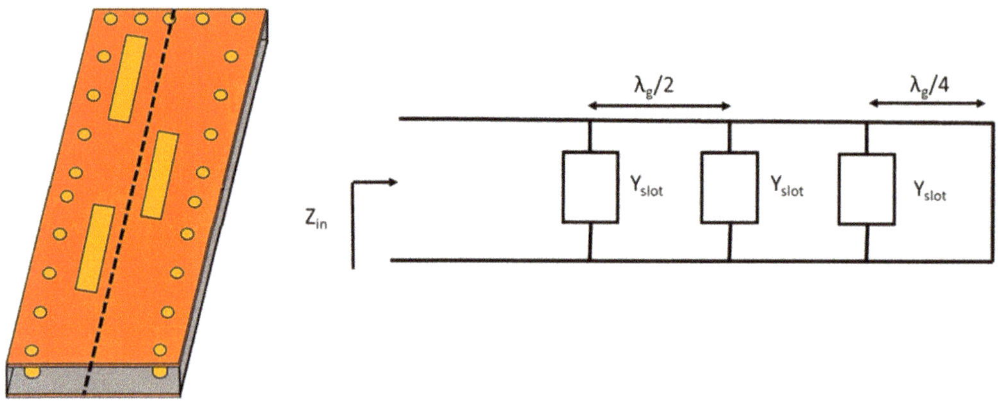

Figure 4.43 Slot antenna array excitation by a short-circuited SIW section: (a) design; and (b) equivalent transmission line model.

The slot is placed slightly shifted from the center line of the SIW such that it disturbs the *X* component of the surface current (refer to (2.4)). The surface current distribution varies as the sinusoid function of position x. Therefore, the more the slot is shifted from the center line of the SIW, the amount of EM energy coupling due to the surface current path disturbance also varies as the square of the sinusoid. This phenomenon can be seen in two ways:

1. The slot antenna is seen as a load in the waveguide, which is a function of its position. Therefore, the equivalent admittance of the slot antenna changes in proportion to the square of the sinusoid of the position.
2. The amount of energy coupled to the slot antenna from the waveguide is also dependent on its position.

To ensure impedance matching at the input end of the SIW, the equivalent impedance of the slot antenna should match the impedance of the line. This can be achieved by choosing the proper slot position. A typical configuration of an SIW-fed slot requires a short-circuited line with the slot placed at a quarter-wavelength from the short-circuited end. The equivalent transmission line representation is shown in Figure 4.42. The input impedance seen by the line should match the characteristic impedance of the line, which can be ensured by choosing proper value of equivalent admittance of the slot, which, in effect, is dependent on the slot position. At the same time, the amount of energy coupled to the slot also depends on the slot position and, as a result, radiation is achieved from the slot antenna into free space.

The working principle can be extended to design a slot array antenna fed by the SIW section. The first slot element is placed at a quarter-wavelength from the short-circuited end and all other slot element is placed at a distance of a half-wavelength from each other. By seeing the surface current distribution in Figure 2.11 in Section 2.2, it is evident that to maintain the same polarity of the excitation field, the slot elements need to be placed on the alternate side of the center line of the waveguide.

The effective conductance of the slot can be calculated by the Stevenson method [13, 14], which takes the form as follows:

$$g(x) = g_0 \sin^2\left(\frac{\pi x}{a}\right) \tag{4.25}$$

where g_0 is the normalized conductance, which depends on the slot dimension and waveguide dimension. For an initial starting point, each slot can be assumed to be a half-wavelength long. The equivalent admittance of each slot element adds in phase due to the placement of the slot at a half-wavelength and in an alternate fashion with respect to the center line of the waveguide. The input impedance of the line is the sum of all slot admittance. However, the slot conductance is proportional to the slot excitation voltage. As a result, we can have the design equations shown below to determine the slot position:

$$g_n(x) = kv_n^2 \tag{4.26}$$

$$\sum_n g_n(x) = Y_0 \tag{4.27}$$

where Y_0 is the characteristic admittance of the equivalent transmission line. The slot excitation voltage is chosen to maintain predefined amplitude distribution (as per the design strategy of uniform or nonuniform excitation of the array antenna), which can be used to determine the optimum position of the slot antenna element.

The above calculation does not include the effect of mutual coupling between different slots, which is a significant phenomenon present in the SIW slot array antenna. The mutual coupling affects the slot admittance by adding additional susceptance terms, which can be compensated by changing the slot length and position. An iterative algorithm may be applied to find out the optimized length and position of each slot, which can give a better result [15].

2.3.1.4 SIW Section-Based, Dual-Frequency Slot Array

The performance of the SIW section-based slot array antenna can be improved by modifying the design to have multiple resonant frequencies while utilizing a similar working principle. Here, a specific example of a dual-frequency antenna array utilizing a similar working principle is explained [16]. The design is shown in Figure 4.44. The array contains two sets of slot arrays corresponding to two resonant frequencies. The slot array *A* corresponding to a lower frequency is placed such a way that the first slot of group A starts at $3 \times \lambda_{g1}/4$ distance where λ_{g1} is a guided wavelength at the higher frequency. All other slots are placed at a distance of $\lambda_{g1}/2$ from each other and placed in an alternate fashion of the center line of the SIW due to the same working principle as described before.

For the slot array *B* corresponding to a higher resonant frequency, the first slot is placed at a distance of $n \times \lambda_{g2}/4$ where n is an odd integer and λ_{g2} is a guided wavelength at a lower resonant frequency. The integer n is chosen such that the first slot of array *B* is placed after the last slot of the array *A*. In addition, a small H-plane step is added in between two slot array groups and after the slot array *B*, which helps to reduce the loading effect of one array on the other as well as to improve impedance matching. The slot array *B* is designed to have impedance matching at the lower frequency, which satisfies (4.27) and, as a result, almost all

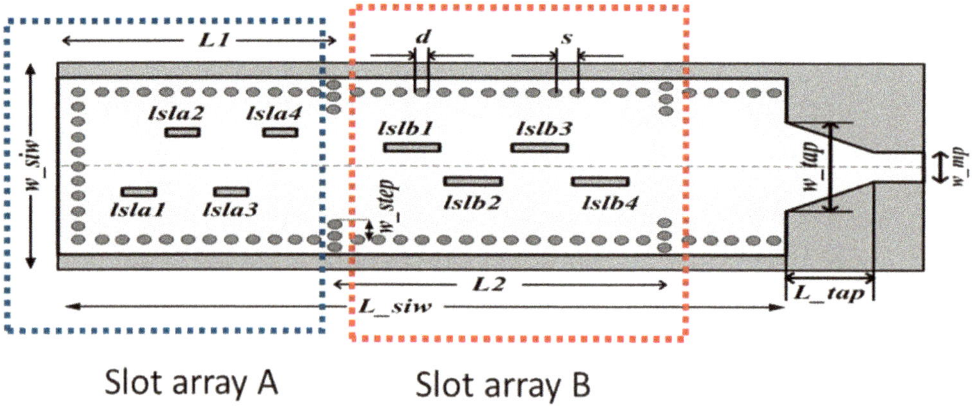

(a) (b)

Figure 4.44 SIW section-based, dual-frequency slot array antenna. (© 2013 IEEE [16].)

the EM wave is radiated into free space, leaving a negligible field passing through the slot array *A*. Uniform excitation is maintained by choosing the slot position properly as explained in (4.26). At a higher operating frequency, the slot array *B* does not act as resonant slot sections and, as a result, the effect of slot array *B* is negligible at this frequency. However, the slot array *A* is impedance-matched and therefore the EM wave is radiated when it reaches to slot array *A*. However, some small effect of slot array *B* is seen in the radiation pattern of a higher frequency. The performance of the array antenna can be seen in Table 4.3. The side lobes are around –13 dB down at 8.6 GHz as shown in Figure 4.47 and around –10 dB down

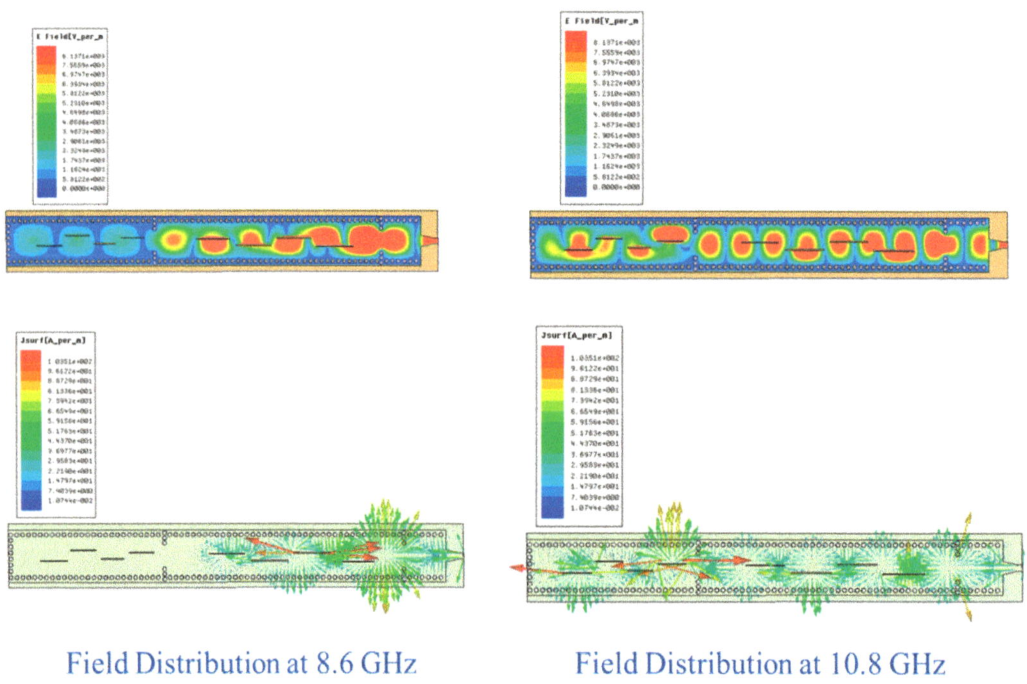

(a) (b)

Figure 4.45 Field distribution at both resonant frequencies of the antenna array. (© 2013 IEEE [16].)

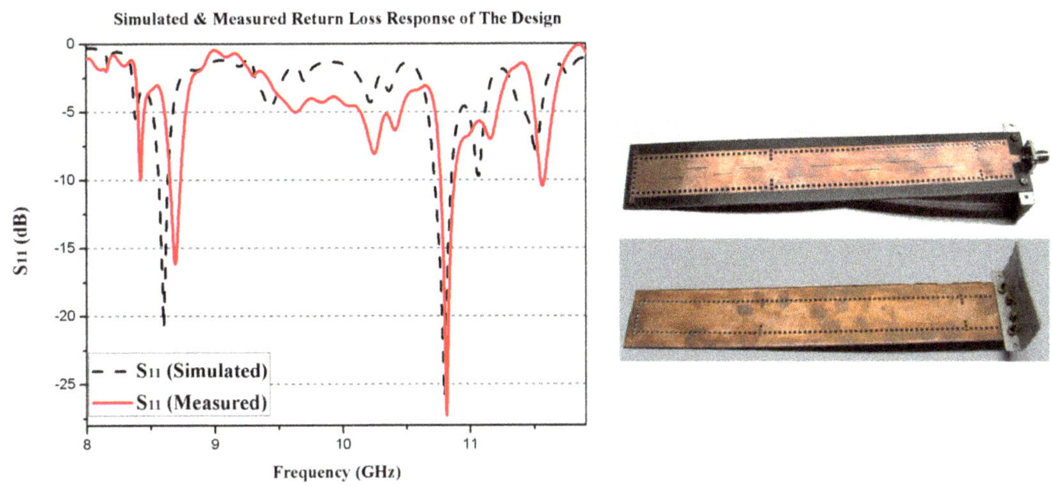

(a) (b)

Figure 4.46 Response of the SIW dual-frequency slot array antenna. (© 2013 IEEE [16].)

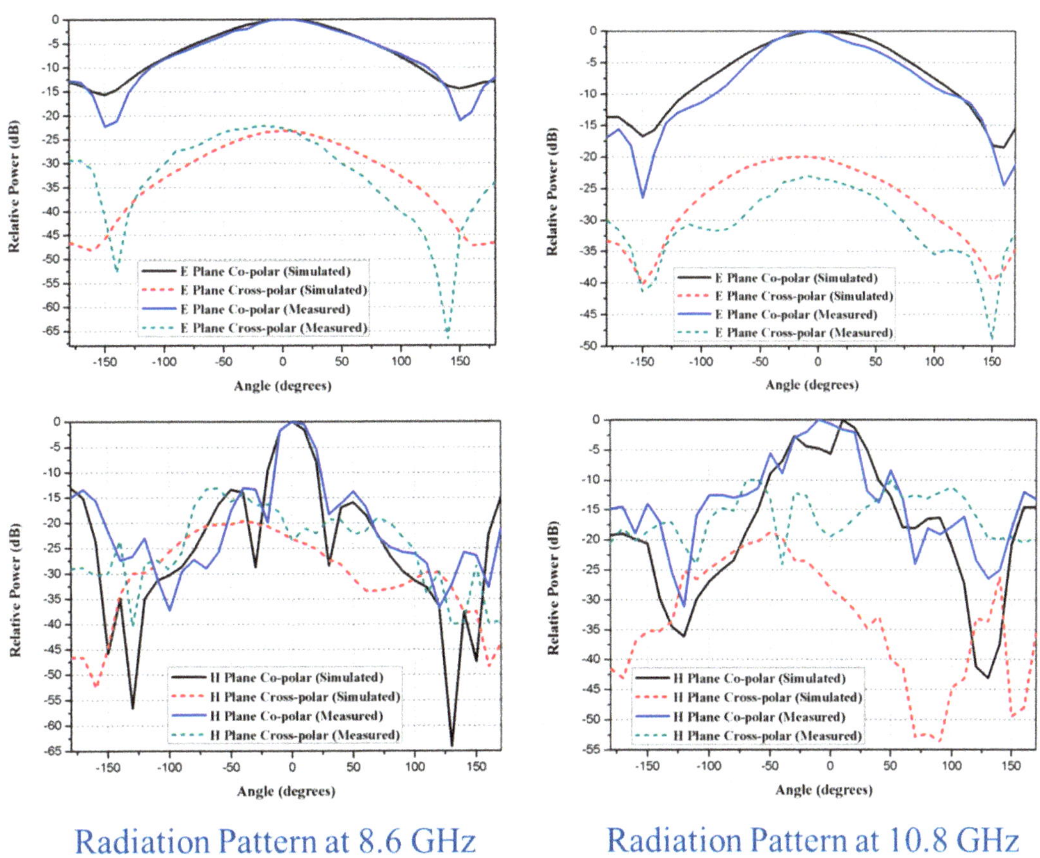

(a) (b)

Figure 4.47 E-plane and H-plane radiation patterns of the antenna array. (© 2013 IEEE [16].)

Table 4.3 Comparative Study of the Performance of the Antenna

Parameter		*Simulated*	*Measured*
First Resonance	Frequency (GHz)	8.6	8.68
	Bandwidth	73.6 MHz (0.85%)	80 MHz (0.92%)
	Gain (dB)	9.99	7.71
Second Resonance	Frequency (GHz)	10.8	10.8
	Bandwidth	135.5 MHz (1.25%)	170 MHz (1.57%)
	Gain (dB)	10.05	9.08

at 10.8 GHz. Both arrays use the same feeding SIW section, reducing the complexity of the feed network.

Several other designs of the SIW slot array antenna implementing high-gain planar array, low side-lobe array, and dual-polarized and circular-polarized arrays have been reported in the literature. The readers are encouraged to review the existing literature to know more details in this regard.

References

[1] Mukherjee, S., and A. Biswas, "Design of Elliptical SIW Cavity Backed High Gain Slot Antenna," *IEEE-INAE Workshop on Electromagnetics*, Trivandrum, 2018.

[2] Kretzschmar, J. G., "Mode Charts for Elliptical Resonant Cavities," *Electron. Lett.*, Vol. 6, 1970, pp. 432–434.

[3] Kretzschmar, J. G., "Wave Propagation in Hollow Conducting Elliptical Waveguides," *IEEE Transactions on Microwave Theory and Techniques*, Vols. MTT-18, 1970, pp. 547–5530.

[4] Mukherjee, S., and A. Biswas, "Design of Planar High-Gain Antenna Using SIW Cavity Hybrid Mode," *IEEE Transactions on Antennas and Propagation*, Vol. 66, No. 2, 2018, pp. 972–977.

[5] Balanis, C. A., *Antenna Theory Analysis and Design*, New York: Wiley, 2005.

[6] Mukherjee, S., and A. Biswas, "Compact High Gain Half-Mode Substrate Integrated Waveguide Cavity Antenna Using Hybrid Mode," *Int. J. RF Microw. Comput. Aided Eng.*, 2019.

[7] Park, D. K., et al., "Self-Diplexed Integrated Antenna Transceiver for Wireless Application," *IEEE Ant. and Propag. Society Symp.*, 2001.

[8] Kurup, D. G., and A. Rydberg, "Amplifying Active Reflect-Antenna Using a Microstrip-T Coupled Patch Design and Measurement," *IEEE Transactions on Microwave Theory and Techniques*, Vol. 51, No. 8, 2003, pp. 1960–1965.

[9] Mukherjee, S., and A. Biswas, "Design of Self-Diplexing Substrate Integrated Waveguide Cavity-Backed Slot Antenna," *IEEE Antennas and Wireless Propagation Letters*, Vol. 15, 2016, pp. 1775–1778.

[10] Mukherjee, S., S. Ghosh, and A. Biswas, "Design of Compact SIW Cavity Backed Self-Triplexing Planar Slot Antenna for Triple Band Application," *2021 15th European Conference on Antennas and Propagation (EuCAP)*, 2021.

[11] Dutta, S., S. Mukherjee and A. Biswas, "Design of Broadband Power Divider Based on Substrate Integrated Waveguide Technology," *IEEE Applied Electromagnetics Conference (AEMC)*, Bhubaneswar, 2013.

[12] Mukherjee, S., and A. Biswas, "Design of Dual-Frequency Substrate Integrated Waveguide (SIW) Cavity Backed Slot Array Antenna," *IEEE Int. Symp. on Antennas and Propag. and North American Radio Science Meeting (APS/URSI) 2015*, Vancouver, Canada, 2015.

[13] Farrall, A., and P. R. Young, "Integrated Waveguide Slot Antennas," *Electronics Letters*, Vol. 40, 2004.

[14] Weng, Z. -B., R. Guo, and Y. -C. Jiao, "Design and Experiment on Substrate Integrated Waveguide Resonant Slot Array Antenna at Ku-Band," *7th International Symposium on Antennas, Propagation & EM Theory*, Guilin, China, 2006.

[15] Chen, X. -P., et al., "Low-Cost High Gain Planar Antenna Array for 60-GHz Band Applications," *IEEE Transactions on Antennas and Propagation*, Vol. 58, No. 6, 2010, pp. 2126–2129.

[16] Mukherjee, S., K. V. Srivastava, and A. Biswas, "Implementation of Dual-Frequency Longitudinal Slot Array Antenna on Substrate Integrated Waveguide at X-Band," *2013 European Microwave Conference*, Nuremberg, Germany, 2013.

CHAPTER 5

SIW Filters

In Chapter 4, the detailed analysis and design of various types of SIW-based cavity-backed slot antennas were discussed.

In this chapter, bandpass filter synthesis technique, SIW-based single-band and dual-band bandpass filters, and bandstop filter design is discussed. Besides that, miniaturized SIW-based bandpass filters design are also demonstrated. All the filters are designed using the filter synthesis procedure and the coupling theory analysis. The use of SIW technology helps to implement high *Q* filter response while maintaining planar configuration.

5.1 Bandpass Filter Synthesis and the Coupling Matrix

A bandpass filter passes the desired frequencies and rejects the undesired one. In almost all RF front-end systems, the microwave filter is an integral component such as mobile radio, satellite communications, and radar systems. The physical realization of microwave filters can differ but the circuit network theory is common to all. The current section explains the fundamental of the filter synthesis technique, which is applicable to the narrowband microwave bandpass filter design.

For any two-port lossless filtering network composed of *N* number of coupled resonators, the reflection and transfer functions can be expressed as a ratio of two *N*th-degree complex polynomials [1].

$$S_{11}(s) = \frac{F(s)}{E(s)} \tag{5.1}$$

$$S_{21}(s) = \frac{P(s)}{\varepsilon E(s)} \tag{5.2}$$

where s is the complex frequency $s = \sigma + j\omega$ and ε is equi-ripple constant. $P(s)$, $F(s)$, and $E(s)$ are the normalized polynomials whose highest degree coefficients are unity. The polynomial $P(s)$ is constructed from design-specified transmission zeros. In this case, the Nth-order filter has finite frequency transmission zeros less than N. Roots of $F(s)$ correspond to the reflection zeros. $E(s)$ is a Hurwitz polynomial of degree N whose roots all lie in the left half of the s-plane [2]. ε can be expressed in terms of prescribed return loss (RL) in dB in the passband given by (5.3).

$$\varepsilon = \frac{1}{\sqrt{10^{RL/10} - 1}} \left| \frac{P(s)}{F(s)} \right|_{s=\pm j} \tag{5.3}$$

The polynomials $F(s)$, $P(s)$, and $E(s)$ must satisfy the following conditions to have a realizable lossless two-port network [3]:

1. $E(s)$ is a strict Hurwitz polynomial. So all its roots must lie in the left half of s-plane.
2. The polynomial $P(s)$ contains prescribed transmission zeros of degree $Z \leq N - 1$ and satisfies the condition $P = (-1)^{N+1} P^*$ (i.e., its roots must be symmetric with respect to the imaginary axis of s-plane).
3. $F(s)$ is a polynomial of degree N.
4. Since the filter is assumed to be a lossless two-port network, it follows the conservation of energy as given in (5.4).

$$E(s)E^*(s) = F(s)F^*(s) + \frac{1}{\varepsilon^2} P(s)P^*(s) \tag{5.4}$$

The discussed polynomials are Nth-order filter with finite frequency transmission zeros less than N. In general, the Nth-order filter can have maximum N finite frequency transmission zeros such as the fully canonical filter where the number of transmission zeros are equal to the degree of the filter. In this case, the transmission and reflection coefficients are in the form:

$$S_{11}(s) = \frac{F(s)}{\varepsilon_R E(s)} \tag{5.5}$$

$$S_{21}(s) = \frac{P(s)}{\varepsilon E(s)} \tag{5.6}$$

Comparing with (5.1), the ripple factor ε_R is added in the denominator of (5.5). The polynomials $E(s)$, $F(s)$, and $P(s)$ are normalized polynomials with the highest coefficients unity and ε is the ripple constant. Using conservation of energy for a two-port lossless system, the filtering polynomial can be expressed as

$$E(s)E^*(s) = \frac{1}{\varepsilon_R^{\,2}} F(s)F^*(s) + \frac{1}{\varepsilon^2} P(s)P^*(s) \tag{5.7}$$

When the total number of transmission zeros is less than the total number of poles (roots of $F(s)$), ε_R is unity. Otherwise, both ε_R and ε can be evaluated [1] using two conditions given in (5.8) and (5.10). The first one is at $\omega = \pm\infty$ and second one is at $\omega = \pm 1$ with the conservation of energy for a lossless network. For N-degree polynomial with N transmission zeros:

$$\frac{1}{\varepsilon_{R^2}}\left|\frac{F(\pm j\infty)}{E(\pm j\infty)}\right|^2 + \frac{1}{\varepsilon^2}\left|\frac{P(\pm j\infty)}{E(\pm j\infty)}\right|^2 = 1 \tag{5.8}$$

$$\frac{1}{\varepsilon_R^2} + \frac{1}{\varepsilon^2} = 1 \tag{5.9}$$

The ripple constant can also be expressed using return loss of the filter at $\omega = \pm 1$.

$$\varepsilon = \varepsilon_R \frac{1}{\sqrt{10^{RL/10} - 1}}\left|\frac{P(\pm j1)}{F(\pm j1)}\right|_{\omega=\pm 1} \tag{5.10}$$

The ripple factor ε_R is obtained using (5.8) and (5.9):

$$\varepsilon_R = \sqrt{1 + \left((10^{RL/10} - 1\right)}\left|\frac{F(\pm j1)}{P(\pm j1)}\right|_{\omega=\pm 1} \tag{5.11}$$

Using (5.7), the transfer functions can be expressed as

$$\left|s_{21}(j\omega)\right|^2 = \frac{1}{1 + \frac{\varepsilon^2}{\varepsilon_R^2}\left|C_N^2(j\omega)\right|} \tag{5.12}$$

where $C_N(s) = F(s)/P(s)$.

$C_N(\omega)$ is known as the filter characteristic function of the Nth-degree filter and its pole and zeros are the roots of $P(\omega)$ and $F(\omega)$, respectively.

5.1.1 Filter Characteristic Functions of the Polynomial

For the general Chebyshev characteristic, the filter characteristic function of the Nth degree is expressed as [1]

$$C_N(\omega) = \cosh\left[\sum_{n=1}^{N}\cosh^{-1}\left(x_n(\omega)\right)\right] \tag{5.13}$$

where $x_n(\omega) = (\omega - 1/\omega)/(1 - \omega/\omega_n)$ and ω_n is the position of the nth transmission zero in the ω-plane.

General Chebyshev approximation ensures equi-ripple insertion loss in the passband and arbitrary placed transmission zeros at the stopband. Thus, highly selective filters with asymmetrical frequency response can be realized, which is an essential requirement for modern wireless communication systems. The necessary conditions for Chebyshev response are as follows:

$$C_N(\omega) \equiv \begin{cases} = 0, & \text{if } |\omega| = 1 \\ < 1, & \text{if } |\omega| < 1 \\ > 1, & \text{if } |\omega| > 1 \end{cases} \tag{5.14}$$

When the transmission zeros are approached to infinity, $C_N(\omega)$ becomes pure Chebyshev response (5.15).

$$C_N(\omega)\Big|_{\omega_n \to \infty} = \cosh\left[N \cosh^{-1}(\omega)\right] \tag{5.15}$$

Since

$$x_n(\omega)\Big|_{\omega_n \to \infty} = \omega - \frac{1}{\omega}\Big/ 1 - \frac{\omega}{\omega_n} = \omega$$

5.1.2 Coupling Matrix

5.1.2.1 Circuit Model

In the early 1970s, Atia and Williams [4] proposed the concept of the coupling matrix, which is applied to dual-mode narrowband waveguide filters. The coupling matrix was constructed from the voltage-current relationship of the equivalent circuit model. The proposed equivalent circuit model was a narrowband bandpass prototype as shown in Figure 5.1. The circuit is driven by a source of open-circuit voltages e_S with resistance R_s and terminated by a load R_L. The circuit model is composed of lumped-element series resonators, which are coupled by transformers. i is the loop current flowing in each series resonator given in (5.16). Each resonator is represented by a capacitor of 1 F in series with the self-inductance of 1H within each loop. So the resonant frequency of each resonator is 1 rad/s and the bandwidth is 1 rad/sec. Each loop is coupled to every other loop through mutual couplings and these couplings are assumed to be frequency-invariant due to the narrowband approximation. The matrix M is called the coupling matrix in (5.17). In the coupling

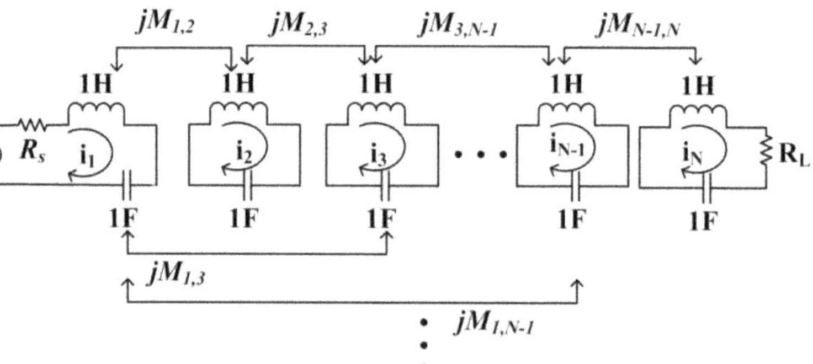

Figure 5.1 Prototype of the general *N*th-order coupled bandpass filter.

matrix M, all the diagonal elements are zero because the self-resonant frequency of each series resonator is equal.

$$\begin{bmatrix} e_S \\ 0 \\ \cdot \\ \cdot \\ \cdot \\ 0 \\ 0 \end{bmatrix} = \begin{bmatrix} R_S + j\left(\omega - \frac{1}{\omega}\right) & jM_{12} & \cdots & jM_{1,N-1} & jM_{1,N} \\ jM_{12} & j\left(\omega - \frac{1}{\omega}\right) & \cdots & jM_{2,N-1} & jM_{2,N} \\ \vdots & \vdots & \ddots & \vdots & \vdots \\ jM_{N-1,1} & jM_{N-1,2} & \cdots & j\left(\omega - \frac{1}{\omega}\right) & jM_{N-1,N} \\ jM_{N,1} & jM_{N,2} & \cdots & jM_{N,N-1} & R_L + j\left(\omega - \frac{1}{\omega}\right) \end{bmatrix} \begin{bmatrix} i_1 \\ i_2 \\ \cdot \\ \cdot \\ \cdot \\ i_{N-1} \\ i_N \end{bmatrix} \quad (5.16)$$

$$e_S \begin{bmatrix} 1 \\ 0 \\ \cdot \\ \cdot \\ \cdot \\ 0 \\ 0 \end{bmatrix} = \left[R + j\left(\omega - \frac{1}{\omega}\right) U + jM \right] \begin{bmatrix} i_1 \\ i_2 \\ \cdot \\ \cdot \\ \cdot \\ i_{N-1} \\ i_N \end{bmatrix} \quad (5.17)$$

$$E = e_S \begin{bmatrix} 1 \\ 0 \\ \cdot \\ \cdot \\ \cdot \\ 0 \\ 0 \end{bmatrix} \text{ and } I = \begin{bmatrix} i_1 \\ i_2 \\ \cdot \\ \cdot \\ \cdot \\ i_{N-1} \\ i_N \end{bmatrix} \quad (5.18)$$

Lowpass to bandpass transformation is given by

$$s = j\left(\frac{\omega_0}{\Delta\omega}\right)\left(\omega - \frac{1}{\omega}\right) \quad (5.19)$$

Since $\omega_0 = 1$ rad/sec and $\Delta\omega = 1$ rad/sec, the above equations result in

$$s = j\left(\omega - \frac{1}{\omega}\right) \quad (5.20)$$

$$E = AI \quad (5.21)$$

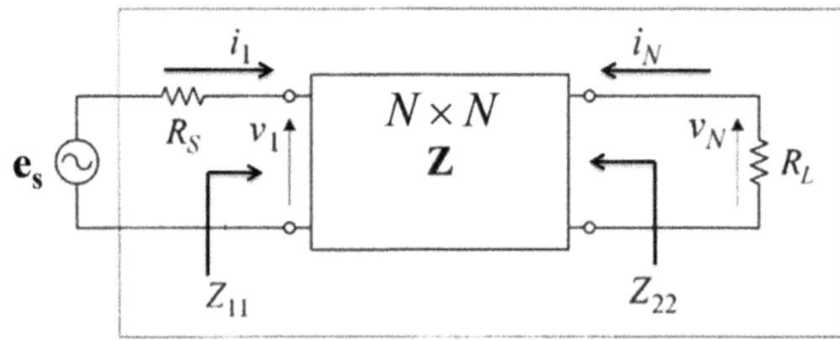

Figure 5.4 An *N*th-order filter representing in impedance matrix including source and load terminations. The impedance matrix *Z* represents a purely reactive network operating between a voltage source with internal impedance and a load.

$$E = AI \tag{5.27}$$

$$A = R + sI + jM; \tag{5.28}$$

$$I = A^{-1}E \tag{5.29}$$

$$A^{-1} = \frac{adj(A)}{\det(A)} \tag{5.30}$$

$$i_1 = A_{1,1}{}^{-1}E \tag{5.31}$$

$$i_N = A_{N,1}{}^{-1}E = \frac{v_N}{R_L} \tag{5.32}$$

$$\text{The transmission coefficient } S_{21} = 2\sqrt{\frac{R_S}{R_L}} \bullet \frac{v_N}{e_S} = 2\sqrt{R_S R_L} A_{N,1}{}^{-1} \tag{5.33}$$

$$\text{The input reflection coefficient } S_{11} = \frac{Z_{11} - R_S}{Z_{11} + R_S} = 1 - \frac{2R_S}{Z_{11} + R_S} \tag{5.34}$$

$$i_1 = \frac{e_S}{Z_{11} + R_S} = A_{1,1}{}^{-1}e_S \tag{5.35}$$

Therefore,

$$A_{1,1}{}^{-1} = \frac{1}{Z_{11} + R_S} \tag{5.36}$$

$$S_{11} = 1 - 2R_S A_{1,1}^{-1} \tag{5.37}$$

Similarly, the output reflection coefficient can be calculated as

$$S_{22} = 1 - 2R_L A_{N,N}^{-1} \tag{5.38}$$

Filter synthesis determines the coupling matrix, R_S and R_L, such that the S-parameters, given by (5.33) to (5.38), reproduce the predetermined insertion loss and return loss. It can be performed by using either analytic methods or optimization methods.

Figure 5.5(a) shows that source and load terminations are unity by introducing impedance inverters $M_{S,1} = \sqrt{R_S}$ and $M_{N,L} = \sqrt{R_L}$. Both these inverters convert the $N \times N$ impedance matrix into $(N + 2) \times (N + 2)$ impedance matrix as shown in Figure 5.5(b). In this matrix, generally $M_{S,S} = 0$ and $M_{L,L} = 0$ and it is also possible to exist coupling between $M_{S,1}$, $M_{N,L}$ and all internal nodes within the core $N \times N$ matrix. In addition to that, direct source and load ($M_{S,L}$) coupling can also be considered for realizing a fully canonical filter functions (5.39). Here, all the diagonal elements ($M_{i,i}$) are self-resonant frequencies. Since the filter is a passive and reciprocal component so $M_{i,j} = M_{j,i}$.

$$M = \begin{bmatrix} 0 & M_{S,1} & M_{S,2} & \cdots & M_{S,N-1} & M_{S,N} & M_{S,L} \\ M_{S,1} & M_{1,1} & M_{1,2} & \cdots & M_{1,N-1} & M_{1,N} & M_{1,L} \\ M_{S,2} & M_{1,2} & M_{2,2} & \cdots & \cdot & \cdot & \cdot \\ \cdot & \cdot & \cdot & \cdots & \cdot & \cdot & \cdot \\ \cdot & \cdot & \cdot & \cdots & \cdot & \cdot & \cdot \\ \cdot & \cdot & \cdot & \cdots & \cdot & \cdot & \cdot \\ M_{S,N} & M_{N,1} & \cdot & \cdots & M_{N,N-1} & M_{N,N} & M_{N,L} \\ M_{S,L} & M_{L,1} & \cdot & \cdots & M_{L,N-1} & M_{L,N} & 0 \end{bmatrix} \tag{5.39}$$

The transmission and refection coefficients are calculated using (5.40):

$$S_{11} = 1 - 2A_{1,1}^{-1} \text{ and } S_{21} = 2A_{N+2,1}^{-1} \tag{5.40}$$

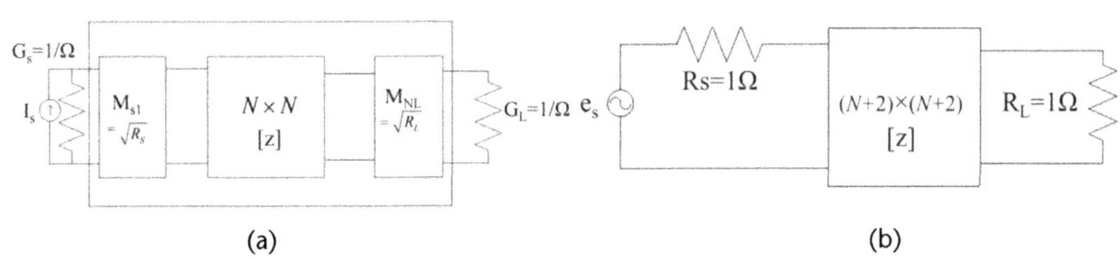

Figure 5.5 (a) $N \times N$ impedance matrix with inverters to normalize the termination to unity and (b) $(N + 2) \times (N + 2)$ impedance matrix with unity normalized terminating source (R_s) and load (R_L) resistances.

$$K = \sum_{i=1}^{NP} \left|S_{11}\left(\omega_{Pi}\right)\right|^2 + \sum_{i=1}^{NZ} \left|S_{21}\left(\omega_{Zi}\right)\right|^2 + + \left|S_{11}(\omega = 1) - \frac{\epsilon}{\sqrt{1+\epsilon^2}}\right|^2 + \left|S_{11}(\omega = -1) - \frac{\epsilon}{\sqrt{1+\epsilon^2}}\right|^2 \tag{5.48}$$

where ω_{Pi} is the location of ith pole of the filtering function and ω_{Zi} is the location of ith zero of the filtering function. The gradient components of the cost function are as follows:

$$\frac{\partial S_{11}}{\partial M_{pq}} = -4jP_{pq}\left[A^{-1}\right]_{1p}\left[A^{-1}\right]_{q1}, \quad p \neq q \tag{5.49}$$

$$\frac{\partial S_{21}}{\partial M_{pq}} = 2jP_{pq} \cdot \left(\left[A^{-1}\right]_{N+2,p}\left[A^{-1}\right]_{q1} + \left[A^{-1}\right]_{N+2,q}\left[A^{-1}\right]_{p1}\right), \quad p \neq q \tag{5.50}$$

Here, $P_{pq} = 1$, when $M_{pq} \neq 0$; if $M_{pq} = 0$, then $P_{pq} = 0$.

$$\frac{\partial S_{11}}{\partial M_{pp}} = -2jP_{pp}\left[A^{-1}\right]_{p1}\left[A^{-1}\right]_{p1} \tag{5.51}$$

$$\frac{\partial S_{21}}{\partial M_{pp}} = 2jP_{pp}\left[A^{-1}\right]_{N+2,P}\left[A^{-1}\right]_{p1} \tag{5.52}$$

Here, $P_{pp} = 1$, when $M_{pp} \neq 0$; if $M_{pp} = 0$, then $P_{pp} = 0$.

The only difference between the diagonal and off-diagonal coupling gradient is that diagonal coupling coefficient's gradient is divided by a factor of 2 because the diagonal elements of a symmetric matrix occur only once. Thus, the optimization technique using a cost function as well as a gradient of the cost function brings faster convergence and makes the optimization process much more robust.

5.2 Bandpass Filter Design Using a Single-Mode Rectangular SIW Cavity Loaded with a Metallic Inductive Post and Iris

Waveguide-based bandpass filter can be realized using E-plane or H-plane irises and metallic inductive posts. In SIW technology, E-plane-based filter is difficult to realize due to fabrication complexity, whereas H-plane filter design is very popular because of the ease of fabrication. Here, H-plane-based bandpass filter is realized using two different techniques. One is an inductive iris and another is an inductive metallic post. The general filter theory method is carried out to realize both the filters. The schematics of inductive metallic post and inductive iris-based filters are shown in Figure 5.9(a, b), respectively, along with the dimensions. Both of the filters

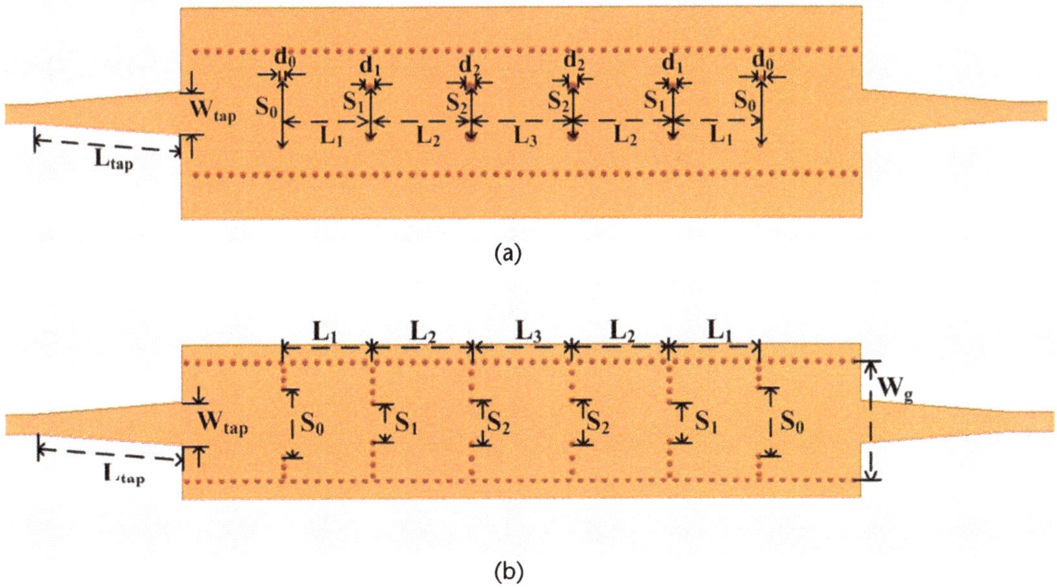

Figure 5.9 Top view of H-plane SIW based bandpass filter. (a) Inductive metallic post filter along with its dimensions as follows: S_0 = 8.62 mm; S_1 = 6.32 mm; S_2 = 6.27 mm; L_1 = 11.8 mm; L_2 = 13.32 mm; L_3 = 13.625; d_0 = 0.8 mm; d_1 = 1.2 mm; d_2 = 1.5 mm; W_{tap} = 5.8 mm; and L_{tap} = 20 mm. (b) Inductive iris filter along with its dimensions as follows: W_{tap} = 5.6 mm; S_0 = 9 mm; S_1 = 5.16 mm; S_2 = 5.72 mm; L_1 = 12 mm; L_2 = 13.15 mm; L_3 = 13.5 mm; L_{tap} = 20 mm; and W_g = 15.4 mm.

are realized using Rogers RT/Duroid 5880 substrate with dielectric constant (ε_r) of 2.2, a thickness of 0.787 mm, and a loss tangent of 0.0009. Here, the dominant mode of SIW TE_{10} is utilized to design the filter. Figure 5.10(a, b) represents metallic inductive post and iris coupling in SIW, respectively. The corresponding lumped equivalent circuit for the shunt inductive discontinuity is shown in Figure 5.10(c). Since the iris diameter (d) is very small compared to a, X_a is negligible. Similarly, for inductive post coupling, X_a is also considered zero, which reduces the design complexity. When the diameter of inductive post or thick inductive iris is comparable to a, X_a is not negligible and should be considered for design process. Shunt inductive coupling can be considered as a normalized impedance inverter along with transmission line (φ), which is attached to both sides of the impedance inverter (K) as shown in Figure 5.10(d) [7]. The relation between normalized impedance inverter and normalized shunt reactance can be expressed as given in (5.53) and (5.54):

$$\frac{K}{Z_0} = \left|\tan\left(\frac{\phi}{2} + \tan^{-1}\frac{X_a}{Z_0}\right)\right| \tag{5.53}$$

$$\phi = \tan^{-1}\left(\frac{2X_b}{Z_0} + \frac{X_a}{Z_0}\right) - \tan^{-1}\frac{X_a}{Z_0} \tag{5.54}$$

The shunt reactance can be expressed in terms of impedance inverter in more simplifying manner as:

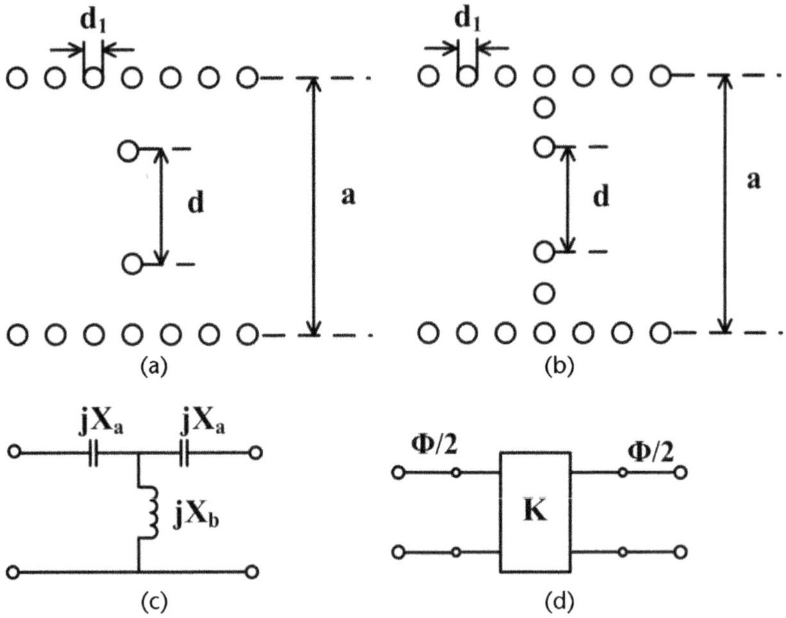

Figure 5.10 Design process of the H-plane shunt inductive coupling discontinuity: (a) metallic inductive post, (b) inductive iris, (c) equivalent circuit for the shunt inductive discontinuity, and (d) equivalent *K*-inverter network.

$$\frac{X_{j,j+1}}{Z_0} = \frac{\dfrac{K_{j,j+1}}{Z_0}}{1-\left(\dfrac{K_{j,j+1}}{Z_0}\right)^2} \tag{5.55}$$

The design specifications are the same for both filters. The center frequency is at 9.75 GHz, and the passband ripple and return loss are 0.1 dB and 25 dB, respectively. Stopband attenuation is a minimum of 25 dB at 9.3 GHz and the bandwidth of the filter (Δf) is 0.5 GHz. Based on design specifications, the minimum order of the filter is 4 ($N = 4$). In the design, the fifth-order filter using a Chebyshev polynomial is chosen. For $N = 5$ and 0.1-dB equi-ripple passband, the value of elements for the Chebyshev filter are: $g_0 = g_6 = 1$; $g_1 = g_5 = 1.1468$; $g_2 = g_4 = 1.3712$; and $g_3 = 1.9750$.

The normalized impedance inverters are calculated using filter's prototype values and design specifications as given in (5.56) to (5.58):

$$\frac{K_{01}}{Z_0} = \sqrt{\frac{\pi}{2}\frac{\omega_\lambda}{g_0 g_1}} = \frac{K_{56}}{Z_0} = 0.411 \tag{5.56}$$

$$\frac{K_{12}}{Z_0} = \frac{\pi\omega_\lambda}{2}\frac{1}{\sqrt{g_1 g_2}} = \frac{K_{45}}{Z_0} = 0.108 \tag{5.57}$$

$$\frac{K_{23}}{Z_0} = \frac{\pi\omega_\lambda}{2}\frac{1}{\sqrt{g_1 g_2}} = \frac{K_{34}}{Z_0} = 0.117 \tag{5.58}$$

where ω_λ is derived from

$$\omega_\lambda = (\lambda_{g0}/\lambda_0)^2 \cdot FBW$$

Here, λ_{g0} and λ_0 are guided and free-space wavelengths at design frequency (f_0), respectively. FBW is the fractional bandwidth, which corresponds to $\Delta f/f_0$. Based on the calculated values of impedance inverter, corresponding shunt reactances are computed using (5.55). The filter has five resonators and the length of each resonating section is calculated using (5.59).

$$\theta_j = \pi - \frac{1}{2}\left(\tan^{-1}\frac{2X_{j-1,j}}{Z_0} + \tan^{-1}\frac{2X_{j,j+1}}{Z_0}\right) \tag{5.59}$$

where θ_j is in radians, j varies from 1 to 5, and $\theta_1 = \theta_5$; $\theta_2 = \theta_4$. Figure 5.9 shows the schematic of the filter with spacing between irises. Based on the calculated coupling coefficients and electrical length of each resonator, the filters are designed. The simulated responses of the filters are presented in Figure 5.11.

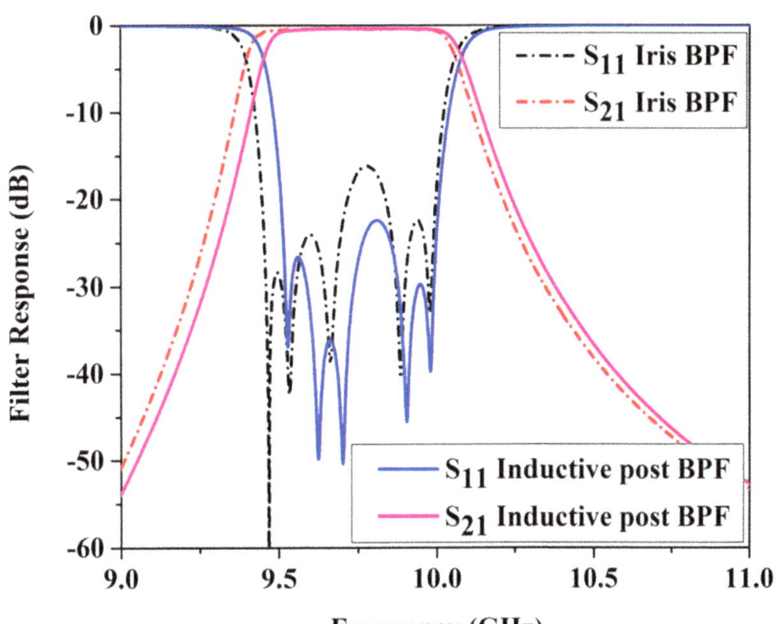

Figure 5.11 Simulated responses of inductive iris and inductive metallic post bandpass filter.

5.3 Bandpass Filter Design Using Substrate Integrated Hexagonal Cavity

In this section, the dual-mode substrate integrated hexagonal cavity (SIHC) filters have been utilized to realize transmission zeros without introducing negative coupling [8]. Figure 5.12(a, b) shows the electric field distributions of two orthogonal modes at 10 GHz. Different filter responses can be realized with the help of the perturbation technique applied to the side walls of the hexagonal cavity. Four filters have been designed using the perturbation technique in a single-layer substrate. Out of these, a pair of two poles filters using a single SIHC are demonstrated with transmission zero on either side of the passband. The third filter is realized using a single hexagonal cavity with two transmission zeros on both side of the passband with less design complexity. The fourth designed bandpass filter is realized using two cascaded dual-mode SIHCs with enhanced bandwidth by increasing the number of poles and symmetric responses due to the presence of two transmission zeros on both sides of the passband. The bandwidth of the filter, number, and position of the transmission zeros can be controlled by the perturbation of the side wall as well as a feeding position of the source and load, which are presented in this section. Utilizing six side walls of the SIHC, a large degree of design flexibility is achieved in the designed design.

5.4 Perturbation of Dual-Mode Hexagonal Cavity and Its Application in Filter Design

The resonant frequencies of the hexagonal cavity are estimated using full-wave electromagnetic simulator (HFSS eigenmode solver). The first two higher-order modes are utilized to implement a dual-mode response. When all the side-wall lengths of the hexagonal cavity are kept the same, it is observed that both of the first higher-order modes exhibit the same resonant frequency. This is because of symmetry in the hexagonal structure. The perturbation using the side wall of the hexagonal cavity disturbs the symmetry of the structure. As a consequence, splitting of degenerate modes occurs. This splitting of modes brings the transmission

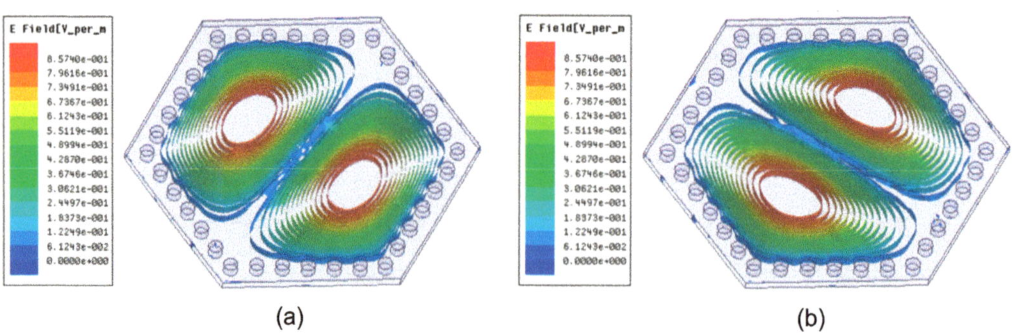

Figure 5.12 Electric field distributions of resonant modes in hexagonal SIW cavity. (a) and (b): two orthogonal higher-order modes at 10 GHz. (© 2014 Wiley [8].)

zero at above or below or even both sides of the passband. Figure 5.13 shows the schematic design of a single dual-mode hexagonal cavity filter along with the direction of perturbation. The side walls of the hexagonal cavity are denoted as: {1}, {2}, {3}, {4}, {5}, and {6}. The perturbed length is termed by *inset* along the direction of perturbation, which is normal to the side wall {5}. When all the hexagonal sidewalls are of the same length, the *inset* is considered 0 mm, which means no perturbation. For inward perturbation, inset is considered as the +*ve* value, and for outward perturbation, inset is considered as the −*ve* value. The side wall {5} provides the symmetrical perturbation with respect to the input and output ports. This perturbation affects only one particular mode, whereas the other mode remains insensitive. By controlling this perturbation, an asymmetric filter response can easily be achieved. Figure 5.14 provides the electric field distribution for odd mode (mode 1) and even mode (mode 2) with respect to the excitation. It can also be inferred from this figure that both modes are orthogonal to each other. Figure 5.15(a) shows the variations of self-resonant frequencies of both odd (mode 1) and even (mode 2) modes with respect to the *inset* variation from −2.5 mm to +0.5 mm. The analysis is carried out using weak coupling at both ports of the cavity. It is observed that by varying *inset*, only the even-mode frequency is shifting, while the odd-mode frequency remains insensitive to it as shown in Figure 5.15(a). This shifting of even mode from lower frequency to higher frequency with respect to the odd mode leads to switching of the

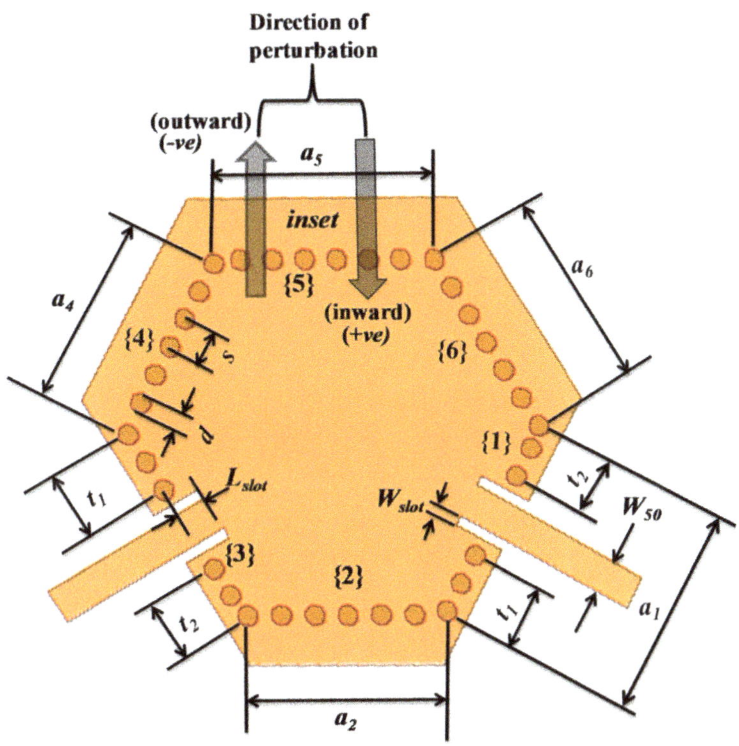

Figure 5.13 A schematic design of a single hexagonal cavity filter with the direction of the perturbation inset using side wall {5}. (© 2014 Wiley [8].)

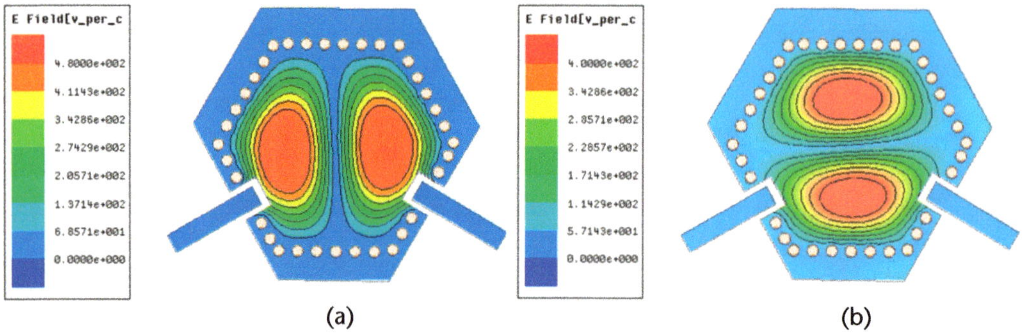

Figure 5.14 Simulated electric field distribution using weak coupling at the source and load end. (a) Odd mode (mode 1). (b) Even mode (mode 2). (© 2014 Wiley [8].)

transmission zero from the lower side to the higher side of the passband as shown in Figure 5.15(b). It provides the design flexibility and more control over the transmission zero on either side of the passband as compared to a circular cavity. The dual-mode hexagonal cavity is used to design four different filters by only varying the perturbation length and changing the position of the source and load. Maintaining design specifications, all these filters are realized theoretically with the help of a coupling matrix. Due to the perturbation, both the first higher-order modes split. However, it does not introduce any intra-cavity coupling between both the modes [9, 10]. The microstrip lines with coupling slots are used to feed planar hexagonal cavity. The designed hexagonal dual-mode filters are fabricated using Rogers RT/Duroid 5880 with a relative dielectric constant of 2.2 and a thickness of 0.787 mm. The simulated unloaded quality factor of the degenerate modes in the hexagonal cavity is around 485 at 10 GHz considering the same substrate with a loss tangent of 0.0009, metal conductivity of 5.8×10^7 S/m, and metal thickness of 0.036 mm.

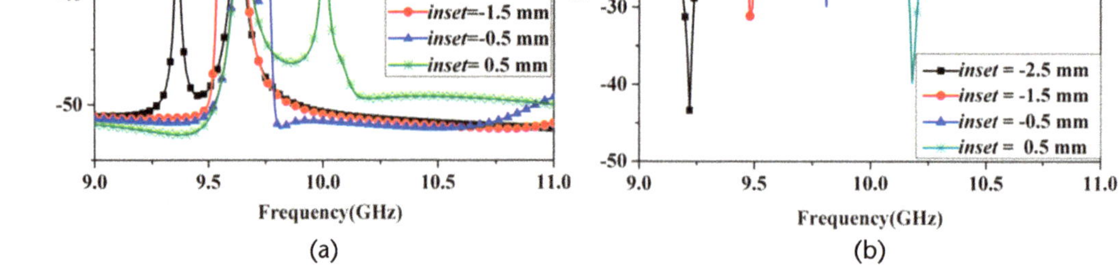

Figure 5.15 The *inset* dependence on dual-mode filter characteristics. (a) Modal resonant characteristic of a designed dual-mode filter with respect to *inset*. (b) Transmission zero characteristic of a designed dual-mode filter with respect to *inset*. (© 2014 Wiley [8].)

5.5 Single-Band SIHC Filter Design Using the Dual-Mode Cavity

As we discussed in the previous section, the perturbation technique splits the dual-mode degenerate resonant frequencies and therefore one or two transmission zeros appear on either or both sides of the passband, thereby improving the selectivity of the filter. In this section, the SIHC is exploited to realize the compact single-band bandpass filters with the change in perturbation.

5.5.1 Dual-Mode SIHC Filter with a Transmission Zero Above the Passband Using a Single Cavity

Figure 5.13 shows the schematic design of a single dual-mode hexagonal cavity filter. To ensure the split of degenerate modes and the transmission zero above the passband, perturbation is introduced along the inward direction. The filter coupling topology is provided in Figure 5.16(a) where it is shown that no intra-cavity coupling is available between two split modes. The coupling matrix of this designed filter is obtained using design specifications of center frequency at 9.8 GHz, fractional bandwidth of 2.1% with return loss of 25 dB in the passband and one transmission zero at 10.1 GHz. The calculated coupling matrix is given in (5.60). Only side wall *{5}* is perturbed. The length of *inset* is kept +0.3 mm for this designed filter. With the help of weak coupling at both source and load ends, the self-resonant frequencies of both odd (mode 1) and even (mode 2) modes are found at 9.64 GHz and 10 GHz, respectively. The even mode is at a higher frequency than the odd mode, which can be observed from the coupling matrix as M_{22} is negative as given in (5.60). From coupling matrix, it is observed that source to odd-mode coupling (M_{S1}) is higher than source to even-mode coupling (M_{S2}).

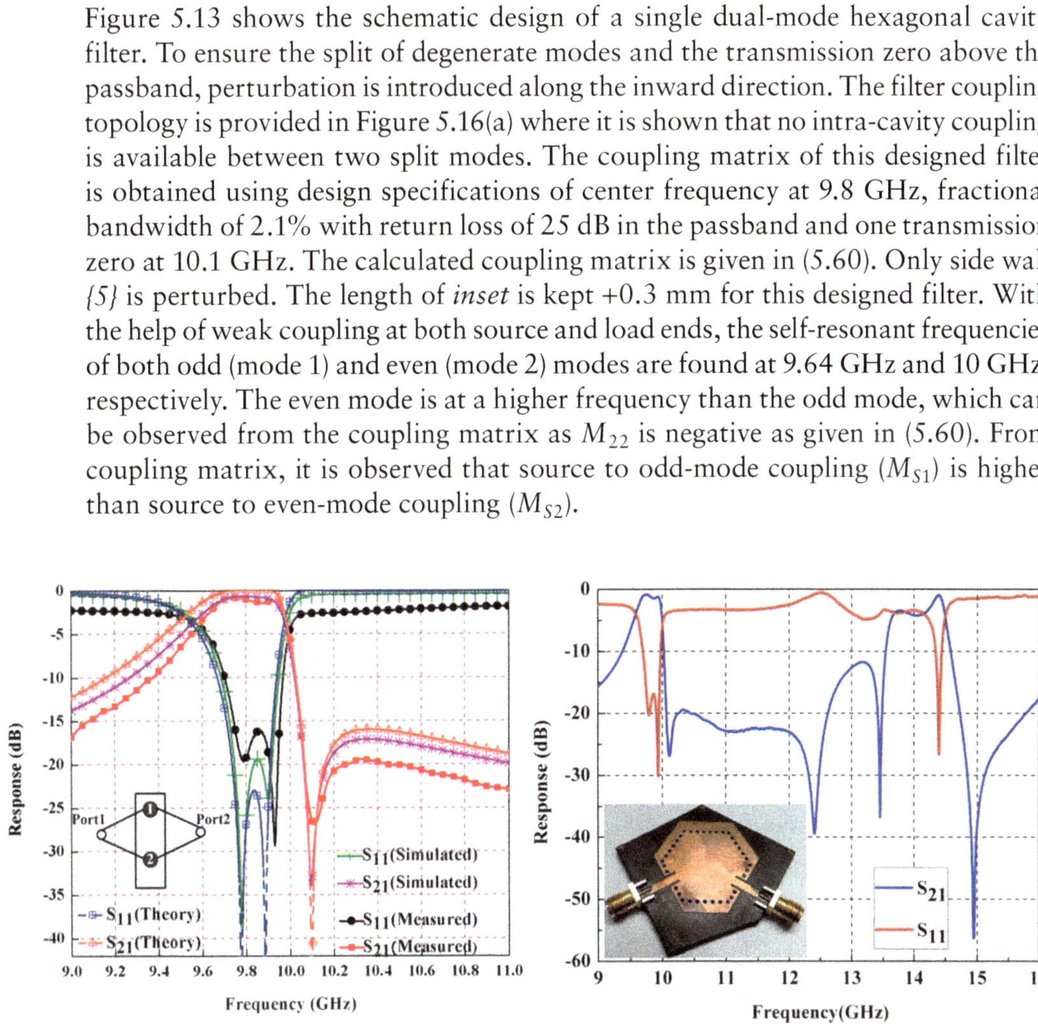

(a) (b)

Figure 5.16 (a) Filter responses using the coupling matrix and simulated and measured results of a single hexagonal dual-mode filter with one transmission zero at the right side of the passband. (Dimension of the designed filter: $a_1 = a_2 = a_3 = 13.8$ mm; $a_4 = 13.8$ mm; $a_5 = 14.7$ mm; $a_6 = 13.8$ mm; $t_1 = 4.54$ mm; $t_2 = 3.8$ mm; $L_{slot} = 2.3$ mm; $W_{slot} = 0.75$ mm; $d = 1.4$ mm; and $s = 2.3$ mm; and $W_{50} = 2.4$ mm.) (b) Fabricated prototype and measured spurious responses. (© 2014 Wiley [8].)

$$M = \begin{bmatrix} & S & 1 & 2 & L \\ S & 0 & 1.15 & 0.63 & 0 \\ 1 & 1.15 & 1.8 & 0 & -1.15 \\ 2 & 0.63 & 0 & -1.35 & 0.63 \\ L & 0 & -1.15 & 0.63 & 0 \end{bmatrix} \quad (5.60)$$

The nonzero diagonal elements of the coupling matrix are due to the split of modes, which represent self-resonant frequencies of the split modes. The self-resonant frequency can be calculated using diagonal elements of a coupling matrix [10] as given in (5.61):

$$f_{0/e} = f_0\left(1 - \frac{M_{ii} FBW}{2}\right) \quad (5.61)$$

Here, M_{ii} corresponds to diagonal elements, FBW is fractional bandwidth, $f_{0/e}$ is self-resonance frequency of the odd/even mode, and f_0 is center frequency of the filter. Since source to odd-mode (mode 1) coupling is higher than source to even-mode (mode 2) coupling and also the even-mode frequency is higher than the odd-mode frequency, the transmission zero appears at the right side of the passband. Figure 5.16(a) shows the comparison of measured, simulated, and ideal filter responses based on the filter theory [10]. The passband insertion loss for simulated and measured filter are 0.74 dB and 1.7 dB, respectively, and the measured transmission zero appears at 10.1 GHz. The fabricated prototype of the filter with the measured wideband spurious frequency response from 9 to 16 GHz is shown in Figure 5.16(b), where five higher-order modes are observed.

5.5.2 Dual-Mode SIHC Filter with a Transmission Zero Below the Passband Using a Single Cavity

In the previous design, only one hexagonal side wall *{5}* is used for perturbation. In this section, the same side wall *{5}* is used for perturbation, but the main difference lies in the direction of perturbation, which is the reverse direction (outward) to the previous one as shown in Figure 5.13. Here, the *inset* value is *–ve* because of outward perturbation and is kept at –2.3 mm. Therefore, the self-resonance frequency of the odd mode is higher than the even mode and one transmission zero appears at the lower side of the passband, which is shown in Figure 5.17(a). The same coupling topology, as mentioned in previous design, is followed here for designing the filter. The design specifications of the designed filter are: center frequency at 10 GHz, fractional bandwidth of 2.2% with a return loss of 20 dB in the passband, and one transmission zero at 9.6 GHz. On the basis of coupling topology and design specifications, the coupling matrix is calculated, which is given in (5.62). The odd-mode (mode 1) and even-mode (mode 2) frequencies are found at 9.96 GHz and 9.72 GHz, respectively, by applying weak coupling as an excitation at both input and output ports.

$$M = \begin{bmatrix} & S & 1 & 2 & L \\ S & 0 & 0.946 & 0.52 & 0 \\ 1 & 0.946 & -0.4 & 0 & -0.946 \\ 2 & 0.52 & 0 & -1.8 & 0.52 \\ L & 0 & -946 & 0.52 & 0 \end{bmatrix} \tag{5.62}$$

The filter has been designed and fabricated based on coupling topology and coupling coefficients. The measured, simulated, and ideal filter responses are calculated from the filter theory, as shown in Figure 5.17(a). The passband insertion loss for the simulated and measured filters are 0.64 dB and 1.42 dB, respectively, and the measured transmission zero appears at 9.6 GHz. Figure 5.17(b) shows the fabricated prototype of the filter along with measured wideband spurious frequency response from 9 to 16 GHz where five higher-order modes are observed and identical spurious responses are found like in the previous case.

5.5.3 Dual-Mode SIHC Filter with a Pair of Transmission Zeros at Both Sides of the Passband Using a Single Cavity

In the previous two designs, the single hexagonal cavity is used for designing bandpass filter with an asymmetric frequency response (i.e., the transmission zero appears either at the higher or lower sides of passband). In this section, the designed filter is designed using a single hexagonal cavity with two poles and two transmission zeros where the transmission zeros appear at both sides of the passband, thereby improving the selectivity of the filter. Here, in the dual-mode hexagonal cavity, perturbation helps only to split the modes without introducing intra-cavity coupling between the modes. The schematic of the designed filter is shown in Figure 5.18. This design

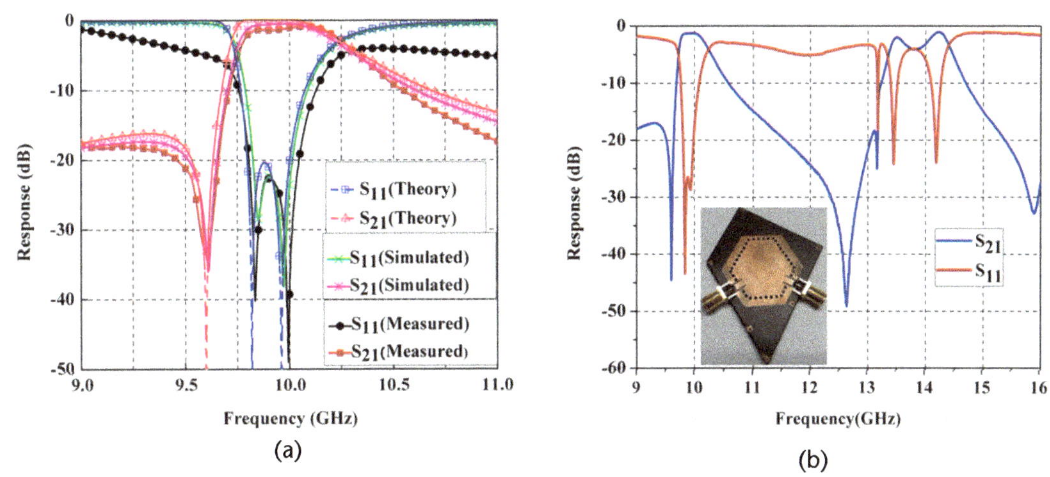

Figure 5.17 (a) Filter responses using coupling matrix, simulated, and measured S-parameters of the single SIHC hexagonal dual-mode filter with one transmission zero below the passband (dimension of the designed filter: $a_1 = a_2 = a_3 = 13.8$ mm; $a_4 = 16.1$ mm; $a_5 = 11.28$ mm; $a_6 = 16.1$ mm; $t_1 = 4.17$ mm; $t_2 = 3.94$ mm; $L_{slot} = 2.3$ mm; $W_{slot} = 0.75$ mm; $d = 1.4$ mm; $s = 2.3$ mm; and $W_{50} = 2.4$ mm). (b) Photograph and measured spurious response of the fabricated filter of single SIHC with a transmission zero below the passband. (© 2014 Wiley [8].)

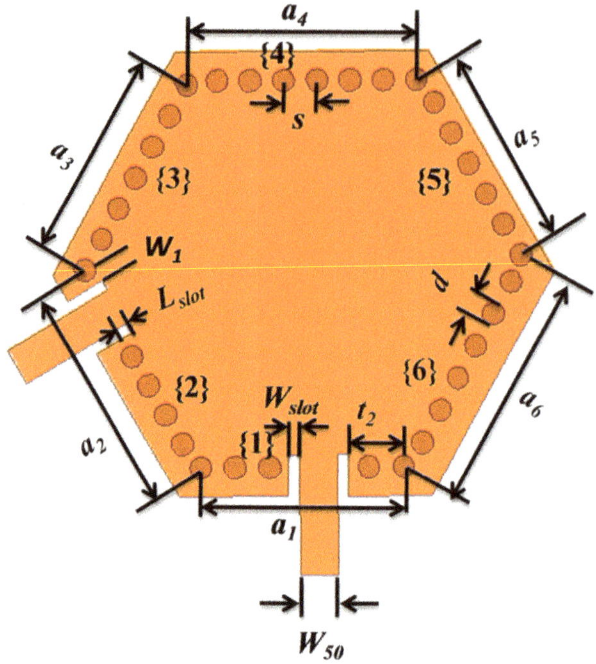

Figure 5.18 Schematic of the designed filter of single hexagonal cavity with two transmission zeros at both side of the passband (dimensions of the designed filter: a_1 = 14.86 mm; a_2 = 14.32 mm; a_3 = 13.2 mm; a_4 = 14.53 mm; a_5 = 12.8 mm; a_6 = 15.13 mm; W_1 = 1.1 mm; t_2 = 3.48 mm; L_{slot} = 0.8 mm; W_{slot} = 0.7 mm; d = 1.4 mm; s = 2.3 mm; and W_{50} = 2.4 mm). (© 2014 Wiley [8].)

provides two transmission zeros: onc appears due to the dual-mode hexagonal cavity and the other appears due to weak coupling between the source and load. The filter topology depicted in Figure 5.19(a) shows this weak coupling as a dashed line between the source and the load. This is practically possible in the design by placing the source and load at the consecutive side walls. The coupling matrix of this designed filter is obtained using coupling topology. The design specifications are as follows: center frequency at 10.1 GHz, fractional bandwidth of 1.3% with 20-dB return loss at the passband and transmission zeros at 9.7 GHz and 10.7 GHz. The coupling matrix corresponding to the design specification is given in (5.63).

$$M = \begin{bmatrix} & S & 1 & 2 & L \\ S & 0 & 0.74 & 0.69 & -0.01 \\ 1 & 0.74 & -0.8 & 0 & -0.78 \\ 2 & 0.69 & 0 & 1.7 & 0.74 \\ L & -0.01 & -0.78 & 0.74 & 0 \end{bmatrix} \tag{5.63}$$

To realize this calculated coupling matrix, it is required to adjust a proper feeding position as well as a perturbation length of the side walls of the hexagonal cavity. Unlike previous designs, three side walls *{3}*, *{5}*, and *{6}* have been used for perturbation to achieve a calculated coupling matrix. The frequency of a normalized

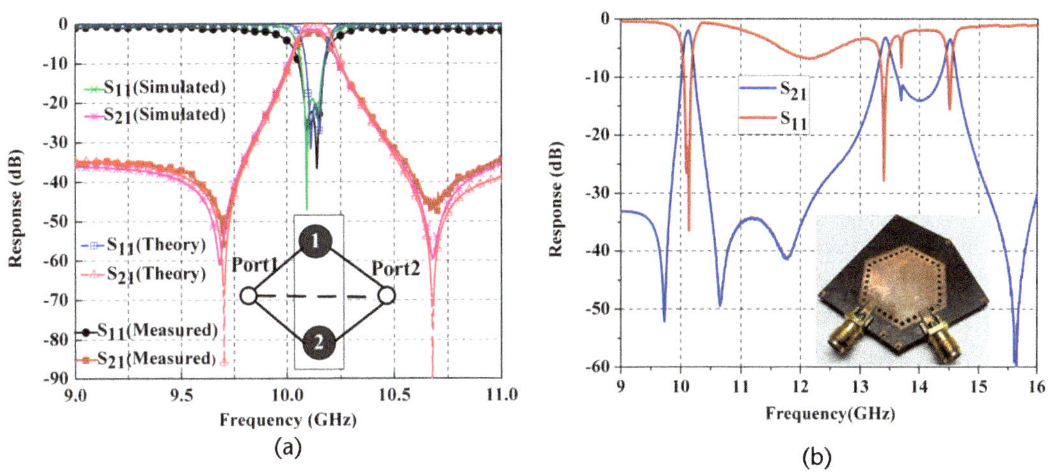

Figure 5.19 Coupling topology and filter response using the coupling matrix and simulated and measured results of a dual-mode single hexagonal cavity filter with two transmission zeros at both ends of the passband. (b) Photograph and measured spurious response of the fabricated filter of a single SIHC with two transmission zeros at the both side of the passband. (© 2014 Wiley [8].)

transmission zero can be calculated from the coupling matrix, which is given in (5.64):

$$M_{SL}\omega^2 + M_{SL}\left(M_{11} + M_{22}\right) - \left(M_{1L}M_{S1} + M_{2L}M_{S2}\right)\omega$$
$$- \left(M_{11}M_{2L}M_{S2} - M_{11}M_{22}M_{SL} + M_{1L}M_{22}M_{S1}\right) = 0 \tag{5.64}$$

where ω is a complex lowpass frequency variable. This quadratic equation implies two transmission zeros, which are due to source-load coupling. The measured, simulated, and ideal filter responses are shown in Figure 5.19(a). The simulated and measured passband insertion loss are 1.67 dB and 2.1 dB, respectively, and the measured transmission zeros appear at 9.7 GHz and 10.68 GHz on both sides of the passband. The fabricated prototype and the measured wideband spurious frequency response from 9 to 16 GHz are shown in Figure 5.19(b) where five higher-order modes are observed and identical spurious responses are found like previous cases.

5.5.4 Dual-Mode SIHC Filter with Two Transmission Zeros at Both Sides of the Passband Using Double Cavities

Two cascaded dual-mode hexagonal cavities are used to design dual-mode filters with two transmission zeros and enhanced bandwidth. The perturbations are mainly carried out by the side wall *{5}* of the left cavity and the side wall *{5′}* of right cavity as shown in Figure 5.20(a). The coupling topology is provided in Figure 5.20(b) and the design specifications are given as follows: center frequency at 10 GHz, fractional bandwidth of 6.5% with 23-dB return loss at the passband, and two transmission zeros at 9.4 GHz and 10.6 GHz. Based on the coupling topology and design specifications, the coupling matrix of this designed design is calculated

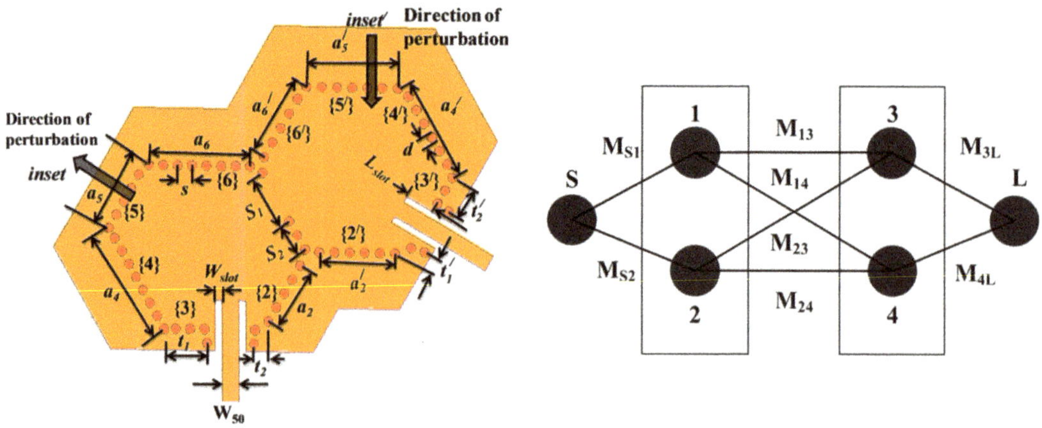

(a) (b)

Figure 5.20 (a) A schematic of the designed double hexagonal cavity based dual-mode filter with two transmission zeros at both side of the passband. (Dimension of the designed filter: a_2 = 9.2 mm; a_4 = 16.4 mm; a_5 = 10.8 mm; a_6 = 14.7 mm; a'_2 = 11 mm; a'_4 = 16.4 mm; a'_5 = 13.2 mm; a'_6 = 11.72 mm; t_1 = 5.86 mm; t_2 = 1.95 mm; t'_1 = 2.36 mm; t'_2 = 4.2 mm; S_1 = 7.85 mm; S_2 = 4.6 mm; L_{slot} = 4.125 mm; W_{slot} = 1.1 mm; d = 1.3 mm; s = 2.3 mm; and W_{50} = 2.4 mm.) (b) Coupling matrix topology of the designed filter. (© 2014 Wiley [8].)

and given in (5.65). In the left cavity, perturbation is done in an outward direction, whereas for the right cavity, perturbation is done in an inward direction using side walls *{5}* and *{5′}*, respectively. Hence, the odd mode (mode 1) is at a higher frequency than the even mode (mode 2) for the left cavity and the even mode (mode 4) is at a higher frequency than the odd mode (mode 3) for the right cavity, which can be justified by $-ve$ value of M_{11} and $-ve$ value of M_{44} in diagonal elements of the coupling matrix. Figure 5.21(a) shows the measured, simulated, and ideal filter responses together. The passband insertion loss for both simulated and measured

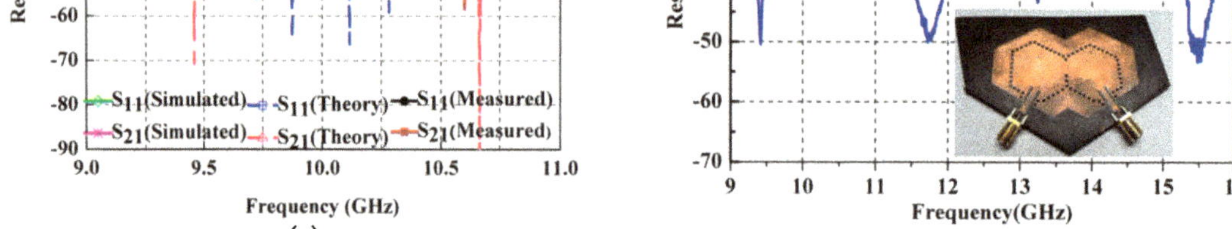

(a) (b)

Figure 5.21 Filter response using a coupling matrix and the simulated and measured results of a dual-mode double hexagonal cavity filter with two transmission zeros at both sides of the passband. (b) Photograph and measured spurious response of the of the fabricated filter of double SIHCs with two transmission zeros at both sides of the passband. (© 2014 Wiley [8].)

filters are 0.85 dB and 1.62 dB, respectively, and the measured transmission zeros appear at 9.4 GHz and 10.6 GHz.

$$M = \begin{bmatrix} & S & 1 & 2 & 3 & 4 & L \\ S & 0 & 0.91 & 0.60 & 0 & 0 & 0 \\ 1 & 0.91 & -0.527 & 0 & 0.247 & 0.704 & 0 \\ 2 & 0.6 & 0 & 1.238 & 0.128 & -0.299 & 0 \\ 3 & 0 & 0.247 & 0.128 & 1.238 & 0 & -0.572 \\ 4 & 0 & 0.704 & -0.299 & 0 & -0.481 & 0.928 \\ L & 0 & 0 & 0 & -0.572 & 0.928 & 0 \end{bmatrix} \quad (5.65)$$

The fabricated prototype and the measured wideband spurious frequency response of the filter from 9 to 16 GHz are shown in Figure 5.21(b). It is observed that the out-of-band spurious responses of this filter are identical with the previous one. First, four higher-order modes are utilized to design the passband of the desired bandpass filter. Besides this, three higher-order modes exist for each hexagonal cavity. These spurious higher-order modes appear far above the desired passband of the filter.

5.6 Dual-Band SIHC Filter Design Using a Dual-Mode Cavity

With the advancement of modern multiband wireless communication system, there is a huge demand for multiband filters. Dual-band or multiband filters are an integral part of the RF dual-band or multiband front-end systems. Dual-band filters can be realized using single-mode or dual-mode resonators. A dual-band filter-based on a dual-mode resonator ensures the design to be more compact with a lesser volume. In this section, a compact dual-mode-based dual-bandpass filter is designed. The design specified transmission zeros and wide stopband improve the filter selectivity. A single hexagonal cavity is used to design dual-band filter. The first two higher-order degenerate modes are exploited to obtain lower-band and the next three higher-order orthogonal modes are employed to obtain a higher band. The center frequency of the first and second bands are at 10.0 GHz and 14.0 GHz, respectively. The SIW-based feeding technique is used to design an SIHC.

The iris coupling window is used to couple power from the SIW to SIHC. The perturbation using side wall a_5 allows to split the orthogonal modes in such a way that the first two orthogonal modes form the lower band and next three higher-order modes form the next higher band. The schematic of the designed dual-band filter is shown in Figure 5.22(a) along with the dimensions of the filter. The coupling topology of this designed filter is shown in Figure 5.22(b) where mode 1 and mode 2 help in designing lower band and mode 3, mode 4, and mode 5 are forming the higher band. The electric field distributions of all these modes are depicted in Figures 5.23 and 5.24. From these figures, it is observed that the mode 1 and mode 2 are orthogonal to each other and the remaining three modes (mode 3, mode 4, and mode 5) are also orthogonal to each other. The perturbation is termed as *inset* and it is kept at −2.25 mm. The −*ve* value of *inset* means the perturbation is in outward direction. The effect of this perturbation on all the modes is shown in

Figure 5.25(a). From the electric field distribution of the mode 1 and mode 2, it is observed that mode 2 does not get affected with the perturbation, whereas mode 1 resonant frequency is sensitive to these inset variations. Mode 3, mode 4, and mode 5 are also sensitive to this perturbation. The modal resonant characteristic of the SIHC is carried out by applying weak coupling to the cavity in order to avoid the loading effect of the source and load. The *inset* variations affect lower and higher-band transmission zeros as shown in Figure 5.25(b).

While *inset* is varying from −2.5 mm to 0 mm, transmission zero is shifting from the lower to higher side of both the passbands. If the *inset* is kept at 0 mm, the transmission zero disturbs the higher-band response; therefore, we need to choose the *inset* properly in order to keep the transmission zero to the left side of the higher passband. The SIW feeding (using iris coupling) allows one to bring all

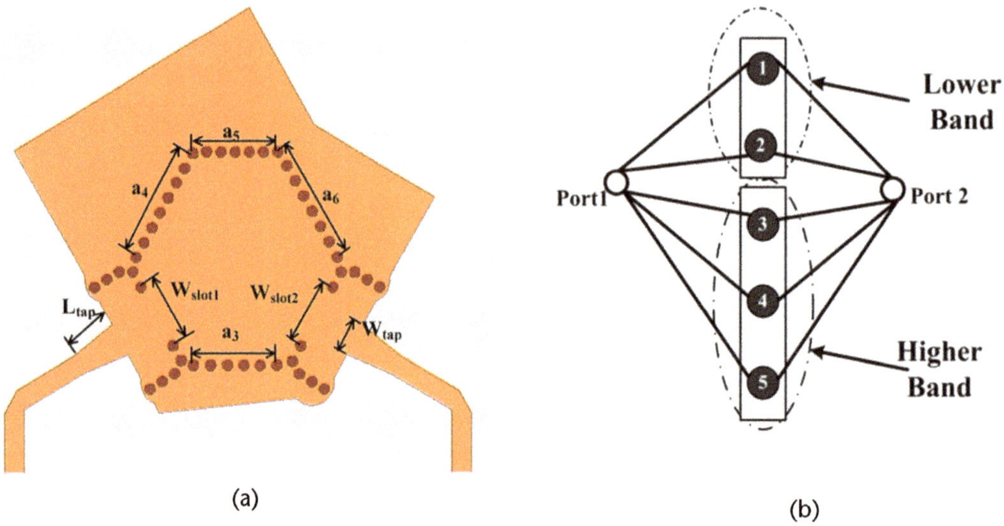

Figure 5.22 (a) Schematic of dual-band bandpass filter with dimensions as follows: $a_3 = 10.5$ mm, $a_4 = 14.35$ mm, $a_5 = 9.0$ mm, $a_6 = 12.6$ mm, $L_{tap} = 6.0$ mm, $W_{slot1} = 8.075$ mm, $W_{slot2} = 8.075$ mm, and $W_{tap} = 4$ mm. Design based on substrate: Rogers 5880 with height 0.787 mm. (b) Coupling topology of dual-band SIHC-based filter. (© 2019 IEEE [11].)

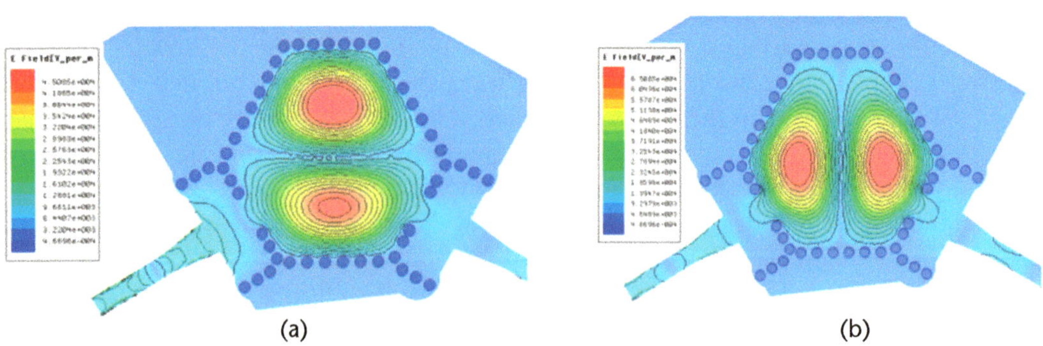

Figure 5.23 Electric field distribution of first two higher-order orthogonal modes in the lower band while applying weak coupling to both the source and the load. (a) The even mode at 10.13 GHz (mode 1). (b) The odd mode at 10.56 GHz (mode 2). (© 2019 IEEE [11].)

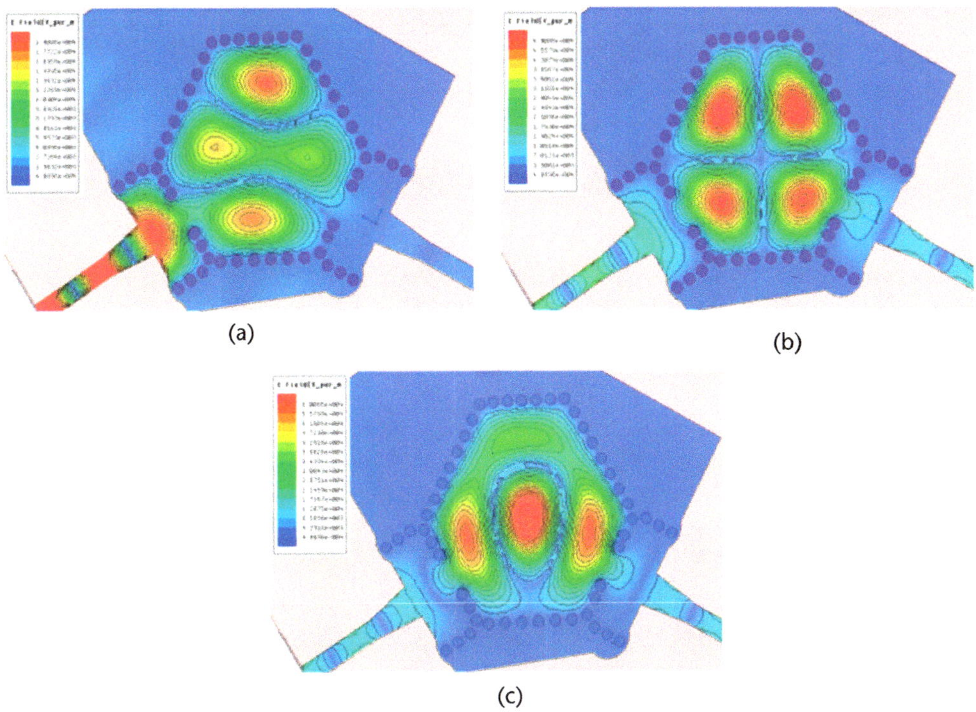

Figure 5.24 Electric field distribution of the next three higher-order orthogonal modes in the higher band while applying weak coupling to both the source and the load. (a) Mode 3 at 13.70 GHz. (b) Mode 4 at 14.0 GHz. (c) Mode 5 at 14.92 GHz. (© 2019 IEEE [11].)

three higher-order modes of the SIHC close to each other that ensures a dual-band response of a single cavity. In addition, the SIW-based feeding technique helps the designed filter to integrate with other SIW circuits without using any transition.

The lower band of the filter is at X-band with two poles and higher band of the filter is at Ku-band with three poles as shown in Figure 5.26. The center frequencies

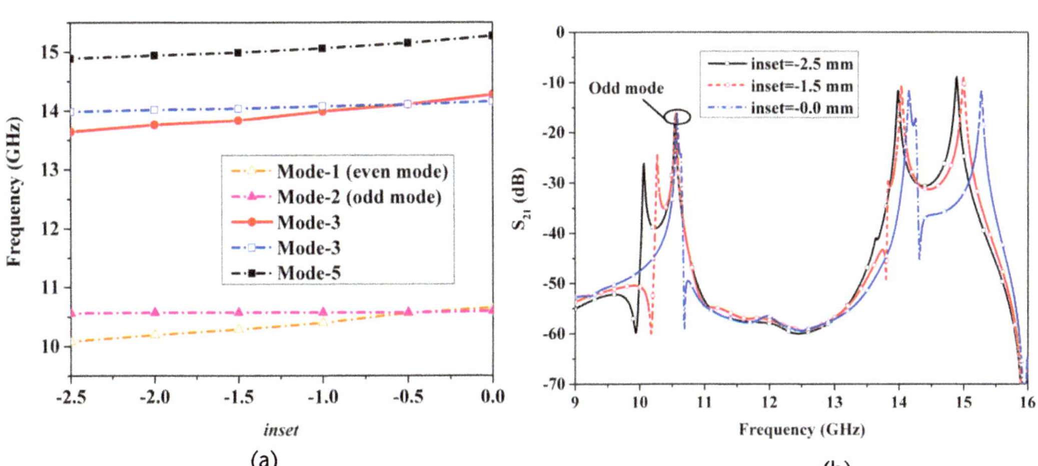

Figure 5.25 (a) Modal resonant characteristic of a dual-band SIHC filter with inset variations. (b) Effects of *inset* variation on insertion loss using weak coupling excitations. (© 2019 IEEE [11].)

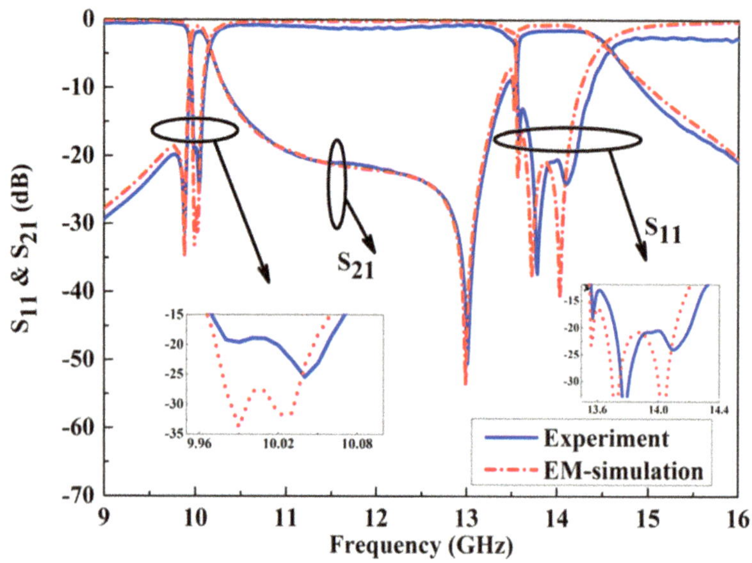

Figure 5.26 Measured and simulated responses of the SIHC dual-band filter using a single cavity. (© 2019 IEEE [11].)

of the lower and higher bands are at 10 GHz and 13.95 GHz, respectively. To achieve the dual-band response of a single SIHC resonator, perturbation is introduced by exploiting only one side wall a_5. The *inset* is kept at 2.25 mm. The fractional bandwidths are 0.7% and 3.9% the in lower and upper bands, respectively. The resonant frequencies of all the modes are given here: Mode 1 = 10.1 GHz, Mode 2 = 9.92 GHz, Mode 3 = 13.42 GHz, Mode 4 = 13.5 GHz, and Mode 5 = 14.4 GHz. The lower and higher passbands' insertion losses are 1.5 dB and 1.2 dB, respectively, and transmission zeros appear at 9.87 GHz and 13.52 GHz, respectively, that are below their respective passbands. The measured passband return losses are about 20 dB and 15 dB for both lower and higher bands, respectively.

5.7 Miniaturized SIW-Based Bandpass Filter Design

There are several techniques to miniaturize an SIW-based filter while retaining its characteristics without disturbing its field patterns such as utilizing its half-mode, quarter-mode, and sixth-mode. This section elaborates the design of a few miniaturized SIW-based filters.

5.7.1 Half-Mode SIW

The concept of half-mode SIW (HMSIW) was designed by Hong et al. [12]. The difference between the SIW and the half-mode SIW is that, in the SIW, both side walls of the structure are shielded by two rows of metallic vias, whereas in the HMSIW, one side wall is shielded by metallic vias and the other is kept open. This is because a fictitious perfect magnetic wall can be considered within the SIW along the central line without affecting the field distribution of the line. As a result, the SIW can be bisected to realize the HMSIW structure without disturbing the original field distribution. At the same time, miniaturization of the SIW structure is implemented by making it almost one-half of the original SIW along the width. The geometry and its dominant mode of electric field distribution at 10 GHz along with its dimensions are provided in Figure 5.27. It is evident from this figure that the HMSIW has the same electric field behavior as the SIW. It is observed that the maximum electric field is along the open edge of the HMSIW, whereas the electric field is terminated at the other side because of rows of metallic vias placed along that side wall.

Like SIW, similar microstrip transition is used here to excite HMSIW. The simulated response of HMSIW-to-microstrip transition is demonstrated in Figure 5.28.

One side wall of the HMSIW should be ideally the perfect magnetic wall, but, in practice, it is not so. As a result, it may introduce radiation loss along the open edge of the HMSIW. Therefore, the HMSIW has degraded the insertion loss response as compared to the SIW.

5.7.2 Sixth-Mode SIW-Based Highly Selective Bandpass Filter Design

In this section, a novel bandpass filter using the sixth mode of an SIHC is designed [13]. The sixth-mode SIHC is obtained by symmetrically cutting the SIHC into six parts, as shown in Figure 5.29. As a result, we get an equilateral triangle-shaped resonator in which two edges are open-ended and one end is confined by vias as shown in Figure 5.30.

Four sixth-mode SIHCs are employed to realize the fourth-order bandpass filter. The iris coupling mechanism (inductive coupling) is applied to the couple power between two resonators. In addition, interdigital capacitor (IDC) coupling is used between the resonators, which provides electrical coupling between them. The field

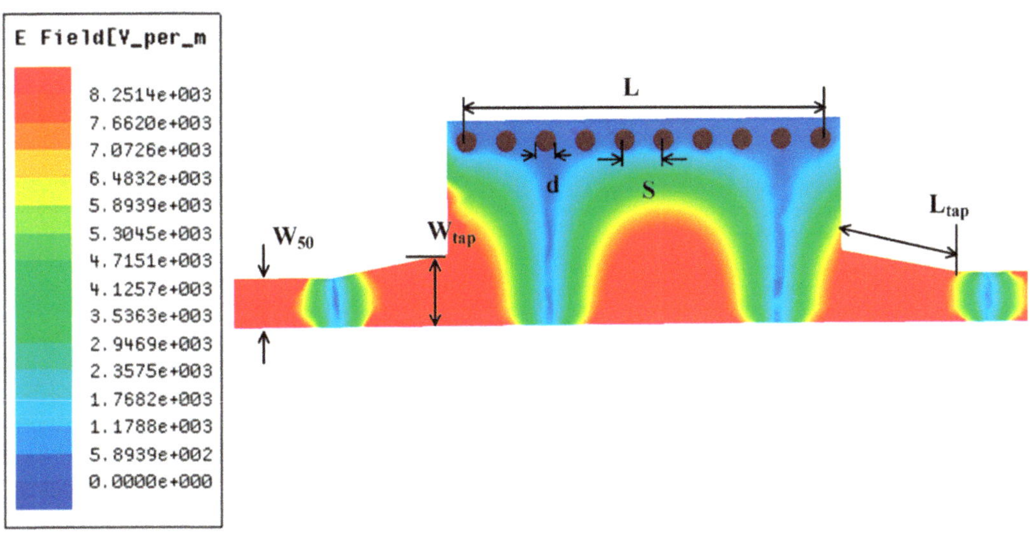

Figure 5.27 Schematic of the half-mode SIW to the microstrip transition with electric field distribution. The dimensions of the schematic are as follows: $d = 1$ mm, $S = 2$ mm, $L = 18$ mm, $W_{tap} = 3.5$ mm, $L_{tap} = 6$ mm, and $W_{50} = 2.4$ mm.

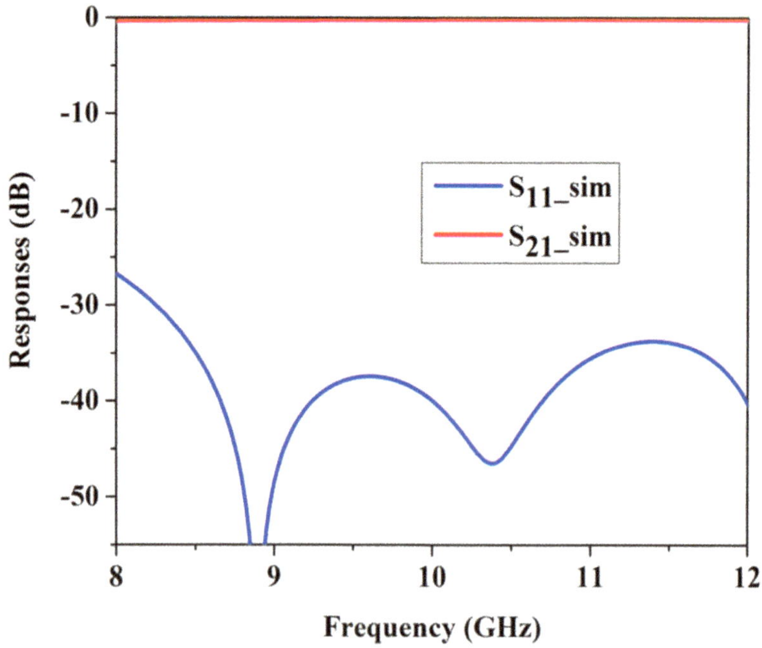

Figure 5.28 Simulated S-parameters of the microstrip-to-HMSIW transition.

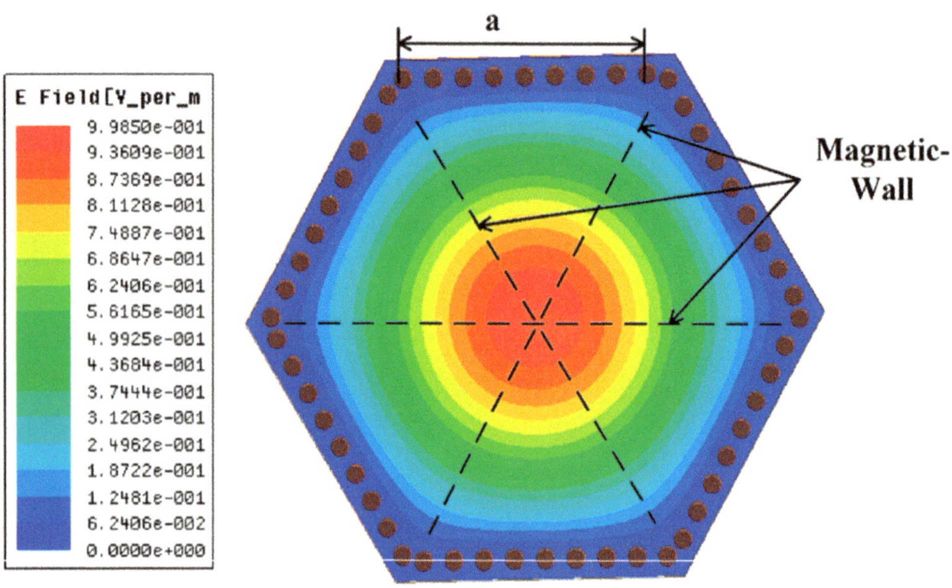

Figure 5.29 Schematic of an SIHC with three fictitious perfect magnetic walls with electric field distribution at 4.7 GHz. Each side-wall dimension of the SIHC: a = 18.5 mm. (© 2016 IEEE [13].)

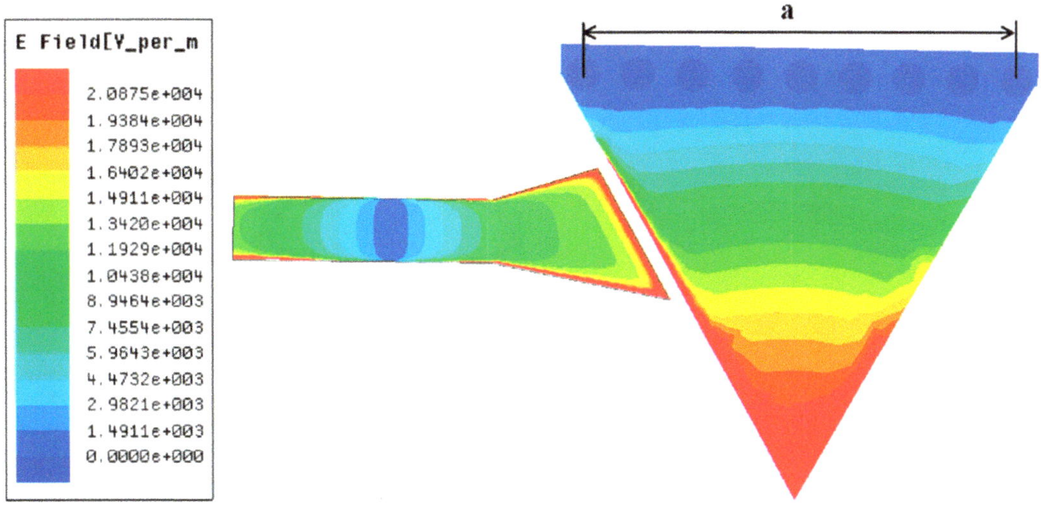

Figure 5.30 Electric field distribution of a sixth-mode SIHC using weak coupling when each side-wall dimension of the sixth-mode SIHC: a = 18.5 mm. (© 2016 IEEE [13].)

distributions of electrically and magnetically coupled resonators are also presented in this section. The variations of coupling coefficients with the change in design parameters are also demonstrated in this section. The operating frequency of the single-mode SIHC is at 4.7 GHz. The dimension of each side wall of the hexagonal cavity is 18.5 mm. Here, the dimension of each edge of the sixth-mode SIHC is also 18.5 mm. The electric field distribution of a sixth-mode cavity is shown in Figure 5.30 while applying weak coupling at the source end. The magnetic field distributions between two sixth-mode resonators for odd and even modes are shown in Figure 5.31 when iris coupling (H-coupling) is applied between two resonators. The even-mode frequency (f_e) varies with change in iris coupling window, while odd-mode (f_0) frequency remains unchanged. The variation of coupling coefficients with a change in the iris aperture is shown in Figure 5.32(a). Larger spacing between f_e

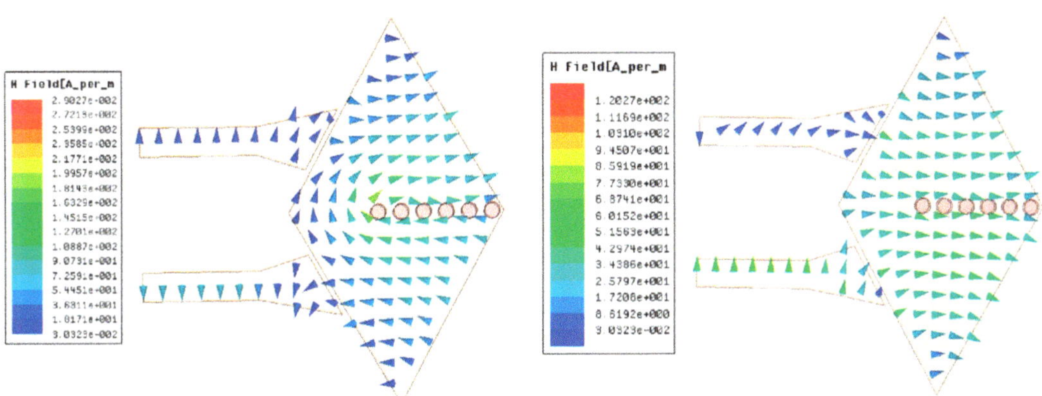

(a) (b)

Figure 5.31 Magnetic vector field distribution through iris coupling for even and odd modes. (a) Even mode. (b) Odd mode. (© 2016 IEEE [13].)

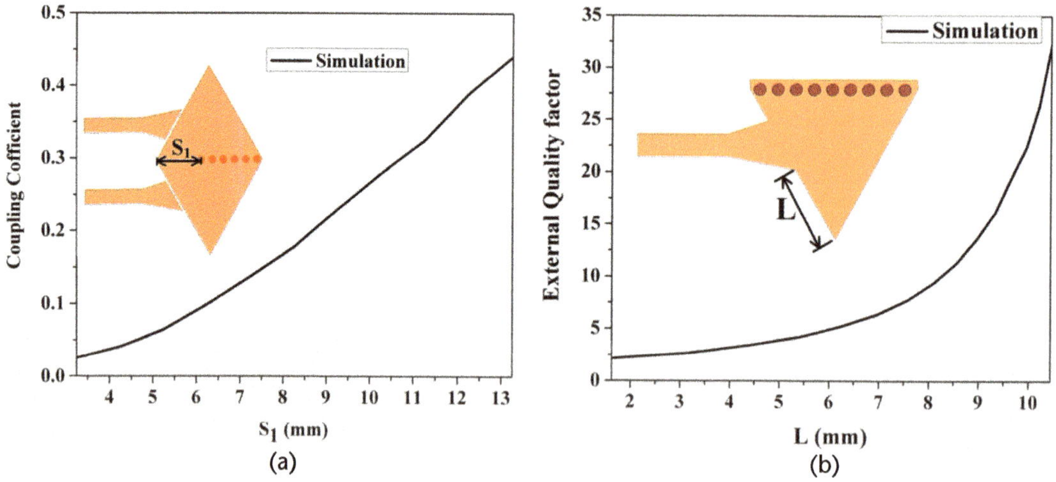

Figure 5.32 (a) Coupling coefficient variations with the changes in the iris window width placed between two sixth-mode resonators. (b) External *Q* variations with the changes in offset loading. (© 2016 IEEE [13].)

and f_0 implies higher coupling between the resonators. The microstrip taper line is used to excite the sixth-mode cavity. The change in external quality factor with the position of offset feed-line is also depicted in Figure 5.32(b). It is observed that if the feed-line position is closer to the open-ended vertex, the coupling becomes higher. As a result, external quality factor gets lower. It is because of inverse relation between external quality factor with coupling coefficient, which is evident from Figure 5.32(b). If the IDC is placed between two sixth-mode resonators as a coupling element, it leads to electrical coupling between the resonators. The electric field distributions for even and odd modes (f_e and f_0) between the resonators are exhibited in Figure 5.33. The electric coupling variation with IDC strip width (W) is shown in Figure 5.34. The fourth-order bandpass filter is realized by adopting

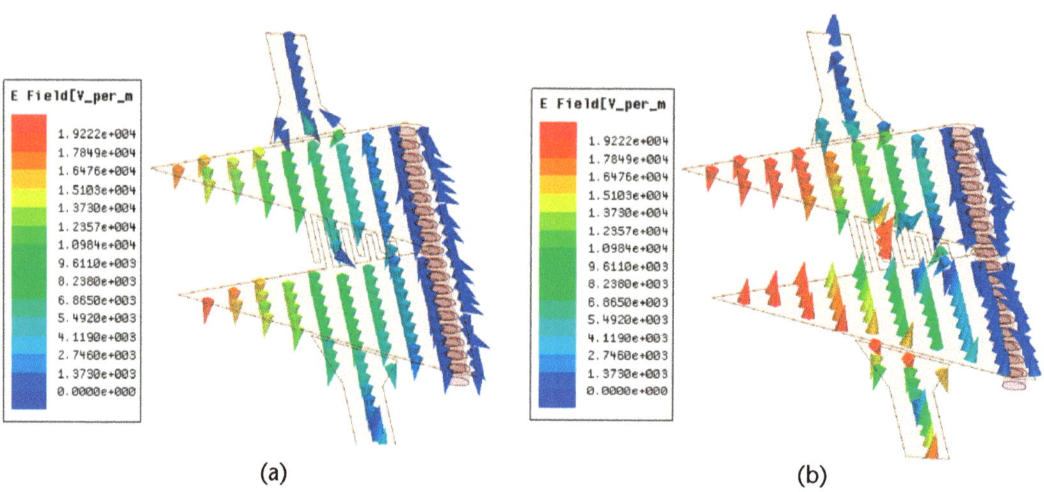

Figure 5.33 Distribution of electric field vector through IDC coupling for both even and odd mode. (a) Even mode. (b) Odd mode.

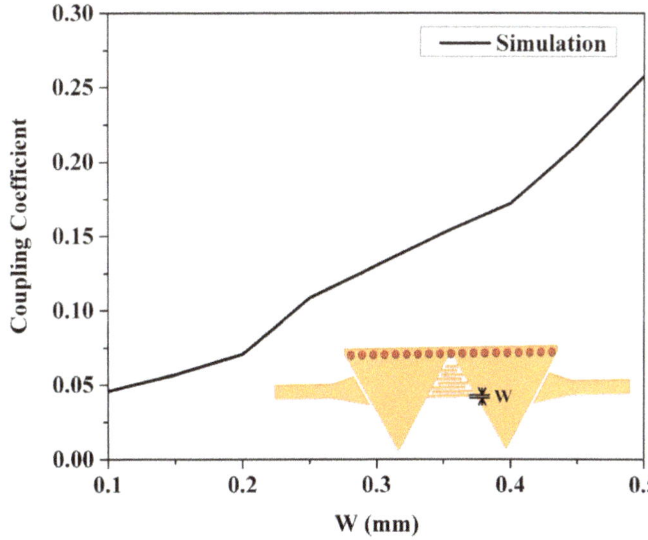

Figure 5.34 Electrical coupling variation with strip width of IDC.

both electric and magnetic coupling. The schematic of the designed sixth-mode bandpass filter with dimensions is provided in Figure 5.35. The diameter of the via is 1.3 mm and the spacing between two vias is 2 mm. The design specifications of the filter are as follows: the center frequency is at 4 GHz with fractional bandwidth (FBW) of 12.5%. All four sixth-mode SIHC have same self-resonant frequency. So the coupling coefficient (k) between two sixth-mode resonators can be calculated using (5.66). The interdigital capacitor provides transmission zero, which improves the selectivity of the filter.

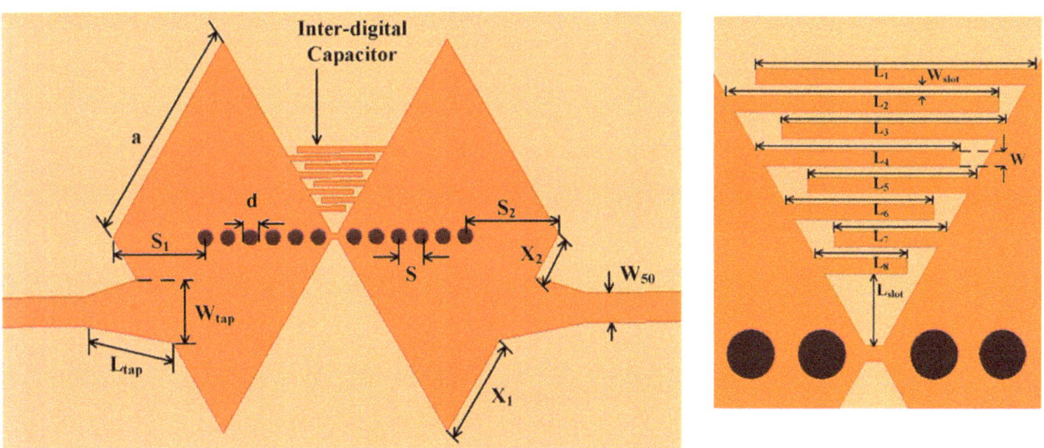

(a) (b)

Figure 5.35 (a) Schematic of sixth-mode SIW-based bandpass filter with dimensions: a = 18.5 mm, S_1 = S_2 = 7.75 mm, d = 1.3 mm, S = 1.85 mm, L_{tap} = 7.35 mm, X_1 = 8.62 mm, X_2 = 4.0 mm, W_{tap} = 5.9 mm, and W_{50} = 2.4 mm. (b) Schematic of interdigital capacitor acting as a coupling element between two resonators with dimensions (L_1 = 7.5 mm, L_2 = 7.3 mm, L_3 = 6 mm, L_4 = 5.45 mm, L_5 = 4.5 mm, L_6 = 3.95 mm, L_7 = 3 mm, L_8 = 2.45 mm, L_{slot} = 2 mm, W = 0.4 mm, and W_{slot} = 0.3 mm).

$$k = \frac{f_U^2 - f_L^2}{f_U^2 + f_L^2} \tag{5.66}$$

where f_U and f_L are the upper and lower resonant frequencies, respectively. When there is a strong coupling, wider separation between two resonant peaks appears. The external quality factor (Q_e) is extracted using (5.67).

$$Q_e = \frac{f_0}{\Delta f_{+90^\circ} - \Delta f_{-90^\circ}} \tag{5.67}$$

where, Δf_{90° and Δf_{-90° are the frequencies at which the S_{11} phase shows +90° and −90° difference with respect to the phase of center frequency (f_0). The prototype is fabricated on Rogers RT/Duroid 5880 with thickness 0.787 mm and it is depicted in Figure 5.36(a) along with the measured and simulated responses. The measured insertion loss and return loss in the passband are 2.0 dB and about 20 dB, respectively. Two transmission zeros appear on both sides of the passband. The fabrication inaccuracy leads to the differences in the measured and simulated responses. As observed from Figure 5.36(b), the measured group delay is flat around the center frequency of the passband.

5.7.3 Second-Order Miniaturized Bandpass Filter Using SIW-Based Rhombic Resonator (Design I)

The schematic of Design I (BPF) along with dimensions is shown in Figure 5.37(a). The design specification of this filter is based on the Chebyshev synthesis method. The passband of the filter is from 9.5 to 10.4 GHz with a return loss below 20 dB. The electric field distribution in this rhombic resonator is shown in Figure 5.37(b).

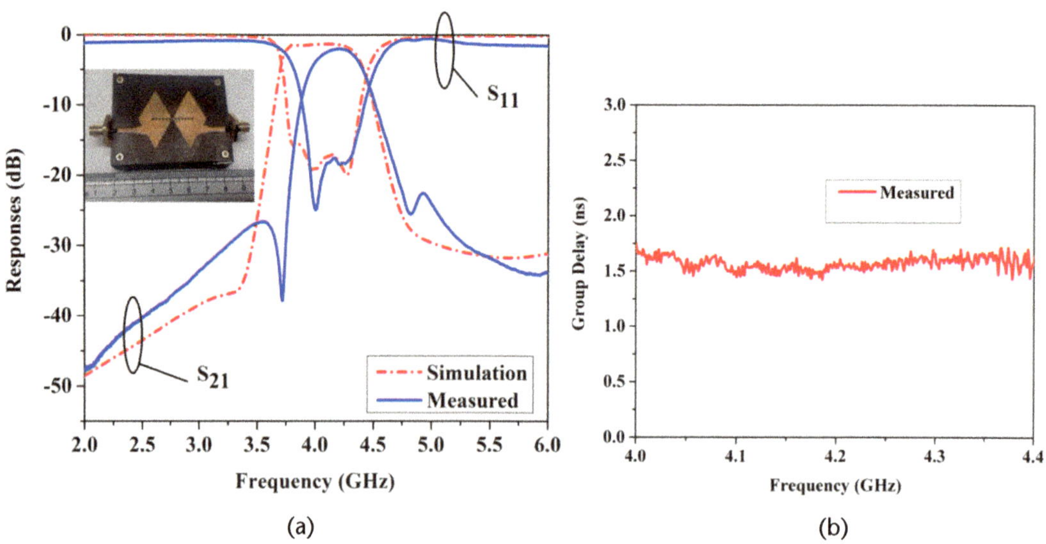

Figure 5.36 (a) Measured and simulated responses of the designed sixth-mode SIHC along with fabricated prototype, and (b) measured group delay.

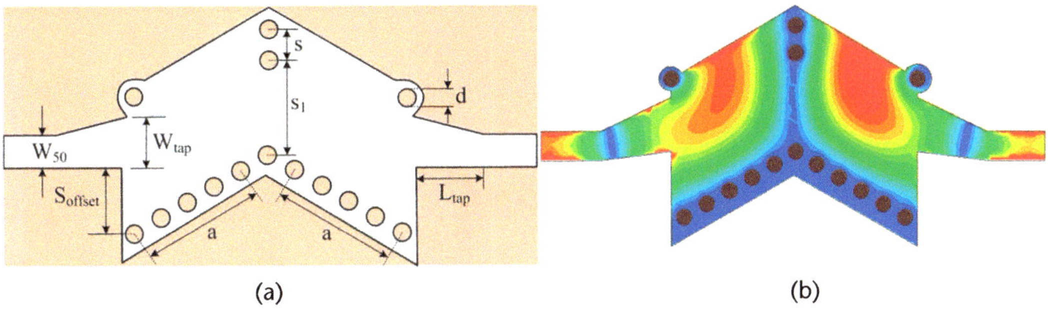

Figure 5.37 (a) Schematic of the second-order miniaturized bandpass filter. (b) Electric field distribution in the miniaturized SIW cavity. The dimension related to the design parameters of the designed second-order BPF are as follows: a = 8.8 mm; S = 2.3 mm; S_1 = 8.2 mm; W_{tap} = 4.5 mm; L_{tap} = 6.2 mm; W_{50} = 2.4 mm; S_{offset} = 5.45 mm; and d = 1.3 mm. (© 2016 IEEE [13].)

A single via is placed at diagonal vertex of the rhombus that helps to improve the matching of the designed filter at the cost of shifting the fundamental resonant frequency slightly to the higher end. The fabricated prototype with simulated and measured results are provided in Figure 5.38.

The simulated and measured insertion losses in the passband are 1.05 dB and 1.6 dB, respectively, and return losses are below 20 dB. The center frequency and fractional bandwidth of the filter are 10 GHz and 9.04%, respectively.

5.7.4 Fourth-Order Miniaturized Bandpass Filter Using SIW-Based Rhombic Resonator (Design II)

Design II is a fourth-order BPF. Higher-order modes of rhombic resonators are exploited to design the fourth-order BPF. The schematic of the fourth-order BPF with dimensions is shown in Figure 5.39(a). Electric field distributions of higher-order

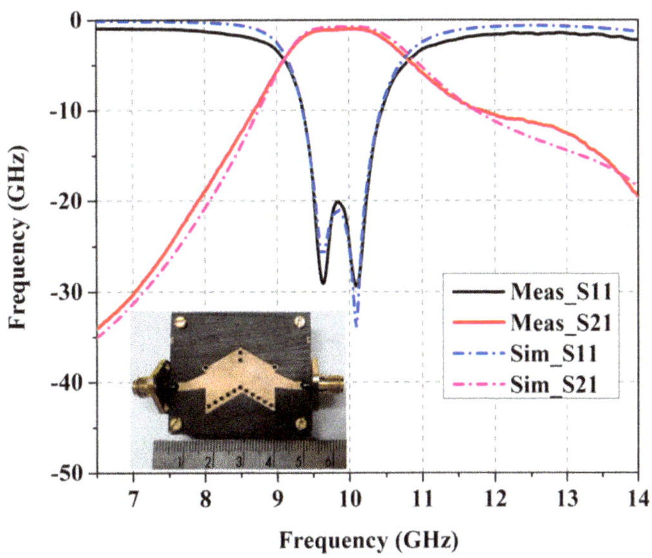

Figure 5.38 Measured and simulated S-parameters of the second-order designed filter along with fabricated prototype of the filter. (© 2016 IEEE [13].)

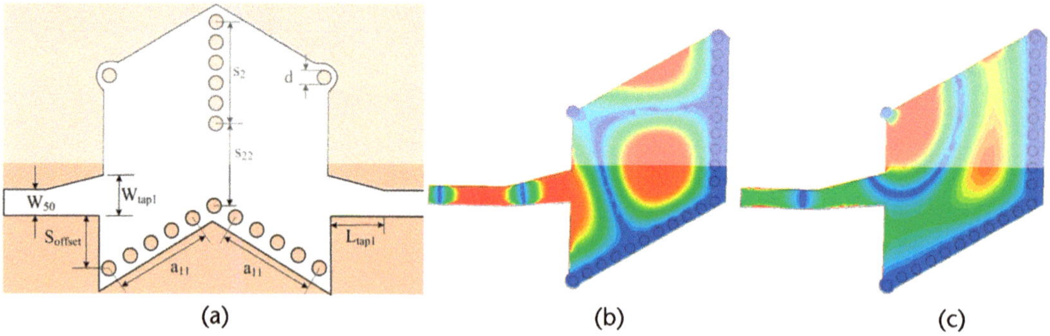

Figure 5.39 (a) Schematic of the fourth-order miniaturized BPF. (b) Electric field distribution in the miniaturized SIW cavity: mode 1 at 9.98 GHz. (c) Electric field distribution in the miniaturized SIW cavity: mode 2 at 10.3 GHz. The dimensions related to the design parameters of the designed fourth-order BPF are as follows: a = 19 mm; S = 12 mm; S_1 = 9 mm; W_{tap} = 4 mm; L_{tap} = 8.5 mm; W_{50} = 2.4 mm; S_{offset} = 9.95 mm; and d = 1.3 mm. (© 2016 IEEE [13].)

orthogonal modes are shown in Figure 5.39(b, c). Two higher-order orthogonal modes are employed in the design to obtain one transmission zero above the higher side of the passband. The design-specified transmission zero improves the selectivity of the filter as compared to Design I. The coupling topology of this filter is illustrated in Figure 5.40. No coupling exists between mode 1 to mode 2 and mode 3 to mode 4 because these modes are orthogonal. The design specifications of this filter are passband from 9.5 to 10.4 GHz with transmission zero at 10.6 GHz. Since higher-order modes are used to design BPF, the side-wall dimension of rhombic resonator is higher than Design I. The dimension of each side arm of this rhombic resonator is 19 mm. In Design II BPF, a single via is placed at a diagonal vertex of the designed resonator, which not only helps to bring two higher-order orthogonal modes close to each other, but also improves the matching of the filter. The simulated and measured results of this filter, along with the fabricated prototype, are shown in Figure 5.41(a, b), respectively. The passband insertion losses of the measured and simulated responses of the designed filter are 1.2 dB and 2.4 dB, respectively. The measured return loss of the designed Design II is about 15 dB in the passband. The center frequency and fractional bandwidth of the filter are at 10 GHz and 8.37%, respectively.

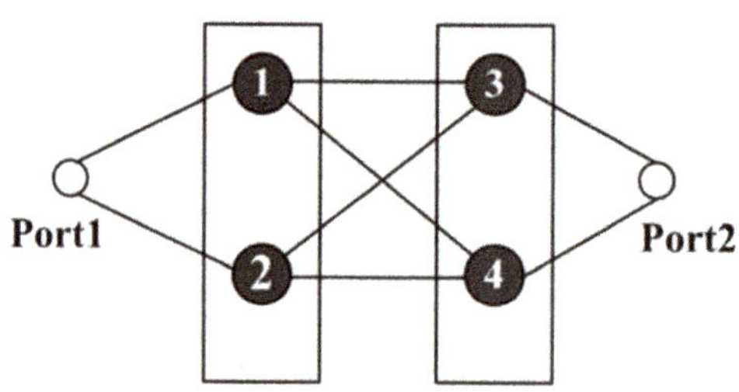

Figure 5.40 Coupling topology of the fourth-order miniaturized bandpass filter. (© 2016 IEEE [13].)

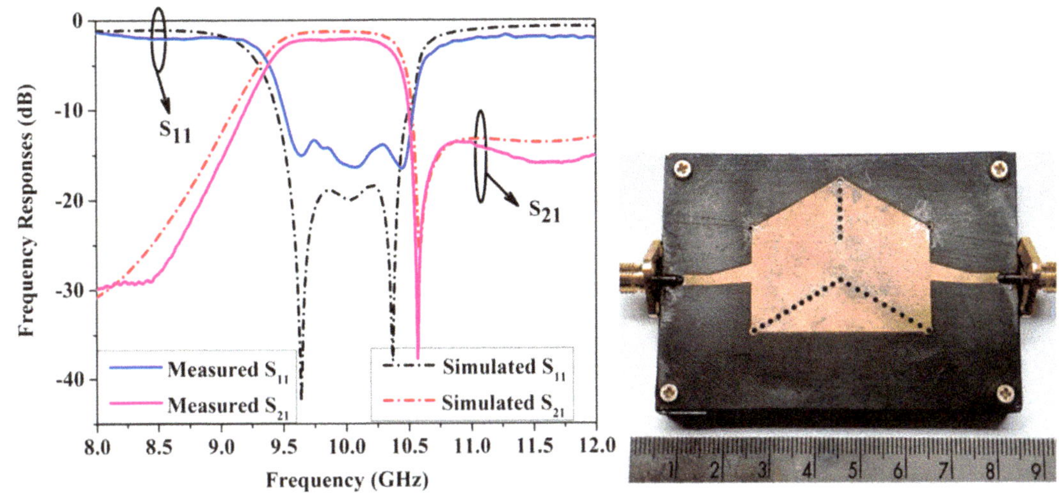

(a) (b)

Figure 5.41 (a) Measured and simulated S-parameters of the fourth-order designed filter. (b) Fabricated prototype of the designed filter. (© 2016 IEEE [13].)

5.8 SIHC-Based Bandstop Filter

In this section, compact dual-mode-based, single-band and dual-band bandstop filters are realized using SIW technology. Three different bandstop filters are implemented using the SIHC with the help of a perturbation in the side wall of the hexagonal cavity that helps in realizing the desired bandstop response. The first two designs are second-order and fourth-order dual-mode-based, single-band bandstop filter. The other design is a compact dual-band bandstop filter using single and dual-mode SIHC. The perturbation effect on reflection zeros of the bandstop filter is investigated. In all three designs, source and load are directly coupled to each other with the help of a microstrip transmission line that generates the required transmission zeros to design the bandstop filter. All the designs are fabricated using 30-mil-thick Rogers RT/Duroid 5880 substrate.

5.8.1 Second-Order SIHC Bandstop Filter (Design I)

The schematic of the second-order bandstop filter with its dimensions is shown in Figure 5.42(a). The hexagonal cavity consists of six side walls. Side wall a_5 is used for perturbation. The direction of this perturbation in side wall a_5 is shown using arrows. Both dual modes exist at same resonant frequency when there is no perturbation. Since the perturbation is applied to the side wall, both degenerate modes are split. Due to the presence of excitations, both the modes are oriented as even and odd modes [14].

The coupling topology of this second-order bandstop filter is provided in Figure 5.42(b). The effect of the *inset* variations using side wall a_5 on reflection zeros and even mode self-resonant frequency is shown in Figure 5.43(a, b), respectively. The

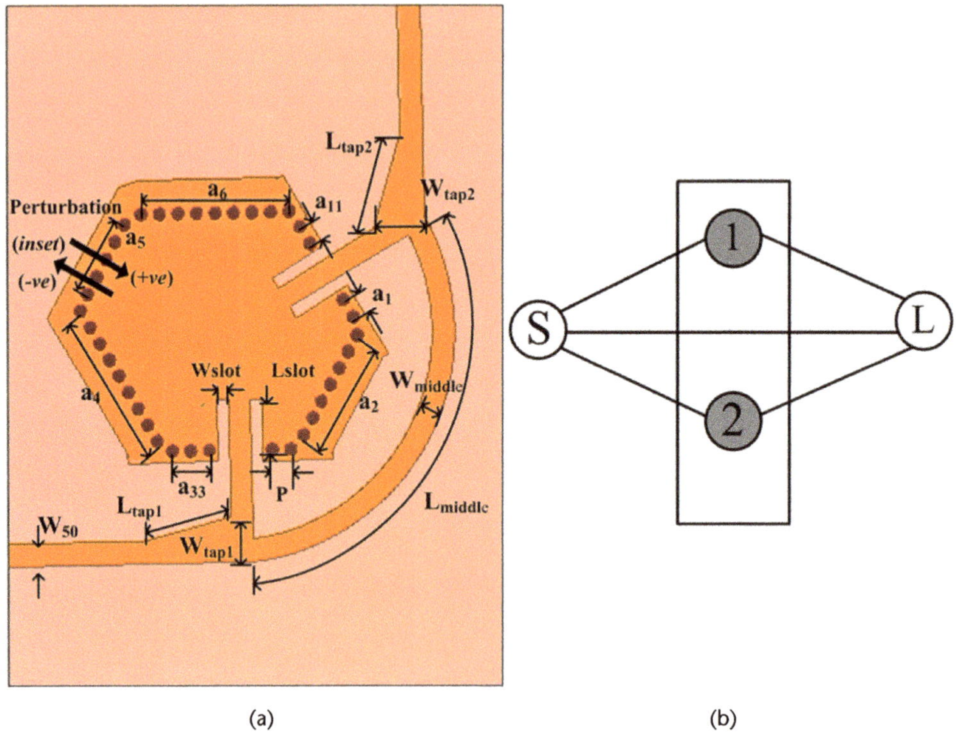

Figure 5.42 (a) Schematic of the bandstop filter Design I. (b) Coupling topology of bandstop filter Design-I dimensions: $L_{slot} = 5.3$ mm, $W_{slot} = 1.25$ mm, $L_{tap1} = L_{tap2} = 10.5$ mm, $W_{tap1} = 4.8$ mm, $W_{tap2} = 5.2$ mm, $a_1 = 2.1$, mm, $a_{11} = 2.1$ mm, $a_2 = 12$ mm, $P = 2$ mm, $a_{33} = 4$ mm, $a_4 = 16$ mm, $a_5 = 8.4$ mm, $a_6 = 16$ mm, and $W_{50} = 2.4$ mm, *inset* = −2.62 mm.

inset is varying from −3 mm to −1 mm, which shifts the design-specified left-side reflection zero towards the right side of the stopband as shown in Figure 5.43(a), but the right-side reflection zero remains unchanged to this perturbation. The effect of this *inset* variations on even-mode self-resonant frequency is also depicted in Figure 5.43(b), whereas the odd-mode self-resonant frequency remains unchanged at 9.86

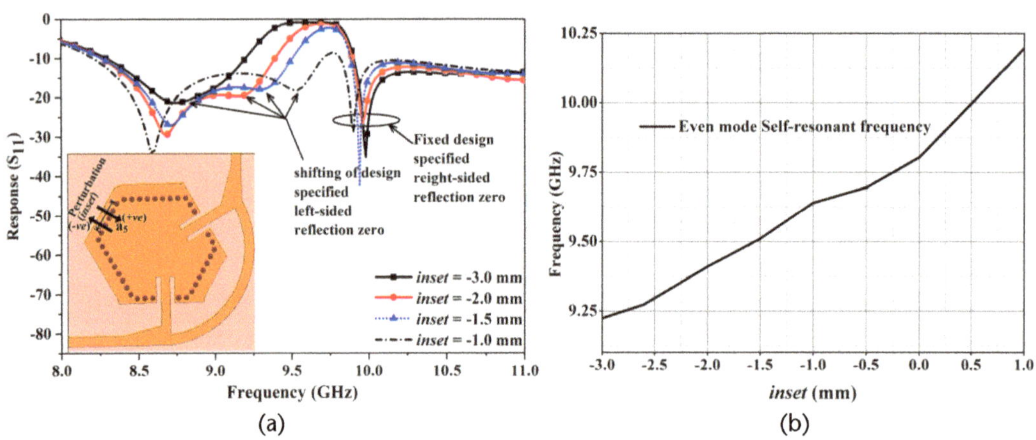

Figure 5.43 Perturbation effect (using side wall a_5) on: (a) reflection zeros of Design I and (b) even-mode self-resonant frequency of Design I.

GHz throughout *inset* variations. However, when side wall a_4 is used for perturbation, both reflection zeros are affected as shown in Figure 5.44(a). It influences both odd-mode and even-mode self-resonant frequencies but significant influence of this perturbation is observed on odd-mode self-resonant frequency as shown in Figure 5.44(b).

The design specifications of this bandstop filter are as follows: the center frequency of the stopband is at 9.7 GHz, the fractional bandwidth of the stopband is 2%, the stopband rejection is 25 dB, and two reflection zeros are at 9 GHz and 10 GHz. The coupling topology of the filter is shown in Figure 5.42(b). The coupling-coefficient values and the self-resonant frequencies are obtained with the help of filter design specifications and coupling topology. In the coupling matrix (5.68), the first and last rows and columns represent source and load, respectively. The second and third rows and columns represent resonators 1 and 2. The coupling coefficients and de-normalized self-resonant frequencies of the resonators [15] are obtained using (5.69) to (5.72).

$$M = \begin{bmatrix} 0 & 0.903 & 1.382 & 0.737 \\ 0.903 & -2.984 & 0 & -0.903 \\ 1.382 & 0 & 4.385 & 1.382 \\ 0.737 & 1.382 & -0.903 & 0 \end{bmatrix} \tag{5.68}$$

$$K_{ij} = FBW \cdot M_{ij} \qquad \text{for } i \neq j, \quad i,j = 1,2,3,\ldots N \tag{5.69}$$

$$f_{0i} = f_0\sqrt{1 + \left(\frac{M_{ii}FBW}{2}\right)^2} - \frac{f_0 M_{ii}}{2} \tag{5.70}$$

$$Q_{e/o} = \frac{1}{FBW \cdot M_{S/L,m}^2} \tag{5.71}$$

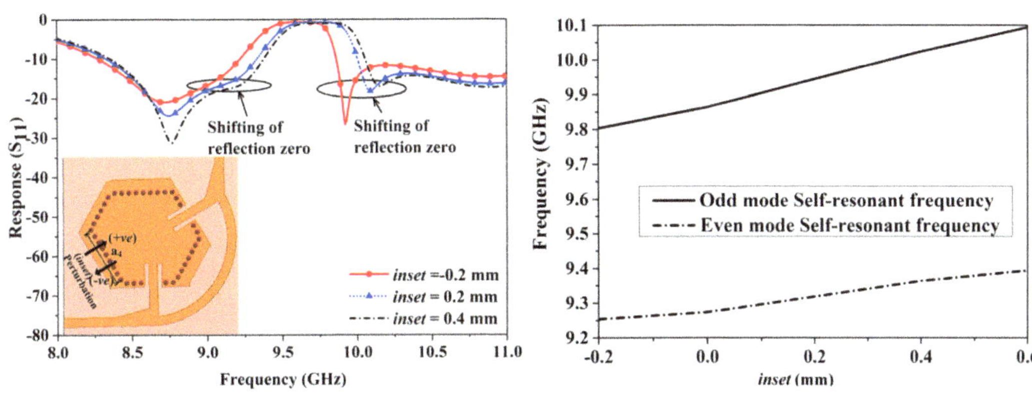

(a) (b)

Figure 5.44 Perturbation effect (using side wall a_4) on: (a) reflection zeros of Design I and (b) even-mode and odd-mode self-resonant frequency of Design I.

$$|M_{SL}| = \sqrt{\frac{1-|S_{11\infty}|}{1+|S_{11\infty}|}} = \frac{1}{|M'_{SL}|} \quad (5.72)$$

Here, M_{ij}, K_{ij} are normalized and de-normalized coupling-coefficient values. M_{ii}, f_{oi}, are normalized and de-normalized self-resonant frequencies, and f_0 is the center frequency of the bandstop filter. $Q_{e/o}$ is the external quality factor of the filter at source or load. $M_{S/L,m}$ is source or load coupling coefficients with the mth resonator. M_{SL} is source-load coupling-coefficient value. To realize a bandstop filter, source and load are directly coupled using the microstrip transmission line of $2\lambda_{eff}$. The microstrip taper line is required to match both the junctions. Figure 5.45 depicts the measured, EM-simulated, and circuit-simulated results along with the fabricated prototype of the filter. The measured return loss in stopband is found to be 1.5 dB.

5.8.2 Fourth-Order Bandstop Filter (Design II)

Design II is a fourth-order bandstop filter using two dual-mode hexagonal cavities. The filter design specifications are as follows: the center frequency of the stopband is at 10.15 GHz; the fractional bandwidth of the stopband is 3.0%; and two reflection zeros are placed at 9.5 GHz and 10.3 GHz. The in-band rejection is 20 dB. The coupling matrix is provided in (5.73). The schematic with dimensions and coupling topology of this filter is shown in Figure 5.46(a, b). To realize the bandstop filter design, source and load are directly coupled using microstrip transmission line of $5\lambda_{eff}/2$ with 50Ω transmission line.

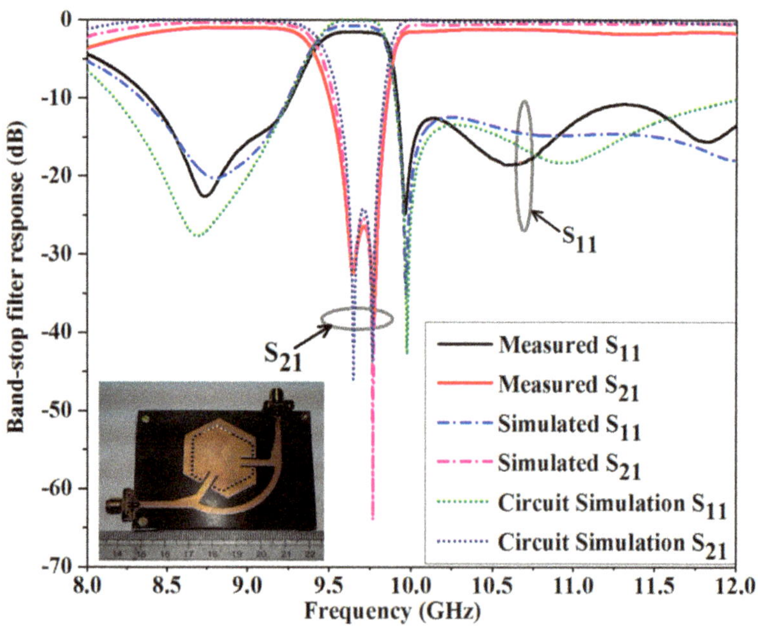

Figure 5.45 Design-I bandstop filter's measured, EM-simulated, and circuit-simulated responses and its fabricated prototype.

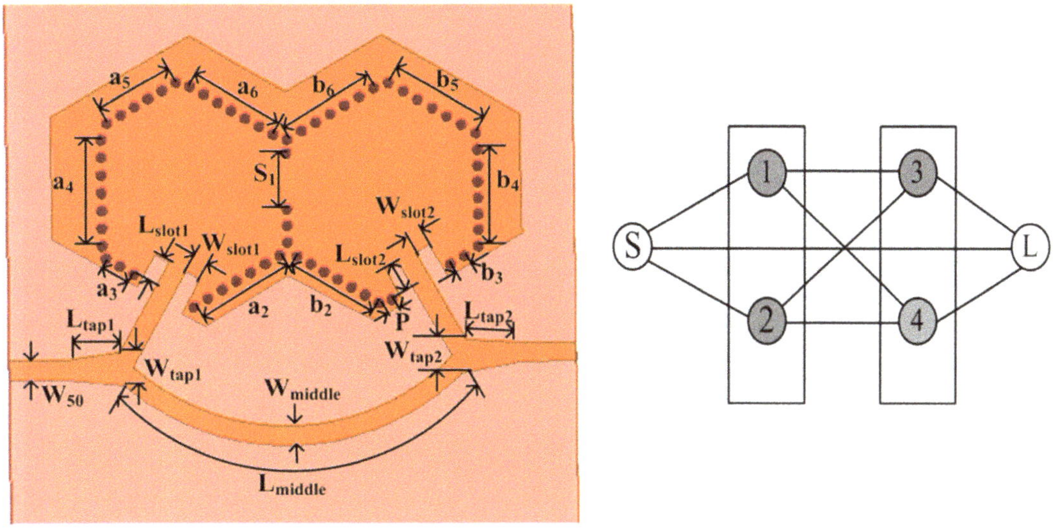

(a) (b)

Figure 5.46 (a) Schematic of the bandstop filter Design II. (b) Coupling topology of the bandstop filter Design II. Dimension of this Design II (bandstop filter): $L_{slot1} = L_{slot2} = 3.5$ mm, $W_{slot1} = W_{slot2} = 2$ mm, $S_1 = 6.975$ mm, $a_2 = 12.6$ mm, $a_3 = 4$ mm, $a_4 = 13.2$ mm, $a_5 = 10$ mm, $a_6 = 12$ mm, $b_2 = 11.7$ mm, $b_3 = 2$ mm, $b_4 = 12$ mm, $b_5 = 12.6$ mm, $b_6 = 12.6$ mm, $L_{tap1} = L_{tap2} = 5.4$ mm, $W_{tap1} = W_{tap2} = 4.0$ mm, $L_{middle} = 49.8$ mm, and $W_{middle} = 2.4$ mm.

$$M = \begin{bmatrix} 0 & 1.298 & 0.566 & 0 & 0 & -1.0 \\ 1.298 & 0.114 & 0 & 0.456 & -1.396 & 0 \\ 0.566 & 0 & 0.536 & 0.350 & 0.566 & 0 \\ 0 & 0.456 & 0.350 & -0.934 & 0 & -0.010 \\ 0 & -1.396 & 0.566 & 0 & -0.085 & 1.416 \\ -1.0 & 0 & 0 & -0.010 & 1.416 & 0 \end{bmatrix} \quad (5.73)$$

Figure 5.46(b) shows that all four modes of both dual-mode cavities are coupled with each other. Coupling between two asynchronous modes are calculated with the help of HFSS eigenmode simulator using (5.74):

$$k = \pm\frac{1}{2}\left(\frac{f_{02}}{f_{01}} + \frac{f_{01}}{f_{02}}\right) \cdot \sqrt{\left(\frac{f_{p2}^2 - f_{p1}^2}{f_{p2}^2 + f_{p1}^2}\right)^2 - \left(\frac{f_{02}^2 - f_{01}^2}{f_{02}^2 + f_{01}^2}\right)^2} \quad (5.74)$$

Here, f_{01} and f_{02} are the self-resonant frequencies of both the modes. f_{p1} and f_{p2} are the lower and higher split resonant frequencies when both the modes are coupled. To calculate the coupling-coefficient between any two modes, the other modes should be suppressed using either perfect electric wall or perfect magnetic wall.

Since source and load both are directly coupled through transmission line, four transmission zeros appear in the design. These four transmission zeros are exploited to realize fourth-order bandstop filter. The pair of hexagonal cavities in the designed

Design II help to realize four reflection zeros. Since the microstrip taper line and Y-junction (both source and load ends) have not been incorporated in the theoretical responses, the discrepancy appears. This is addressed by designing equivalent circuit model of the filter in Keysight Advanced Design System (ADS) software. Both microstrip tapers lines (W_{tap1} and W_{tap2}) are quarter-wavelength transformer, which is used for matching purpose in the junctions.

Figure 5.47 shows the effect of varying microstrip arc length (L_{middle}) on reflection zeros. The proper choosing of microstrip arc length helps in improving bandstop response of the desired bandstop filter. Figure 5.48(a) shows the measured, EM-simulated and circuit-simulated frequency responses of the bandstop filter. Figure 5.48(b) shows the measured spurious responses and fabricated prototype.

5.8.3 Dual-Band Bandstop Filter (Design III)

Here, a pair of dual-mode and a pair of single-mode SIHCs are utilized to realize a dual-band bandstop filter. The schematic of this designed bandstop filter with dimensions is shown in Figure 5.49(a), and the coupling topology is provided in Figure 5.49(b). The design specification of the filter is as follows: the center frequency of the filter is at 9.5 GHz. The filter has two stopbands. The lower stopband is from 9.25 to 9.35 GHz, and band rejection is 25 dB. The higher stopband is from 9.75 to 9.82 GHz, and band rejection is 20 dB. The reflection zeros are placed at 8.9 GHz, 9.15 GHz, and 9.7 GHz. The fractional bandwidth (FBW) of the filter is 5.3%. Based on design specifications and coupling topology, the coupling matrix values are obtained, which are provided in (5.75).

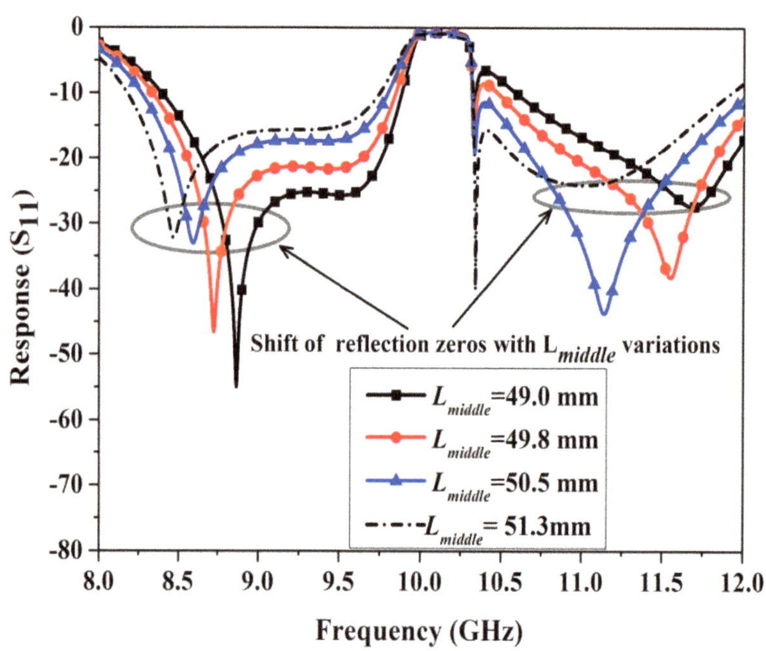

Figure 5.47 Effect of varying microstrip arc length (L_{middle}) on reflection zeros of the Design-II bandstop filter.

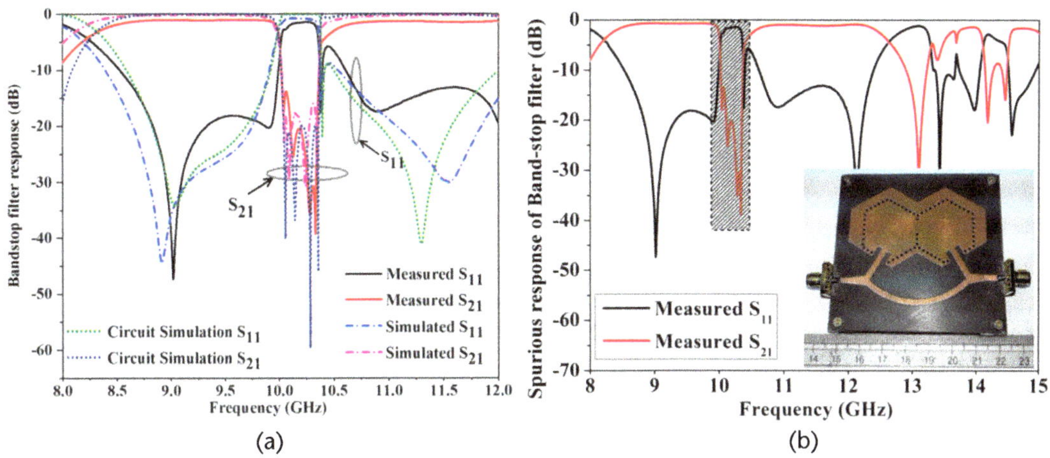

Figure 5.48 (a) Measured, EM-simulated, and circuit-simulated responses. (b) Measured spurious responses of the fabricated prototype of the Design-II bandstop filter.

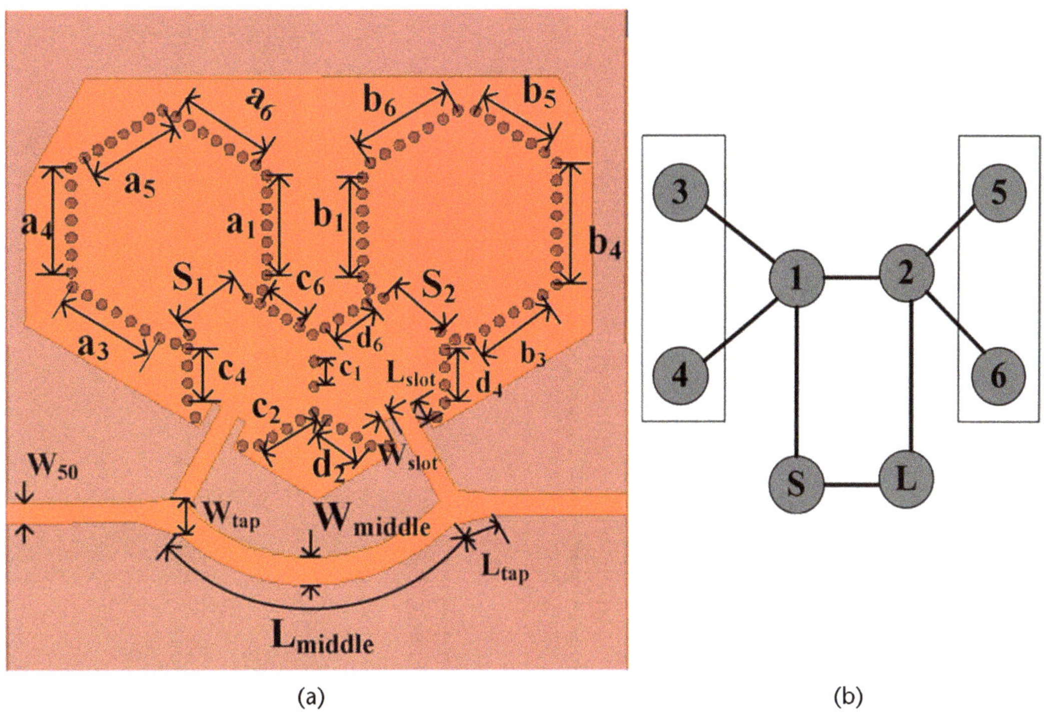

Figure 5.49 (a) Schematic of dual-band bandstop filter Design III. (b) Coupling topology of dual-band bandstop filter Design III. Dimension of this Design III (bandstop filter): W_{tap} = 4.5 mm, C_1 = 2.975 mm, L_{slot} = 2.7 mm, W_{slot} = 1.3 mm, C_2 = 8 mm, C_4 = 6.3 mm, C_6 = 7.6 mm, a_5 = 11.4 mm, a_6 = 11.7, a_1 = 12 mm, a_4 = 12.6 mm, a_3 = 12.6 mm, S_1 = 8.575 mm, d_2 = 6.3 mm, d_4 = 6.3 mm, S_2 = 8.175 mm, d_6 = 6 mm, b_3 = 10.5 mm, b_4 = 14.7 mm, b_5 = 10 mm, b_6 = 12.6 mm, W_{middle} = 3.2 mm, and L_{middle} = 39.0 mm. Substrate: Rogers 5880 with relative dielectric constant 2.2, thickness 0.787 mm, and loss tangent 0.0009.

$$M = \begin{bmatrix} 0 & 1.315 & 0 & 0 & 0 & 0 & 0 & 1.1 \\ 1.315 & 0.379 & 0.609 & 0.498 & 0 & 0 & 0 & 0 \\ 0 & 0.609 & -0.483 & 0 & 0 & 0.01 & 1.118 & 0.927 \\ 0 & 0.498 & 0 & -0.411 & 0 & 0 & 0 & 0 \\ 0 & 0 & 0 & 0 & 1.541 & 0 & 0 & 0 \\ 0 & 0 & 0.01 & 0 & 0 & 0.695 & 0 & 0 \\ 0 & 0 & 1.118 & 0 & 0 & 0 & 0.435 & 0 \\ 1.1 & 0 & 0.927 & 0 & 0 & 0 & 0 & 0 \end{bmatrix} \quad (5.75)$$

Figure 5.50 shows the theoretical and simulated frequency responses of the dual-band bandstop filter. To take care of microstrip taper line and Y-junction, Keysight Advanced Design System (ADS) software is used for circuit simulation. External quality factors, coupling coefficients and self-resonant frequencies are incorporated in the circuit simulation. Equivalent lumped capacitor and inductor of the resonators [10] are dependent on external quality factor and expressed using (5.76) and (5.77)

$$C_{oi} = \frac{Q_{e/o}}{\omega_{0i} Z} \times 10^{12} \text{ (pF)} \quad (5.76)$$

$$L_{oi} = \frac{Z}{\omega_{0i} Q_{e/o}} \times 10^{09} \text{ (nH)} \quad (5.77)$$

$Q_{e/o}$ is external quality factor of a dual-mode resonator's even/odd mode. All the coupling-coefficients can be realized as impedance inverter model. The source and load impedance inverters are calculated using (5.78) to (5.81):

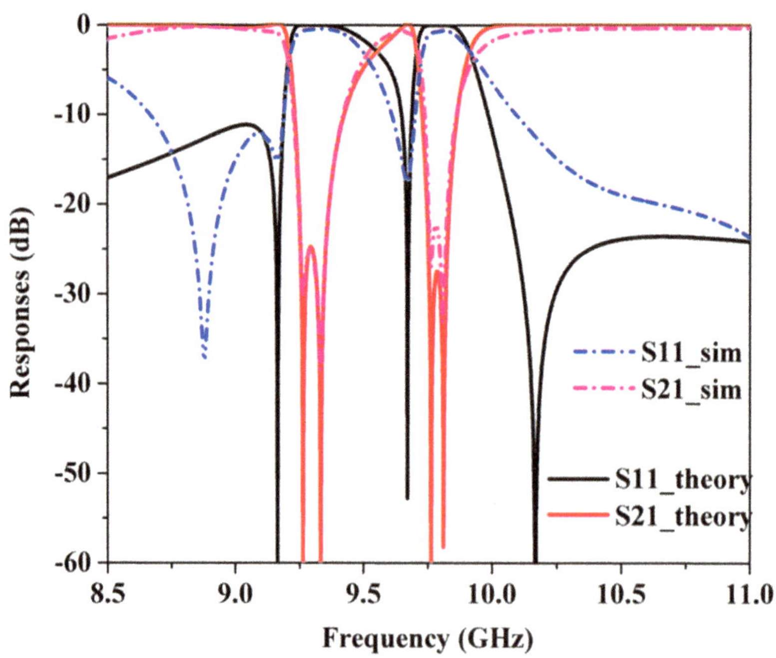

Figure 5.50 Theoretical and simulated responses of Design III (dual-band bandstop filter).

$$C_{\text{in(e/o)}} = \sqrt{\frac{C_{oi}}{ZQ_{e/0}\omega_{0i}}} \tag{5.78}$$

$$Z_{\text{(e/o)}} = \frac{1}{\omega_{0i} C_{\text{in(e/o)}}} \tag{5.79}$$

$$C_{ij} = K_{ij} \cdot \sqrt{C_{oi} C_{oj}} \tag{5.80}$$

$$Z_{ij} = \frac{1}{\omega_0 C_{ij}} \tag{5.81}$$

Here, Z is port impedance (50Ω) and $Z_{\text{(e/o)}}$ corresponds to characteristic impedance of even/odd-mode impedance inverters. C_{oi} and L_{oi} are equivalent lumped models of odd-mode and even-mode resonators. For an odd-mode resonator, the equivalent lumped inductor and capacitor are L_{oo} and C_{oo}. Similarly, for an even-mode resonator, the lumped inductor and capacitor are L_{oe} and C_{oe}. The circuit schematic of the bandstop filter model is shown in Figure 5.51. The microstrip taper line is quarter-wavelength ($L_{\text{tap1}} = L_{\text{tap2}} = 10.6$ mm). In the circuit simulation, the unloaded quality factor (Q_U) is considered infinity. All the design parameters including impedance inverters and lumped inductors and capacitors values are provided as follows:

$L_1 = 0.05$ nH; $C_1 = 6$ pF; $L_2 = 0.043$ nH; $C_2 = 7$ pF; $L_3 = 0.048$ nH; $C_3 = 5.7$ pF; $L_4 = 0.053$ nH; $C_4 = 6.27$ pF; $L_5 = 0.04$ nH; $C_5 = 6.6$ pF; $L_6 = 0.041$ nH; $C_6 = 6.72$ pF; $Z_2 = 50\Omega$; $Z_{13} = 48\Omega$; $Z_{14} = 1{,}570.7\Omega$; $Z_{25} = 1{,}416.7\Omega$; $Z_{26} = 40\Omega$; $Z_{12} = 967\Omega$.

The measured, EM-simulated, and circuit-simulated responses of the dual-band bandstop filter are shown in Figure 5.52(a). Figure 5.52(b) shows a measured spurious response and a fabricated prototype of the filter.

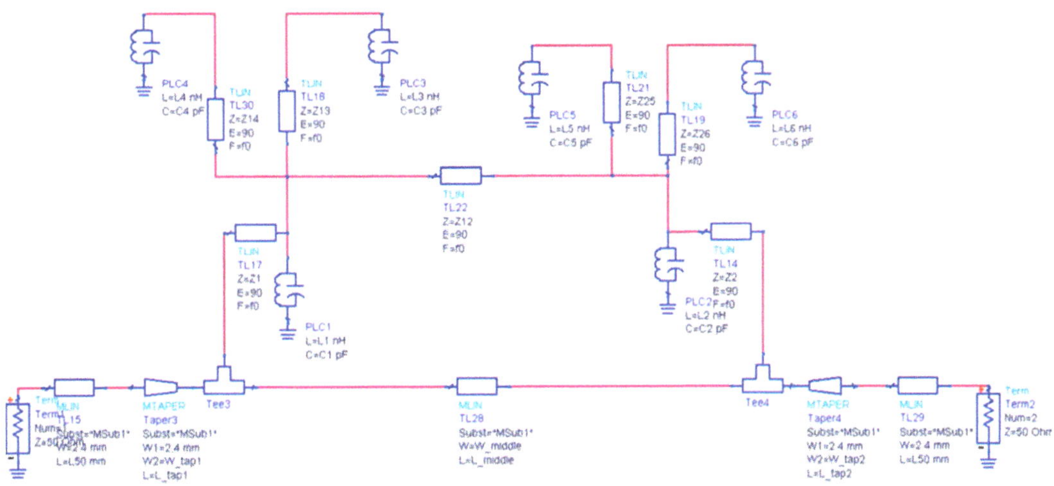

Figure 5.51 Circuit schematic of Design III (dual-band bandstop filter).

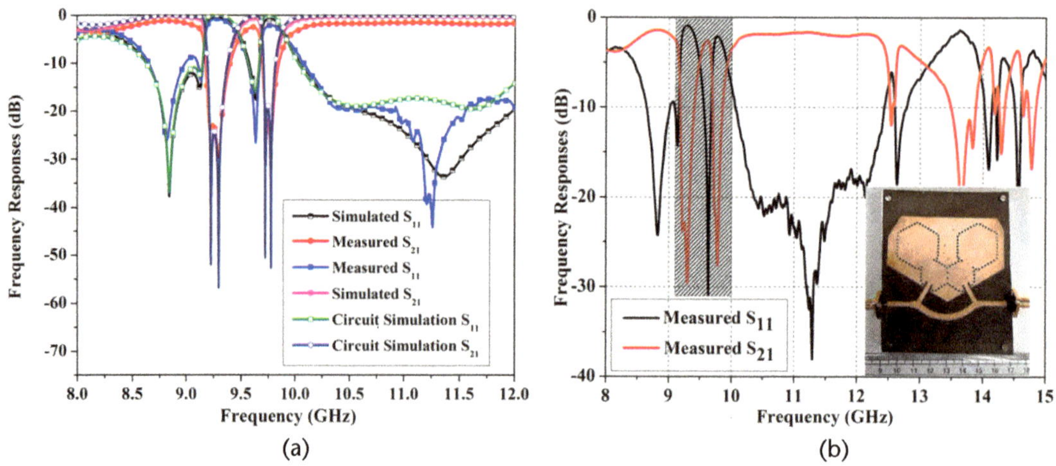

Figure 5.52 (a) Design-III bandstop filter's measured, EM-simulated, and circuit-simulated responses. (b) Spurious response and fabricated prototype of Design-III bandstop filter.

References

[1] Cameron, R. J., R. Mansour, and C. M. Kudsia, *Microwave Filters for Communication Systems: Fundamentals, Design and Applications*, New York: Wiley, 2007.

[2] Zverev, A. I., *Handbook of Filter Synthesis*, New York: Wiley-Interscience, 2005.

[3] Hunter, I., *Theory and Design of Microwave Filters*, London, U.K.: IET, 2001.

[4] Atia, A. E., and A. E. Williams, "Narrow-Bandpass Waveguide Filters," *IEEE Transactions on Microwave Theory and Techniques*, Vol. 20, No. 4, April 1972, pp. 258–265.

[5] Cameron, R. J., "Advanced Coupling Matrix Synthesis Techniques for Microwave Filters," *IEEE Transactions on Microwave Theory and Techniques*, Vol. 51, No. 1, January 2003, pp. 1–10.

[6] Amari, S., "Synthesis of Cross-Coupled Resonator Filters Using an Analytical Gradient-Based Optimization Technique," *IEEE Transactions on Microwave Theory and Techniques*, Vol. 48, No. 9, September 2000, pp. 1559–1564.

[7] Matthaei, G. L., L. Young, and E. M. T. Jones, *Microwave Filters, Impedance Matching Networks, and Coupling Structures*, New York: McGraw-Hill, 1964.

[8] Chongder, P., K. V. Srivastava, and A. Biswas, "Realisation of Controllable Transmission Zeros by Perturbation Technique for Designing Dual-Mode Filter Using Substrate Integrated Hexagonal Cavity," *IET Microwaves, Antennas Propagation*, Vol. 8, No. 6, April 2014, pp. 451–457.

[9] Rosenberg, U., and S. Amari, "Novel Design Possibilities for Dual-Mode Filters Without Intracavity Couplings," *IEEE Microwave and Wireless Components Letters*, Vol. 12, No. 8, August 2002, pp. 296–298.

[10] Hong, J., and M. Lancaster, *Microstrip Filters for RF/Microwave Applications*, Second Edition, New York: Wiley, 2011.

[11] Chongder, P., and A. Biswas, "Novel Single Substrate Integrated Hexagonal Cavity (SIHC) Based Dual-Band Bandpass Filter," *2019 IEEE MTT-S International Microwave and RF Conference (IMARC)*, Mumbai, India, 2019, pp. 1–4.

[12] Hong, W., et al., "Half Mode Substrate Integrated Waveguide: A New Guided Wave Structure for Microwave and Millimeter Wave Application," *2006 Joint 31st International Conference on Infrared Millimeter Waves and 14th International Conference on Terahertz Electronics*, September 2006, p. 219.

[13] Chongder, P., and A. Biswas, "Novel HMSIW Cavity Based Bandpass Filter for X Band Application," *2016 Asia-Pacific Microwave Conference (APMC)*, New Delhi, India, 2016.

[14] Gorur, A., E. Gunturkun, and C. Karpuz, "Dual-Mode Microstrip Bandstop Filters Using Square Loop Resonators," *2014 44th European Microwave Conference (EuMC)*, October 2014, pp. 1060–1063.

[15] Amari, S., and U. Rosenberg, "Direct Synthesis of a New Class of Bandstop Filters," *IEEE Transactions on Microwave Theory and Techniques*, Vol. 52, No. 2, February 2004, pp. 607–616.

CHAPTER 6

SIW Passive Circuits

In Chapter 5, detailed analysis and design of various types of SIW-based bandpass and bandstop filters were discussed.

In this chapter, diplexer and triplexer designs using synthesis technique, dual-band diplexer, reconfigurable diplexer are discussed. Besides these, filtering and diplexing balun and miniaturized hybrid branch line coupler designs are also demonstrated as various passive components in SIW technology.

6.1 Diplexer Design

Microwave diplexer [1] is designed by combining two bandpass filters (BPFs) of different center frequencies and are realized using either single-mode or dual-mode resonators. One of the applications of microwave diplexer is to combine both transmitter and receiver channels using a common junction that is connected to a single antenna. Figure 6.1(a) shows that the junctions attached to Port 1 are generally considered as a Y-junction or T-junction. Generally, in the SIW diplexer, the junction is H-plane T-junction or Y-junction in which fabrication complexity is lesser.

Another way, the common node includes one more extra resonator apart from the Tx and Rx filters resonators and therefore it provides design flexibility as shown in Figure 6.1(b). The resonator makes the design more compact as it occupies less area than the T-junction or Y-junction.

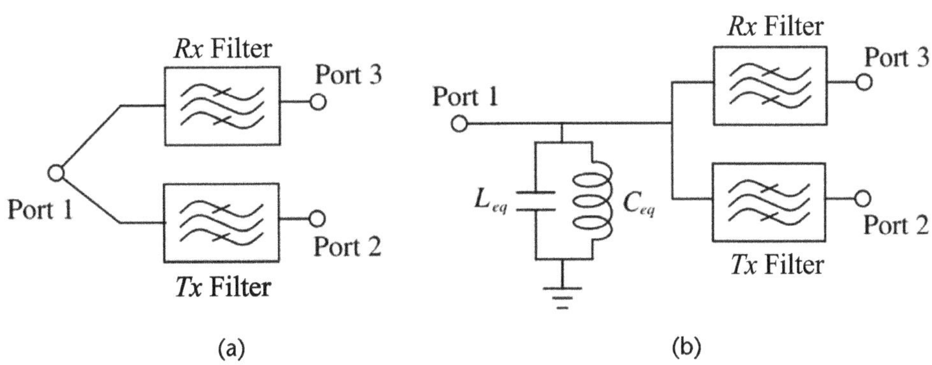

Figure 6.1 Diplexer design with different junction types. (a) Type-I (waveguide H-plane T-junction or Y-junction). (b) Type-II (junction considered as a common resonator) [3].

6.1.1 Diplexer Design Synthesis Technique

The two (Rx and Tx) lowpass filters are separately analyzed in normalized frequency domain through their characteristic polynomials [2].

$$S_{11}^{\mathrm{Tx}} = \frac{F_{\mathrm{Tx}}(s)}{E_{\mathrm{Tx}}(s)},\ S_{21}^{\mathrm{Tx}} = \frac{P_{\mathrm{Tx}}(s)}{E_{\mathrm{Tx}}(s)} = \frac{P_{0\mathrm{Tx}}P_{\mathrm{Tx}n}(s)}{E_{\mathrm{Tx}}(s)}$$
$$S_{11}^{\mathrm{Rx}} = \frac{F_{\mathrm{Rx}}(s)}{E_{\mathrm{Rx}}(s)},\ S_{21}^{\mathrm{Rx}} = \frac{P_{\mathrm{Rx}}(s)}{E_{\mathrm{Rx}}(s)} = \frac{P_{0\mathrm{Rx}}P_{\mathrm{Rx}n}(s)}{E_{\mathrm{Rx}}(s)} \quad (6.1)$$

Both Rx and Tx filters have the degrees of $N_{p\mathrm{Rx}}$ and $N_{p\mathrm{Tx}}$, respectively. These can be found from the characteristic polynomials E_{Rx} and E_{Tx} given in (6.1). Roots of the E_{Rx} and E_{Tx} represent the poles of Rx and Tx filters, respectively. The roots of the polynomial F_{Rx} and F_{Tx} correspond to the reflection zeros of Rx and Tx filters. Similarly, the roots of the polynomial P_{Rx} and P_{Tx} are the transmission zeros of Rx and Tx filters, respectively.

$P_{0\mathrm{Rx}}$ and $P_{0\mathrm{Tx}}$ are the values of highest degree coefficients of P_{Rx} and P_{Tx} polynomials, respectively, when the highest degree of both P_{Rx} and P_{Tx} polynomials are imposed to one. The diplexer is considered as a lossless three-port network that can be defined by four polynomials. These polynomials can be related to scattering parameters in the lowpass normalized domain as [2]:

$$S_{11} = \frac{n_0 N(s)}{D(s)},\ S_{21} = \frac{P_{0t}P_t(s)}{D(s)},\ S_{31} = \frac{P_{0r}P_r(s)}{D(s)} \quad (6.2)$$

The coefficients of highest degree of all four polynomials are normalized to one. Hence, n_0, P_{0t}, and P_{0r} have the highest degree coefficient values of the concerning polynomials of $N(s)$, P_t, and P_r, respectively in (6.2). The roots of $N(s)$ and $D(s)$ represent the reflection zeros and poles of the overall diplexer at Port 1 as shown in Figure 6.1(b). Besides that, the roots of P_t and P_r are the position of transmission zeros in lowpass normalized domain.

The synthesis of the diplexer is carried out in a normalized frequency domain, as shown in Figure 6.2. First, a bandpass to lowpass or vice versa. Frequency transformation is carried out using: $\Omega = (f_0/B)(f/f_0 - f_0/f)$. This is the usual lowpass to bandpass frequency transformation where $B = f_{H2} - f_{L1}$ and $f_0 = \sqrt{f_{L1} \cdot f_{H2}}$. In the design, the Rx band is considered below the Tx band. The Rx passband is from f_{L1} to f_{L2}. The center frequency of Rx band is $\sqrt{f_{L1} \cdot f_{L2}}$ and the corresponding Rx bandwidth is $(f_{L2} - f_{L1})$ and Tx band is from f_{H1} to f_{H2} with center frequency $\sqrt{f_{H1} \cdot f_{H2}}$ and the Tx bandwidth is $(f_{H2} - f_{H1})$. Using the bandpass to lowpass frequency transformation, diplexer response transforms into the Ω-plane. The Ω-plane consists of both Rx and Tx lowpass bands. In the Ω-plane, Rx filter's lower and upper passband limits are -1 and Ω_r, respectively. For the Tx filter upper and lower passband limits are 1 and Ω_t. Next, frequency transformation is carried out from Ω to Ω'-plane. In the Ω'-plane, both Rx and Tx bands are separate lowpass prototype filters. The S-parameters of the two lowpass prototype filters (Rx and Tx) in the normalized frequency domain ($s = j\Omega$) can be expressed through their characteristic polynomials [3]. The frequency transformation from Ω' to Ω is expressed as

$$\Omega = a_1\Omega' + b_1, \quad \text{for Rx filter} \tag{6.3}$$

$$\Omega = a_2\Omega' + b_2, \quad \text{for Tx filter} \tag{6.4}$$

The Rx filter's passband limit is −1 to 1 in the Ω' domain that transform to −1 to Ω_r, respectively, in the Ω domain. Similarly, the Tx filter's passband limit is −1 to 1 in the Ω' domain that transform to Ω_t to 1 in the Ω domain. By enforcing the limit of the Rx and Tx filters, we get [3]:

$$\begin{cases} -1 = -a_1 + b_1 \\ \Omega_r = a_1 + b_1 \end{cases} \quad \begin{cases} 1 = a_2 + b_2 \\ \Omega_t = -a_2 + b_2 \end{cases} \tag{6.5}$$

Solving (6.5):

$$\begin{aligned} a_1 &= \frac{\Omega_r + 1}{2}, \quad b_1 = \frac{\Omega_r - 1}{2}, \quad \text{Rx filter} \\ a_2 &= \frac{1 - \Omega_t}{2}, \quad b_2 = \frac{\Omega_t + 1}{2}, \quad \text{Tx filter} \end{aligned} \tag{6.6}$$

6.1.1.1 Diplexer Design Using Type-I Junction

The input admittance at Port 1 for Type-1 junction [2] shown in Figure 6.3 is defined in (6.7):

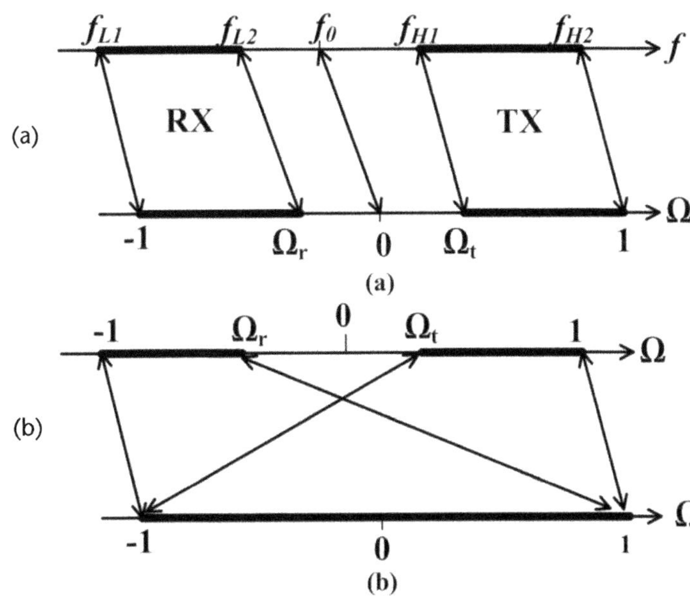

Figure 6.2 Two-stage frequency mapping for diplexer analysis. (a) Frequency mapping from the Ω domain to the f domain. (b) Frequency mapping from Ω' to the Ω domain [3].

$$Y_{in} = n^2\left(jb_0 + Y_{in}^{Tx} + Y_{in}^{Rx}\right) \tag{6.7}$$

where Y_{in}^{Tx} and Y_{in}^{Rx} are input admittances of Tx and Rx filters, respectively, and are defined as:

$$\begin{aligned} Y_{in}^{Tx} &= \frac{1 - S_{11}^{Tx}}{1 + S_{11}^{Tx}} = \frac{E_{Tx} - F_{Tx}}{E_{Tx} + F_{Tx}} = \frac{D_{Tx}}{S_{Tx}} \\ Y_{in}^{Rx} &= \frac{1 - S_{11}^{Rx}}{1 + S_{11}^{Rx}} = \frac{E_{Rx} - F_{Rx}}{E_{Rx} + F_{Rx}} = \frac{D_{Rx}}{S_{Rx}} \end{aligned} \tag{6.8}$$

where

$$\begin{aligned} S_{Tx} &= \frac{E_{Tx} + F_{Tx}}{2} \\ D_{Tx} &= \frac{E_{Tx} - F_{Tx}}{2} \\ S_{Rx} &= \frac{E_{Rx} + F_{Rx}}{2} \\ D_{Rx} &= \frac{E_{Rx} - F_{Rx}}{2} \end{aligned} \tag{6.9}$$

Substituting (6.8) and (6.9) in (6.7), we get

$$Y_{in} = n^2 \frac{jb_0 S_{Tx} S_{Rx} + D_{Tx} S_{Rx} + D_{Rx} S_{Tx}}{S_{Tx} S_{Rx}} \tag{6.10}$$

For Type-I junction, S_{11} can be expressed in terms of Y_{in} as:

$$S_{11} = \frac{1 - Y_{in}}{1 + Y_{in}}$$

where $N(s)$ and $D(s)$ are expressed [2]:

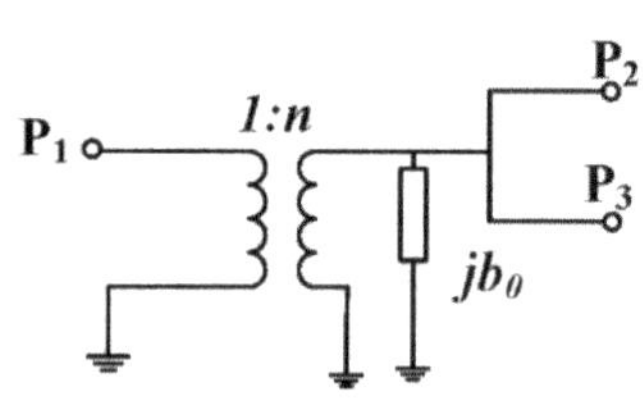

Figure 6.3 Equivalent circuit model of an SIW H-plane T-junction (Type-I junction).

$$
\begin{aligned}
N(s) &= S_{\mathrm{TX}}S_{\mathrm{RX}} - n^2 \frac{D_{\mathrm{TX}}S_{\mathrm{RX}} + D_{\mathrm{RX}}S_{\mathrm{TX}}}{1 - jn^2 b_0} \\
n_0 &= \frac{1 - jn^2 b_0}{1 + jn^2 b_0} \\
D(s) &= S_{\mathrm{TX}}S_{\mathrm{RX}} + n^2 \frac{D_{\mathrm{TX}}S_{\mathrm{RX}} + D_{\mathrm{RX}}S_{\mathrm{TX}}}{1 + jn^2 b_0}
\end{aligned} \tag{6.11}
$$

From the equation given in (6.12), it is evident that the degree of the diplexer's polynomial is $(N_{p\mathrm{Tx}} + N_{p\mathrm{Rx}})$, where $N_{p\mathrm{Tx}}$ and $N_{p\mathrm{Rx}}$ correspond to degree of the individual Tx and Rx filters, respectively. The other important polynomials of the diplexer related to transmission coefficients from Port 1 to Port 2 and Port 3 are expressed [2]:

$$
\begin{aligned}
S_{21} &= \frac{S_{21}^{\mathrm{Tx}}\left(1 + Y_{\mathrm{in}}^{\mathrm{Tx}}\right)}{\dfrac{1}{n} + jnb_0 + n \cdot Y_{\mathrm{in}}^{\mathrm{Tx}} + n \cdot Y_{\mathrm{in}}^{\mathrm{Rx}}} \\
S_{31} &= \frac{S_{21}^{\mathrm{Rx}}\left(1 + Y_{\mathrm{in}}^{\mathrm{Rx}}\right)}{\dfrac{1}{n} + jnb_0 + n \cdot Y_{\mathrm{in}}^{\mathrm{Tx}} + n \cdot Y_{\mathrm{in}}^{\mathrm{Rx}}}
\end{aligned} \tag{6.12}
$$

$P_t(s)$ and $P_r(s)$ can be evaluated by substituting (6.1) and (6.8) in (6.12) and comparing with (6.2). These are given below [2]:

$$
\begin{aligned}
P_t(s) &= P_{\mathrm{Tx}n}S_{\mathrm{Rx}} \\
P_{0t} &= P_{0\mathrm{Tx}}\left(\frac{n}{1 + jn^2 b_0}\right) \\
P_r(s) &= P_{\mathrm{Rx}n}S_{\mathrm{Tx}} \\
P_{0r} &= P_{0\mathrm{Rx}}\left(\frac{n}{1 + jn^2 b_0}\right)
\end{aligned} \tag{6.13}
$$

It is evident from (6.13) that the order of $P_t(s)$ is $(N_{z\mathrm{Tx}} + N_{p\mathrm{Rx}})$, where $N_{z\mathrm{Tx}}$ are the number of transmission zeros of Tx filter and $N_{p\mathrm{Rx}}$ are the number of poles in the Rx filter. $N_{p\mathrm{Rx}}$ are the additional transmission zeros appearing because of S_{Rx}. However, in the Rx section of the diplexer $P_r(s)$, the total number of transmission zeros are $(N_{z\mathrm{Rx}} + N_{p\mathrm{Tx}})$. The additional transmission zeros in the Rx section appear because of S_{Tx}. The degree of S_{Tx} (polynomial) is same as that of the Tx filter.

6.1.1.2 Diplexer Design Using a Type-II Junction

In Type II, the junction is a common resonator. To simplify the design, the resonant frequency of the junction node is kept at f_0, which is the center frequency of the diplexer. In the lowpass equivalent circuit, the shunt resonator is replaced with capacitance c_0. Just like previous case, the diplexer polynomials can be obtained using the following expressions [2]:

$$
\begin{aligned}
N(s) &= sS_{\text{Tx}}S_{\text{Rx}} - \frac{1}{c_0}\left(S_{\text{Tx}}S_{\text{Rx}} - D_{\text{Tx}}S_{\text{Rx}} - S_{\text{Tx}}D_{\text{Rx}}\right) \\
n_0 &= -1 \\
D(s) &= sS_{\text{Tx}}S_{\text{Rx}} + \frac{1}{c_0}\left(S_{\text{Tx}}S_{\text{Rx}} + D_{\text{Tx}}S_{\text{Rx}} + S_{\text{Tx}}D_{\text{Rx}}\right) \\
P_t(s) &= P_{\text{Tx}n}S_{\text{Rx}} \\
P_{0t} &= \frac{P_{0\text{Tx}}}{c_0} \\
P_r(s) &= P_{\text{Rx}n}S_{\text{Tx}} \\
P_{0r} &= \frac{P_{0\text{Rx}}}{c_0}
\end{aligned} \tag{6.14}
$$

Since the junction node provides one more additional pole, which is due to the capacitor c_0, the total number of poles is $N_{p\text{Tx}} + N_{p\text{Rx}} + 1$. Besides that, one more transmission zero appears at infinity in both the Rx and Tx sections. For the Type-II junction, an additional refection zero is imposed by placing capacitance sc_0 in the normalized lowpass prototype domain. In order to ensure that the presence of the sc_0 does not disturb the quasi-equi-ripple diplexer response, the value of sc_0 is chosen to be purely real (i.e., 1.5).

6.1.1.3 Evaluation of the Diplexer Polynomials

Step 1: Rx and Tx filters of the diplexer are synthesized independently in the Ω' plane using a generalized Chebyshev method with the help of design specifications. Design specifications include the number of poles of both the Rx and Tx channels ($N_{p\text{Rx}}$ and $N_{p\text{Tx}}$), transmission zeros of the filters, and return loss in the passband. Based on design specifications, filtering polynomials are generated such as E^0_{Rx}, F^0_{Rx}, P^0_{Rx}, F^0_{Tx}, E^0_{Tx}, and P^0_{Tx}. This general Chebyshev method is a well-established filter synthesis technique. The detailed studies over the method are presented in Chapter 5. Both the filters are linearly transformed to Ω from the Ω'-plane [3].

Step 2: Evaluate both P_t and P_r polynomials using convolution: convolution of ($P_{\text{Tx}n}$, S_{Rx}) and ($P_{\text{Rx}n}$, S_{Tx}), respectively.

Step 3: Evaluate both P_{0t} and P_{0r} of the polynomial coefficients at $\Omega = \pm 1$ using design specified return loss. The diplexer is assumed to be lossless and satisfying a unitary condition.

$$
\begin{aligned}
\frac{|N(j)|^2}{|N(j)|^2 + |P_{0t}|^2 \cdot |P_t(j)|^2 + |P_{0r}|^2 \cdot |P_r(j)|^2} &= 10^{-RL_{\text{Tx}}/10} \\
\frac{|N(-j)|^2}{|N(-j)|^2 + |P_{0t}|^2 \cdot |P_t(-j)|^2 + |P_{0r}|^2 \cdot |P_r(-j)|^2} &= 10^{-RL_{\text{Rx}}/10}
\end{aligned} \tag{6.15}
$$

Using (6.15), calculate $|P_{0t}|^2$ and $|P_{0r}|^2$. After that, place the polynomial coefficients in (6.16) to evaluate the poles of the diplexer $D(s)$ by imposing a unitary condition of the S-parameter. $|n_0|$ is considered unity.

$$D(s)D^*(-s) = |n_0|^2 N(s)N^*(-s) + |P_{0t}|^2 P_t(s)P_t^*(-s) + |P_{0r}|^2 P_r(s)P_r^*(-s) \tag{6.16}$$

Step 4: After an evaluation of the polynomial $D(s)$, the new estimated values of S_{Tx} and S_{Rx} are evaluated by solving the roots of (6.17):

$$mN(s) + nD(s) = S_{\text{Tx}} S_{\text{Rx}} \tag{6.17}$$

For the Type-I junction $(N(s) + D(s))/2 = S_{\text{Tx}}S_{\text{Rx}}$ and for the Type-II junction $(N(s) - D(s))/2 = S_{\text{Tx}}S_{\text{Rx}}$. From the roots of the equation, S_{Tx} and S_{Rx} are evaluated.

Step 5: Compare the roots of S_{Tx} and S_{Rx} with previous data until it satisfies the difference between new and old data less than the specified level using the iteration process. Thus, after few iterations, the desired P_t and P_r are obtained.

6.1.2 Diplexer Design Using the Perturbation Technique

In this section, the substrate integrated hexagonal cavity (SIHC) is used to design the diplexer. Detailed studies on symmetric and asymmetric diplexer responses are performed by shifting transmission zeros and adjusting the number of poles in each Rx and Tx channel. Perturbing the side wall of the hexagonal cavity and adjusting the feeding position leads to a more flexible design and improve the filter selectivity [4]. In Diplexer Designs I to III, both Rx and Tx channels have one dual-mode and single-mode SIHC [5]. The single-mode SIHC is acting as a common resonator. Diplexer Design IV has two diplexers consisting of two dual-mode SIHCs in each Rx and Tx channel [5].

6.1.2.1 The Diplexer Design Principle Using the Perturbation Technique

The dual-mode SIHC diplexer's schematic is shown in Figure 6.4. The input resonator is a single-mode SIHC and both the channels (Rx and Tx) are realized using a dual-mode SIHC. In a dual-mode resonator, both degenerate modes exist at the same frequency but the perturbation applied to the side wall of the cavity results in the mode splitting [4, 6]. With the presence of weak excitation, two orthogonal modes are oriented as odd and even modes and the electric field distributions in the cavity for both modes are depicted in Figure 6.5. The coupling topology corresponding to the schematic is shown in Figure 6.6. Here the resonators 2 and 3 are considered as odd and even resonating modes for the Rx channel. Similarly, resonators 4 and 5 are considered as odd and even resonating modes for the Tx channel.

The side wall $\{a_5\}$ of the Rx channel and $\{b_5\}$ of the Tx channel are used for perturbation as shown in Figure 6.4. The *inset* is mentioned through the arrow as shown in Figure 6.4. The *inset* is shown by the arrow and is considered as $-ve$ and

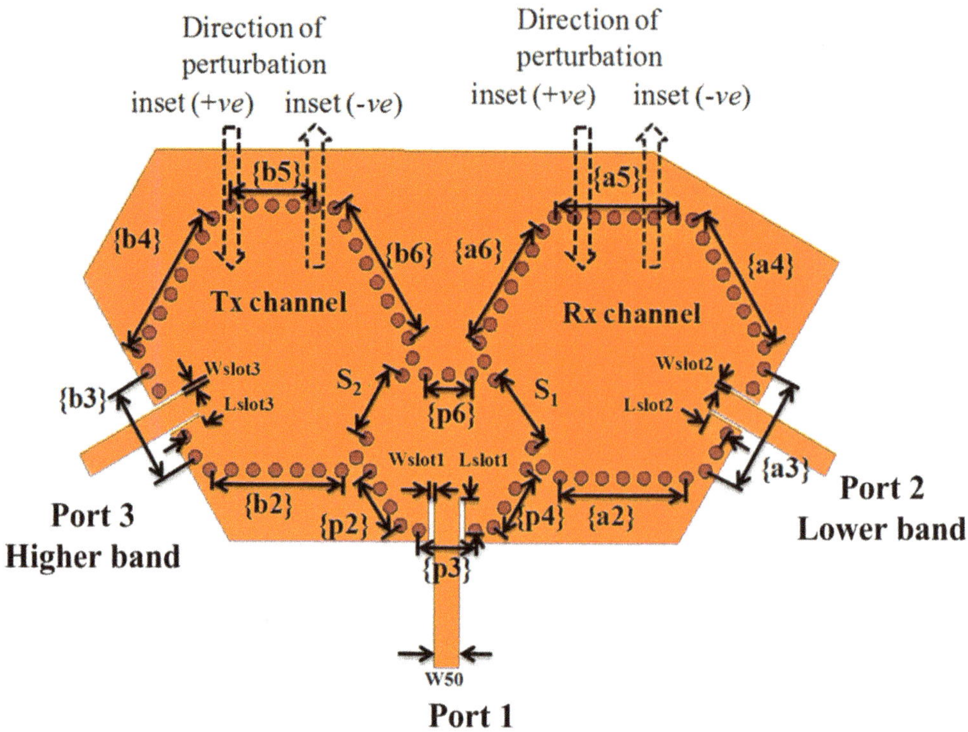

Figure 6.4 Schematic of the designed Diplexer Design I with the dimensions as follows: L_{slot1} = 3.3 mm, W_{slot1} = 0.63 mm, $\{p_2\}$ = 6.15 mm, $\{p_3\}$ = 5.635 mm, $\{p_4\}$ = 6.6 mm, and $\{p_6\}$ = 4.6 mm. Rx channel: $\{a_1\}$ = 11.875 mm, $\{a_2\}$ = 12.6 mm, $\{a_3\}$ = 11.575 mm, $\{a_4\}$ = 14.35 mm, $\{a_5\}$ = 12 mm, $\{a_6\}$ = 12.6 mm, L_{slot2} = 2.6 mm, W_{slot2} = 0.4 mm, and S_1 = 7.575 mm. Tx channel: $\{b_1\}$ = 11.225 mm, $\{b_2\}$ = 13.2 mm, $\{b_3\}$ = 9.575 mm, $\{b_4\}$ = 15.05 mm, $\{b_5\}$ = 8.4 mm, $\{b_6\}$ = 14.7 mm, L_{slot3} = 2.4 mm, W_{slot3} = 0.4 mm, and S_2 = 7.125 mm. (© 2017 Wiley [5].)

$+ve$ for outward and inward perturbation, respectively. This perturbation technique is used to shift the transmission zero on both sides of the passband. In diplexer, the total number of transmission zeros is $(N_{z\text{Rx}} + N_{p\text{Tx}})$ and $(N_{z\text{Tx}} + N_{p\text{Rx}})$ for Rx and Tx channels, respectively. $N_{p\text{Rx}}$ and $N_{p\text{Tx}}$ are the total number of design specified poles of the channel, and $N_{z\text{Rx}}$, $N_{z\text{Tx}}$ are the total number of design-specified transmission zeros appearing in the Rx and Tx channel, respectively [2].

In this design, Rx and Tx channels consist of a single dual-mode SIHC, and each channel has two reflection zeros and three transmission zeros. Out of these, one transmission zero appears due to the design specifications, and the other two transmission zeros appear due to the loading effect of the Tx channel over the Rx channel and vice versa. These two transmission zeros are generally complex in nature in the normalized Ω-plane. If the real part of these transmission zeros becomes negligible, it would act as another ideal transmission zero and contribute to improve the filter selectivity. The normalized frequency of the transmission zeros are $\Omega_{Tz,\text{Rx}}$ and $\Omega_{Tz,\text{Tx}}$ for the Rx and Tx channels, respectively. $\Omega_{Tz,\text{Rx}}$ and $\Omega_{Tz,\text{Tx}}$ can be analytically calculated by solving cubic equations (6.18) and (6.19), respectively.

The equation of the Rx channel's transmission zeros is:

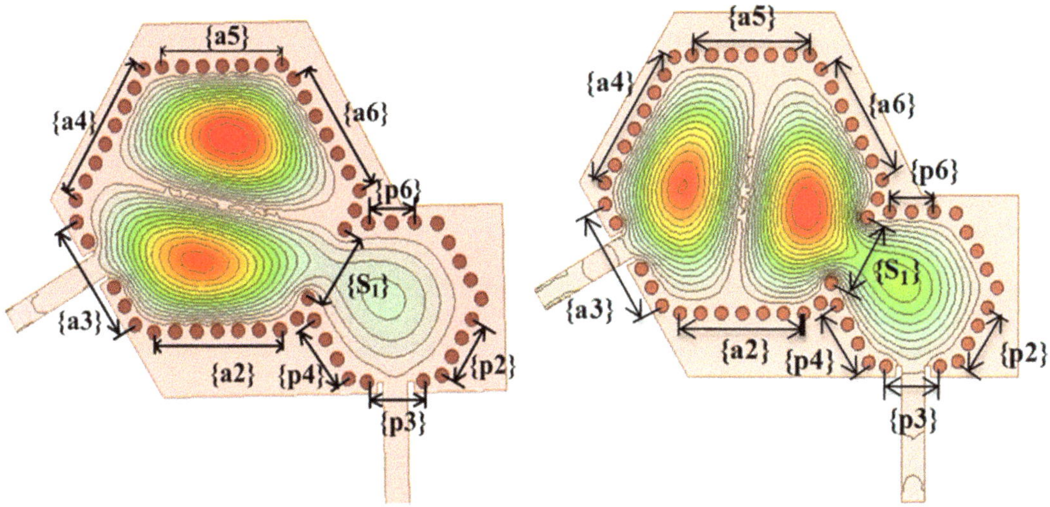

(a) (b)

Figure 6.5 Electric field distribution in a dual-mode hexagonal cavity for: (a) even mode and (b) odd mode. The dimensions are: $\{p_3\} = 5.635$ mm, $\{p_4\} = 6.6$ mm, $\{p_6\} = 4.6$ mm. Rx channel: $\{a_1\} = 11.875$ mm, $\{a_2\} = 12.6$ mm, $\{a_3\} = 11.575$ mm, $\{a_4\} = 14.35$ mm, $\{a_5\} = 12$ mm, $\{a_6\} = 12.6$ mm, $L_{\text{slot2}} = 2.6$ mm, $W_{\text{slot2}} = 0.4$ mm, and $S_1 = 7.575$ mm. Tx channel: $\{b_1\} = 11.225$ mm, $\{b_2\} = 13.2$ mm, $\{b_3\} = 9.575$ mm, $\{b_4\} = 15.05$ mm, $\{b_5\} = 8.4$ mm, $\{b_6\} = 14.7$ mm, $L_{\text{slot3}} = 2.4$ mm, $W_{\text{slot3}} = 0.4$ mm, and $S_2 = 7.125$ mm. (© 2017 Wiley [5].)

$$\begin{aligned}&M_{P_1,1}\left(M_{P_2,3}M_{1,3}M_{2,2} + M_{P_2,2}M_{1,2}M_{3,3} + M_{P_{2,2}}M_{1,2}\Omega + M_{P_2,3}M_{1,3}\Omega\right)\cdot\\&\left(i\Omega^2 + \left(M_{4,4} + M_{5,5} + M_{P_3,4}^2 + M_{P_3,5}^2\right)\Omega + M_{P_3,4}^2 M_{5,5} + M_{P_{3,5}}^2 M_{4,4} + iM_{4,4}M_{5,5}\right) = 0\end{aligned} \tag{6.18}$$

Here, the first part is because of the Rx filter's design specified transmission zero, and the second part is a quadratic equation in which all the coupling terms are associated with the Tx channel.

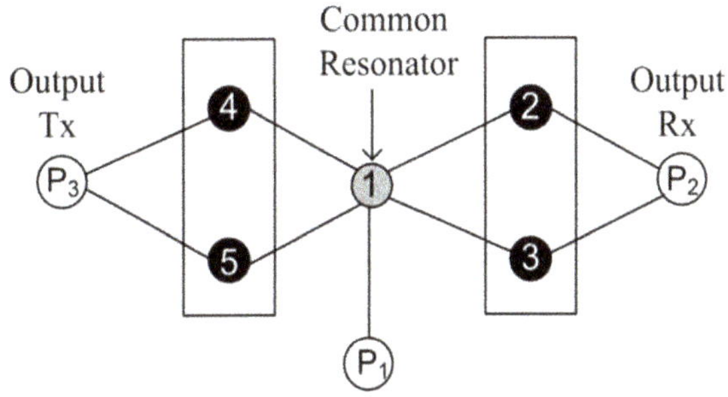

Figure 6.6 Coupling topology of the designed diplexer shown in Figure 6.4. (© 2017 Wiley [5].)

Similarly, the equation of Tx channel's transmission zeros can be written as:

$$M_{P_1,1}\left(M_{P_3,5}M_{1,5}M_{4,4} + M_{P_3,4}M_{1,4}M_{5,5} + M_{P_3,4}M_{1,4}\Omega + M_{P_3,5}M_{1,5}\Omega\right)\cdot \left(i\Omega^2 + \left(M_{2,2} + M_{3,3} + M_{P_2,2}^2 + M_{P_2,3}^2\right)\Omega + M_{P_2,2}^2 M_{3,3} + M_{P_2,3}^2 M_{2,2} + iM_{2,2}M_{3,3}\right) = 0 \qquad (6.19)$$

Here, the first part is due to the Tx filter's design specification and all the terms of the second part of the equation deals with the Rx channel's coupling.

In (6.18) and (6.19), *Mmn* is the normalized coupling coefficients for $m \neq n$, m, $n = 1, 2, \ldots, 5$ where both m, n represent mth and nth resonators, respectively. In this design, the total number of resonators is 5. M_{mm} is the normalized self-resonant frequency of the mth resonator for $m = 1, 2, \ldots, 5$ and $M_{Pr,s}$ is the normalized coupling coefficient between the source and the load with sth resonator where $r = 1, 2, 3$ and $s = 1, 2, \ldots, 5$.

Perturbation is applied to the side wall $\{a_5\}$ of Rx channel using *inset*, which is varied from outward to inward directions (−1.5 mm to 0 mm). Due to perturbation, the even-mode self-resonating frequency of the Rx channel shifts toward a higher frequency, whereas an odd-mode self-resonating frequency remains unaffected as shown in Figure 6.7(a) and this results in the shifting of transmission zero from left to right in its passband. However, the other two transmission zeros of the Rx channel remain unaffected to its own Rx channel perturbation as shown in Figure 6.8(a). As shown in Figure 6.8(b), Tx channel's design specified transmission zero is not sensitive to the perturbation, but other two transmission zeros of Tx channel, appearing due to the loading effect of the Rx channel, are sensitive to this perturbation. Similarly, the effect of perturbation using the side wall $\{b_5\}$ on the self-resonating frequency of the Tx channel is shown in Figure 6.7(b). Figure 6.8(c) shows that perturbation using the side wall $\{b_5\}$ does not affect the Rx filter's design specified transmission zero, while other two transmission zeros are sensitive to the perturbation due to a loading effect of the Tx channel over the Rx channel. This perturbation results in

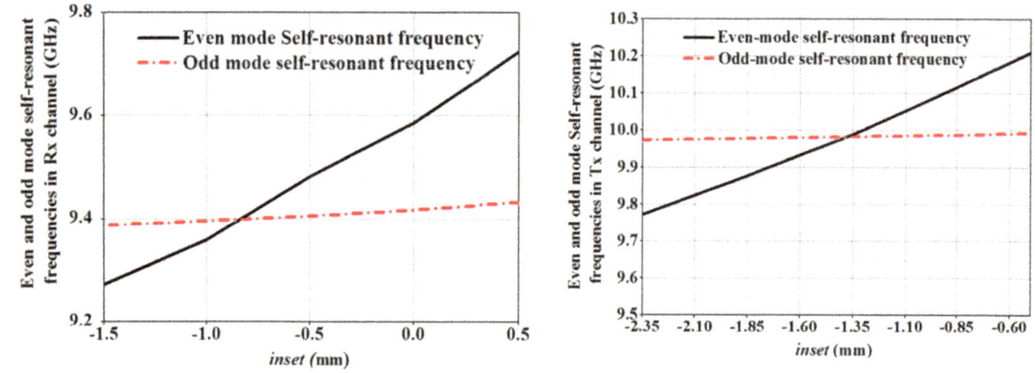

(a) (b)

Figure 6.7 (a) The Rx channel's even and odd-mode self-resonating frequencies' variation with perturbation using the side wall $\{a_5\}$. (b) The Tx channel's even and odd-mode self-resonating frequencies variation with perturbation using the side wall $\{b_5\}$. (© 2017 Wiley [5].)

shifting the Tx channel's design-specified transmission zero from the left to right side of its passband, while other two transmission zeros remain unaffected to this *inset* variations as shown in Figure 6.8(d). To design the diplexer, the Rogers RT/Duroid 5880 substrate with thickness of 0.787 mm is used. The coupling matrix synthesis procedure is followed to design diplexers with a common resonator junction which provides design flexibility since the common port is coupled to the Tx and Rx channels by this common resonator. Once the characteristic polynomials of Tx and Rx channels are obtained according to the method [2], a coupling matrix of the diplexers based on given coupling topology is synthesized by optimization [7].

6.1.3 Diplexer Design I

Design I is an asymmetric diplexer consisting of a single dual-mode SIHC resonator with Rx and Tx channels' design-specified transmission zeros above and below of their respective passband. Figure 6.4 shows the schematic of this diplexer. The design specifications are as follows: center frequency: 9.7 GHz, fractional bandwidth (FBW): 5.36%, passband bandwidth of Rx and Tx channels: 100 MHz, and return loss in

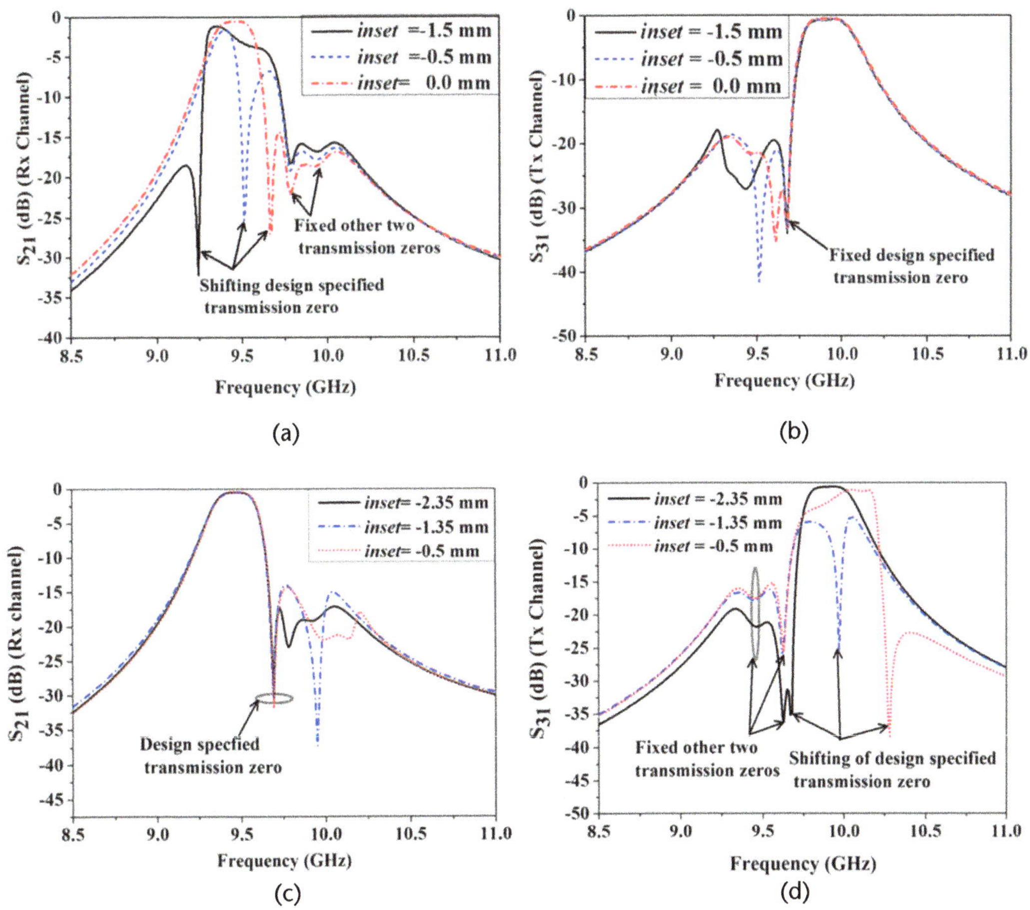

Figure 6.8 (a, b) Effect of perturbation using the side wall $\{a_5\}$. (c, d) Effect of perturbation using the side wall $\{b_5\}$. (© 2017 Wiley [5].)

the passband of Rx and Tx channels: 25 and 23 dB, respectively. The passband of the Tx channel is from 9.86 to 9.96 GHz. In the Rx channel, the transmission zero is kept above the passband at 9.70 GHz, whereas in the Tx channel, transmission zero is kept below the passband at 9.70 GHz. According to the coupling topology as shown in Figure 6.6 and design specifications, coupling coefficients are obtained using the generalized Chebyshev method. The values are as follows: f_{01} = 9.7 GHz, f_{02} = 9.416 GHz, f_{03} = 9.59 GHz, f_{04} = 9.976 GHz, f_{05} = 9.786 GHz, K_{12} = 0.0402, K_{13} = 0.0256, K_{14} = 0.0383, K_{15} = 0.0247, $Q_E(P_1, 1)$ = 8.286, $Q_E(P_2, 2)$ = 65.14, $Q_E(P_2, 3)$ = 176.5, $Q_E(P_3, 4)$ = 72.24, and $Q_E(P_3, 5)$ = 189.1.

Here K_{mn} is the coupling coefficient of the resonators where $m \neq n$, m, $n = 1, 2, \ldots, N$ and N is total number of resonators. $Q_E(P_1, 1)$ represents the external quality factor of the resonating junction 1 loaded with source conductance. $Q_E(P_2, 2)$ and $Q_E(P_2, 3)$ are representing external quality factors determined by loads at P_2 with resonators 2 and 3, respectively. $Q_E(P_3, 4)$ and $Q_E(P_3, 5)$ are representing external quality factors determined by loads at P_3 with resonators 4 and 5, respectively. The coupling coefficients and denormalized self-resonant frequencies of the resonators are obtained using the following equations:

$$
\begin{aligned}
K_{mn} &= FBW \cdot M_{mn} \text{ for } m \neq n, m, n = 1,2,\ldots,N \\
f_{0m} &= f_0\sqrt{1+\left(\frac{M_{mm}FBW}{2}\right)^2} - \frac{f_0 M_{mm}}{2} \quad m = 1,2,..,N \\
Q_E(P_m,n) &= \frac{1}{FBW \cdot M^2_{P_m,n}} \text{ for } m = 1,2,3 \text{ and } n = 1,2,\ldots,N
\end{aligned}
\tag{6.20}
$$

where M_{mm}, f_{0m} are normalized and denormalized self-resonating frequencies of the mth resonator, respectively, and f_0 is the center frequency of the diplexer.

The self-resonant frequency of the common resonator is kept at the center frequency (9.7 GHz) of the diplexer. To design this diplexer, no perturbation is applied to the side wall $\{a_5\}$ of the Rx channel's SIHC (*inset* = 0.0 mm). Even-mode self-resonant frequency (f_{03}) is higher than odd-mode frequency (f_{02}), which is shown in Figure 6.7(a). However, in Tx channel, perturbation is applied in outward direction using the side wall $\{b_5\}$ (*inset* = −2.35 mm). Odd-mode self-resonant frequency (f_{04}) is higher than even-mode self-resonant frequency (f_{05}) in the Tx channel. The theoretical, simulated, and measured diplexer responses are shown together in Figure 6.9(a). The minimum measured insertion loss is 1.42 dB for the Rx channel and 1.38 dB for the Tx channel. Measured spurious response and the fabricated prototype of the diplexer are shown in Figure 6.9(b). The spurious response is measured from 8.5 to 16 GHz, which shows the higher-order modes of both dual-mode hexagonal cavities associated to the Rx and Tx channels.

6.1.4 Diplexer Design II

The above discussion of coupling topology is used to design symmetric and asymmetric diplexers, keeping the design-specified transmission zero above the passband of both channels. The designs are obtained not only by changing the perturbation in

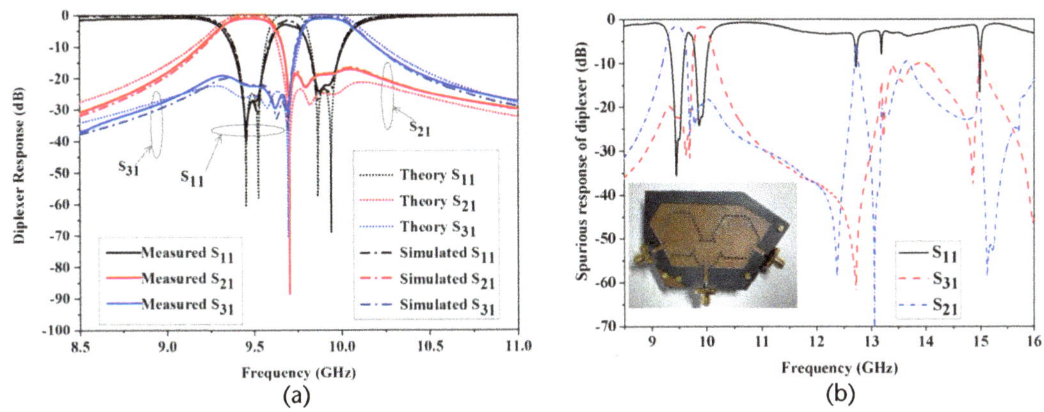

Figure 6.9 (a) Measured, simulated, and theory results of Diplexer Design I. (b) Measured spurious response of the Diplexer Design I with a fabricated prototype. (© 2017 Wiley [5].)

both Rx and Tx channels but also by depending on the dimensions of the common resonator and feeding position. The schematic of this diplexer is shown in Figure 6.10. In the symmetric diplexer, both Rx and Tx channels have two poles and one transmission zero above the passband of its respective channel, whereas in asymmetric diplexer, Rx and Tx channels have three poles and two poles, respectively, and both channels have one transmission zero above the passband.

6.1.4.1 Symmetric Diplexer Design II(a)

The specifications of the diplexer with a symmetric response are as follows. The center frequency is kept at 9.7 GHz with an FBW of 9.6%. Both Rx and Tx channels have 90-MHz bandwidth. Return loss in the passband of both Rx and Tx channels is 20.0 dB. The Rx channel passband is from 9.28 to 9.37 GHz, and the Tx channel passband is from 10.12 to 10.21 GHz.

In the Rx channel, transmission zero is kept at 9.45 GHz, which is above the Rx channel's passband, whereas in the Tx channel, the transmission zero is kept at 10.36 GHz, which is also above the Tx channel's passband. The corresponding coupling coefficients, external quality factor, and de-normalized self-resonating frequencies of this diplexer are as follows: $f_{01} = 9.57$ GHz, $f_{02} = 9.30$ GHz, $f_{03} = 9.40$ GHz, $f_{04} = 9.96$ GHz, $f_{05} = 10.2$ GHz, $K_{12} = 0.0263$, $K_{13} = 0.0150$, $K_{14} = 0.0604$, $K_{15} = 0.0201$, $Q_E(P_1, 1) = 16.1$, $Q_E(P_2, 2) = 88.34$, $Q_E(P_2, 3) = 470.0$, $Q_E(P_3, 4) = 170.6$, and $Q_E(P_3, 5) = 119.9$. The dimensions of this symmetric diplexer is given in Figure 6.10. The simulated and measured diplexer responses are shown in Figure 6.11. For the Rx and Tx channels of this diplexer, the measured minimum insertion losses are 1.6 dB and 1.65 dB, respectively.

6.1.4.2 Asymmetric Diplexer Design II(b)

The asymmetric diplexer has three poles in the Rx channel and two poles in the Tx channel. Rx and Tx channel's design specified transmission zeros are kept above the passband. An asymmetric response is realized by changing the perturbation and dimension of the common resonator. The design specifications are as follows.

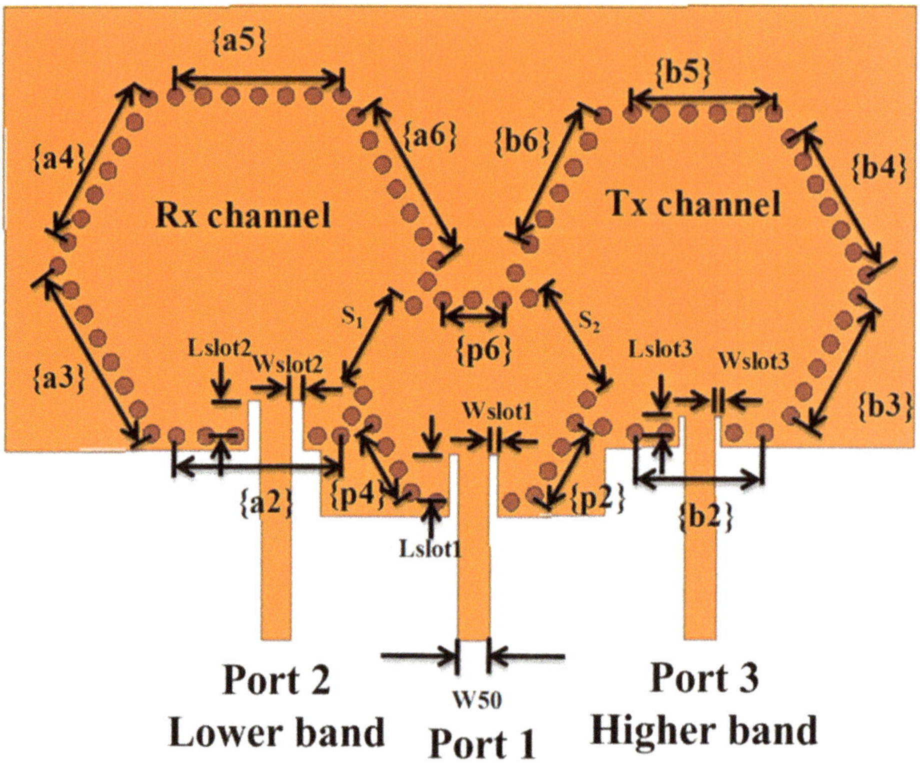

Figure 6.10 Schematic of a diplexer Design II. Dimensions of a symmetric Diplexer Design II(a): L_{slot1} = 3.1 mm, W_{slot1} = 0.6 mm, $\{p_4\}$ = 6.3 mm, $\{p_2\}$ = 6.6 mm, and $\{p_6\}$ = 4.6 mm. Rx channel: $\{a_1\}$ = 11.275 mm,$\{a_2\}$ = 11.475 mm, $\{a_3\}$ = 15.00 mm, $\{a_4\}$ = 13.20 mm, $\{a_5\}$ = 13.2 mm, $\{a_6\}$ = 12.9 mm, L_{slot2} = 2.5 mm, W_{slot2} = 0.4 mm, and S_1 = 6.775 mm. Tx channel: $\{b_1\}$ = 12.675 mm, $\{b_2\}$ = 10.875 mm, $\{b_3\}$ = 11.0 mm, $\{b_4\}$ = 13.2 mm, $\{b_5\}$ = 11.0 mm, $\{b_6\}$ = 10.5 mm, L_{slot3} = 1.8 mm, W_{slot3} = 0.4 mm, and S_2 = 8.875 mm. Dimension of Asymmetric Diplexer Design II(b): L_{slot1} = 3.6 mm, W_{slot1} = 0.7 mm, $\{p_2\}$ = 6.3 mm, $\{p_4\}$ = 6.0 mm, and $\{p_6\}$ = 4.6 mm. Rx channel: $\{a_1\}$ = 11.775 mm, $\{a_2\}$ = 12.775 mm, $\{a_3\}$ = 14.70 mm, $\{a_4\}$ = 12.60 mm, $\{a_5\}$ = 12.9 mm, $\{a_6\}$ = 12.6 mm, L_{slot2} = 2.7 mm, W_{slot2} = 1.0 mm, and S_1 = 7.875 mm. Tx channel: $\{b_1\}$ = 13.475 mm, $\{b_2\}$ = 10.075 mm, $\{b_3\}$ = 10.5 mm, $\{b_4\}$ = 12.0 mm, $\{b_5\}$ = 11.0 mm, $\{b_1\}$ = 11.25 mm, L_{slot3} = 1.3 mm, W_{slot3} = 0.45 mm, and S_2 = 8.875 mm. (© 2017 Wiley [5].)

The center frequency of the diplexer is at 9.7 GHz and the FBW is 9.5%. The Rx channel has 250-MHz bandwidth, and the Tx channel has 55-MHz bandwidth.

Return loss in the passband of Rx and Tx channels is 22.5 dB. Rx channel passband is from 9.22 to 9.47 GHz and Tx channel passband is from 10.1 to 10.155 GHz. The Rx and Tx channels' transmission zeros are at 9.55 GHz and 10.30 GHz, respectively, which are above their respective passbands. The de-normalized self-resonant frequencies, coupling coefficients, and external quality factors values are as follows: f_{01} = 9.5 GHz, f_{02} = 9.25 GHz, f_{03} = 9.504 GHz, f_{04} = 9.926 GHz, f_{05} = 10.153 GHz, K_{12} = 0.0351, K_{13} = 0.0109, K_{14} = 0.0650, K_{15} = 0.0172, $Q_E(P_1, 1)$ = 21.39, $Q_E(P_2, 2)$ = 38.59, $Q_E(P_2, 3)$ = 98.40, $Q_E(P_3, 4)$ = 307.0, and $Q_E(P_3, 5)$ = 135.5.

The resonant frequency of the single-mode hexagonal cavity is kept at 9.5 GHz, which adds one more pole in the Rx channel and therefore enhances the Rx channel

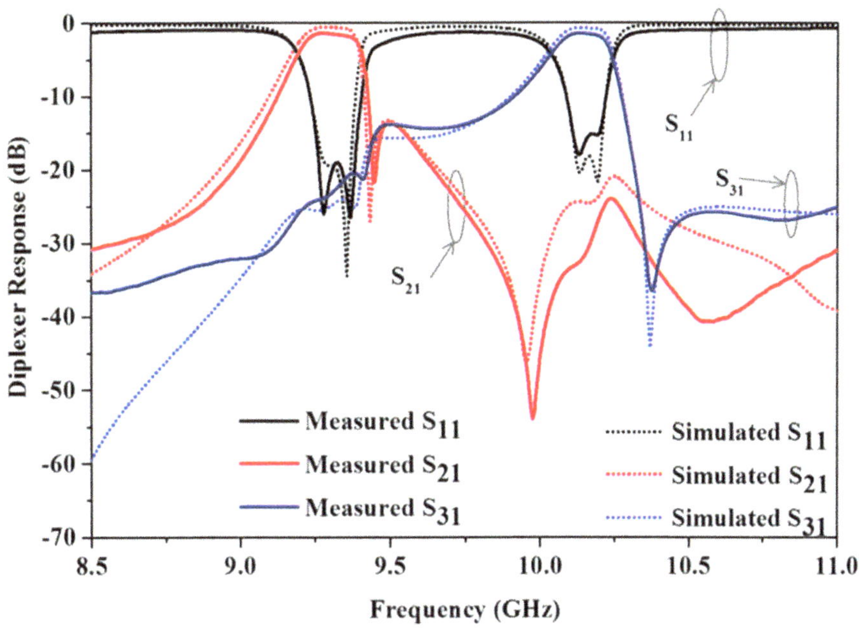

Figure 6.11 Measured and simulated S-parameters of the Symmetric Diplexer Design II(a). (© 2017 Wiley [5].)

bandwidth. The dimensions of this asymmetric diplexer are given in Figure 6.10. The asymmetric diplexer responses using coupling matrix synthesis, simulation (HFSS EM simulation) and measurement are shown in Figure 6.12(a). For Rx and Tx channels of this diplexer, the measured minimum insertion losses are 1.52 dB and 1.81 dB, respectively. The measured spurious response and fabricated prototype of this diplexer are shown in Figure 6.12(b). Both symmetric and asymmetric diplexers have identical spurious responses. The spurious response is measured from 8.5 to

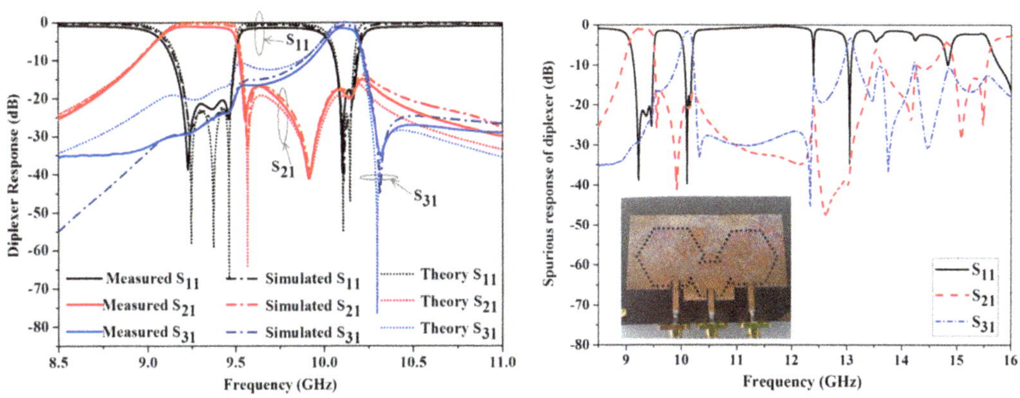

(a) (b)

Figure 6.12 (a) Measured, simulated, and theory results of Asymmetric Diplexer Design II(b), and (b) measured spurious response of Asymmetric Diplexer Design II(b) with a fabricated prototype. (© 2017 Wiley [5].)

16 GHz and shows the higher-order modes of both dual-mode hexagonal cavities attached to the Rx and Tx channels of this diplexer.

6.1.5 Diplexer Design III

This diplexer consists of three poles in the Rx channel and two poles in the Tx channel like a previous asymmetric diplexer design but the Rx channel has two transmission zeros on both sides of the passband. Therefore, it improves the Rx channel's selectivity. The Tx channel has one transmission zero below its passband. This diplexer follows the same coupling topology like all previous designs. The schematic of this diplexer is shown in Figure 6.13. The common resonator is a single-mode hexagonal cavity that contributes one more pole to the Rx channel, thereby enhancing bandwidth. The design specifications are as follows. The center frequency of the diplexer is at 9.70 GHz with an FBW of 6.5%. The Rx and Tx channels' bandwidths are 200 and 85 MHz, respectively, and the return losses in the passband of the Rx and Tx channels are 22 dB and 27.5 dB, respectively. The number of poles of Rx and Tx channels is three and two, respectively. The passbands of Rx and Tx channels are from 9.4 to 9.6 GHz. and 9.94 to 10.025 GHz, respectively. The Rx and Tx channels' design specified transmission zeros are placed at 9.2 GHz and 9.75 GHz, respectively.

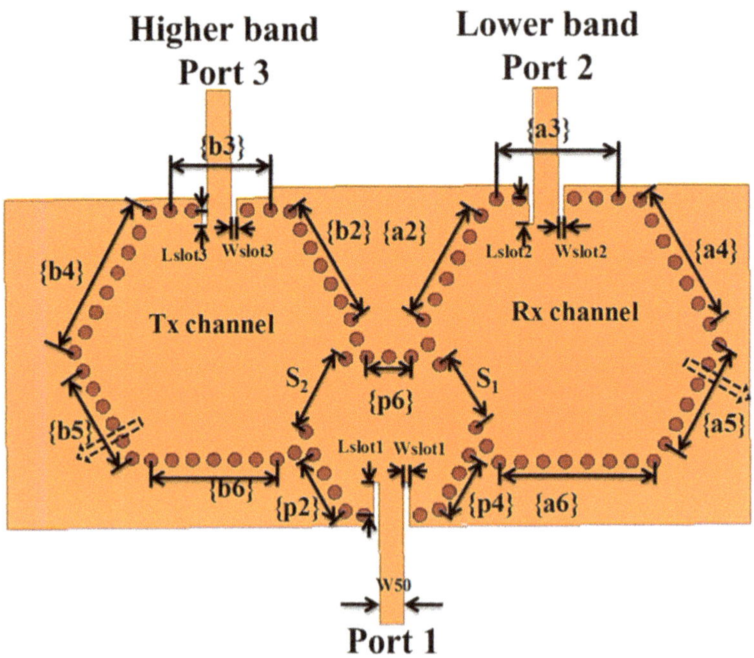

Figure 6.13 Schematic of Diplexer Design III with the dimensions as follows: L_{slot1} = 3.15 mm, W_{slot1} = 0.6 mm, $\{p_2\}$ = 6.3 mm, $\{p_4\}$ = 6.0 mm, and $\{p_6\}$ = 4.4 mm. Rx channel: $\{a_1\}$ = 11.325 mm, $\{a_2\}$ = 11.5 mm, $\{a_3\}$= 12.125 mm, $\{a_4\}$ = 13.80 mm, $\{a_5\}$ = 11.0 mm, $\{a_6\}$ = 15.4 mm, L_{slot2} = 2.4 mm, W_{slot} = 0.55 mm, and S_1 = 6.975 mm. Tx channel: $\{b_1\}$ = 12.575 mm, $\{b_2\}$ = 12.0 mm, $\{b_3\}$ = 9.975 mm, $\{b_4\}$ = 15.4 mm, $\{b_5\}$ = 9.75 mm, $\{b_6\}$ = 12.6 mm, L_{slot3} = 1.5 mm, w_{slot3} = 0.55 mm, and S_2 = 8.075 mm. (© 2017 Wiley [5].)

Both these transmission zeros are appearing below the passband of their respective channels. Based on design specifications and coupling topology, the de-normalized self-resonant frequencies, coupling coefficients, and external quality factors are: f_{01} = 9.70 GHz, f_{02} = 9.587 GHz, f_{03} = 9.378 GHz, f_{04} = 9.989 GHz, f_{05} = 9.782 GHz, K_{12} = 0.0261, K_{13} = 0.0146, K_{14} = 0.0298, K_{15} = 0.0413, $Q_E(P_1, 1)$ = 22.22, $Q_E(P_2, 2)$ = 62.1, $Q_E(P_2, 3)$ = 99.5, $Q_E(P_3, 4)$ = 53.7, and $Q_E(P_3, 5)$ = 5,529.4. The dimensions of this diplexer are provided in Figure 6.13. Here, perturbation is applied outward direction using the Rx and Tx channels' side walls $\{a_5\}$ and $\{b_5\}$ that brings odd-mode self-resonating frequencies (f_{02}, f_{04}) higher than their respective even-mode self-resonant frequencies (f_{03}, f_{05}) for Rx and Tx channels. The self-resonating frequency of the single-mode common resonator is kept at 9.7 GHz, which adds one more pole in the Rx channel and therefore enhances the Rx channel's bandwidth. One more transmission zero appears in the Rx filter at 9.78 GHz due to the loading effect of the Tx channel. This transmission zero is not depending on the perturbation of the Rx channel's filter. Figure 6.14(a) shows the responses using coupling matrix, EM simulation, and measurement. The measured minimum insertion losses in the passbands of the Rx and Tx channels are 1.78 dB and 1.92 dB, respectively. Measured spurious response and fabricated prototype are shown in Figure 6.14(b). This spurious response is measured from 8.5 to 16 GHz and exhibits the generated higher-order modes of dual-mode hexagonal cavities associated with Rx and Tx channels.

6.1.6 Diplexer Design IV

In this section, two diplexers are designed using two dual-mode SIHCs in both Rx and Tx channels using same coupling topology but change in the feeding position and perturbation, which lead to change in diplexer responses. In addition, since both Rx and Tx channels have one more dual-mode hexagonal cavity, it enhances the bandwidth of Rx and Tx channels in comparison to the previous design.

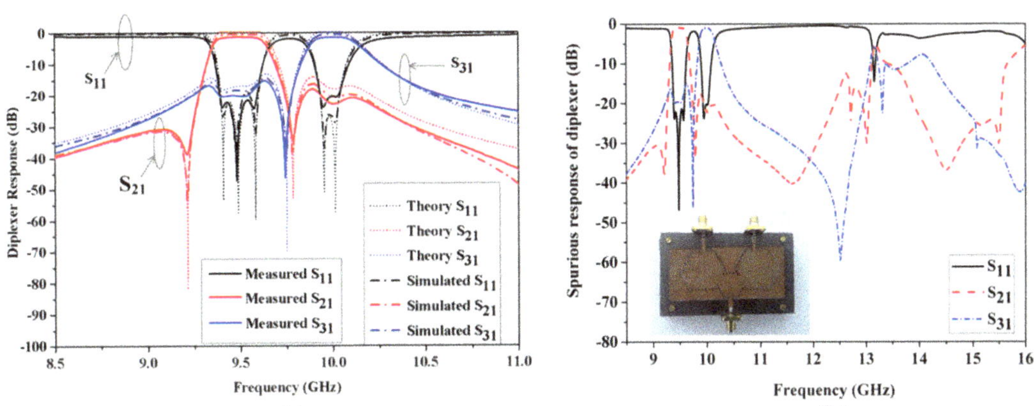

Figure 6.14 (a) Measured, simulated and theoretical results of Diplexer Design III and (b) measured spurious response of the diplexer with fabricated prototype. (© 2017 Wiley [5].)

6.1.6.1 Diplexer Design IV(a)

In this section, the diplexer is designed using a pair of SIHCs with a single design-specified transmission zero on both Rx and Tx channels, and its schematic is shown Figure 6.15. The design specifications are as follows. The center frequency is 9.60 GHz, FBW is 9.7%, the Rx and Tx channels' bandwidth are 323 MHz and 330 MHz, respectively, and the return losses in the passbands of the Rx and Tx channels are 20 dB and 21.5 dB, respectively. The number of poles in the Rx and Tx channels is four in each. The passband of the Rx channel is from 9.147 to 9.47 GHz. The passband of the Tx channel is from 9.75 to 10.08 GHz. Transmission zeros are placed at 9.52 GHz for the Rx channel and 9.60 GHz for the Tx channel.

The coupling topology of this filter is shown in Figure 6.16. Resonators 2, 3, 4, and 5 are associated with the Rx channel, whereas resonators 6, 7, 8, and 9 are associated with Tx channel. Based on the design specifications and given the coupling topology, the coupling coefficients, de-normalized self-resonant frequencies and external quality factors are: f_{01} = 9.6 GHz, f_{02} = 9.28 GHz, f_{03} = 9.371 GHz, f_{04} = 9.273 GHz, f_{05} = 9.373 GHz, f_{06} = 9.944 GHz, f_{07} = 9.932 GHz, f_{08} = 9.972 GHz, f_{09} = 9.938 GHz, K_{12} = 0.0647, K_{13} = 0.0452, K_{16} = 0.0644, K_{17} = 0.0293, K_{24} = 0.0295, K_{25} = 0.0234, K_{34} = −0.0237, K_{35} = 0.0077, K_{68} = 0.0240, K_{69} =

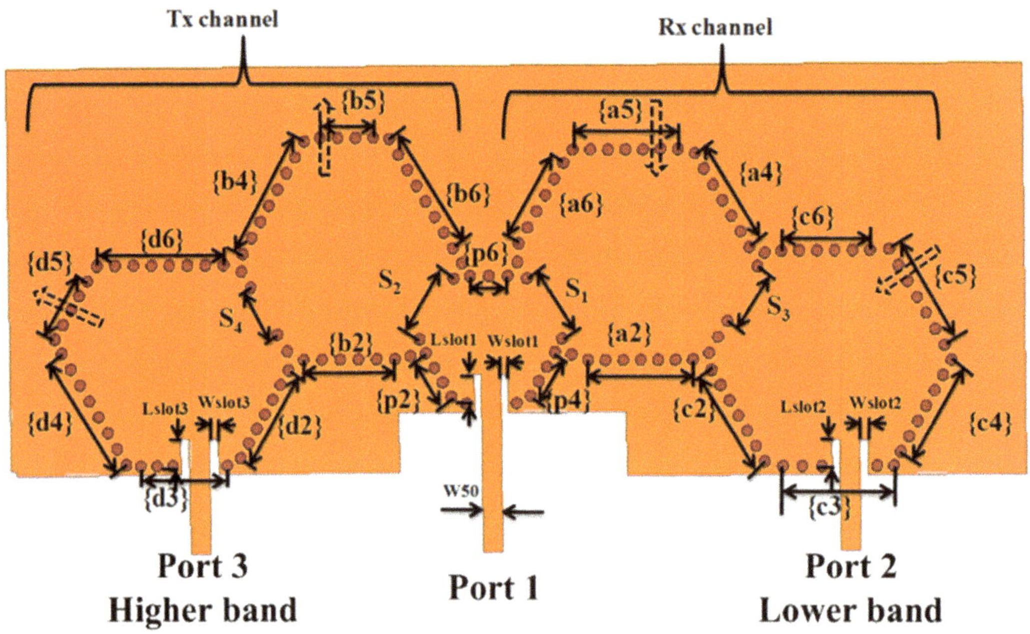

Figure 6.15 Schematic of the Diplexer Design IVa with the dimensions as follows: L_{slot1} = 3.5 mm, W_{slot1} = 0.875 mm, $\{p_2\}$ = 6.0 mm, $\{p_4\}$ = 6.0 mm, and $\{p_6\}$ = 4.6 mm. Rx channel: $\{c_1\}$ = 13.725 mm, $\{c_2\}$ = 12.6 mm, $\{c_3\}$ = 13.425 mm, $\{c_4\}$ = 12.6 mm, $\{c_5\}$ = 13.2 mm, $\{c_6\}$ = 10.75 mm, $\{a_1\}$ = 12.725 mm, $\{a_2\}$ = 12.6 mm, $\{a_3\}$ = 13.725 mm, $\{a_4\}$ = 12.9 mm, $\{a_5\}$ = 12.6 mm, $\{a_6\}$ = 11.0 mm, S_1 = 8.325 mm, S_3 = 7.525 mm, L_{slot2} = 3.1 mm, and W_{slot2} = 1.05 mm. Tx channel: $\{d_1\}$ = 11.025 mm, $\{d_2\}$ = 12 mm, $\{d_3\}$ = 10.625 mm, $\{d_4\}$ = 15.4 mm, $\{d_5\}$ = 8.0 mm, $\{d_6\}$ = 15.54 mm, $\{b_1\}$ = 12.725 mm, $\{b_2\}$ = 11.25 mm, $\{b_3\}$ = 11.025 mm, $\{b_4\}$ = 15.4 mm, $\{b_5\}$ = 6.60 mm, $\{b_6\}$ = 15.05 mm, s_2 = 8.325 mm, S_4 = 6.525 mm, W_{slot3} = 1.125 mm, and L_{slot3} = 3.2 mm. (© 2017 Wiley [5].)

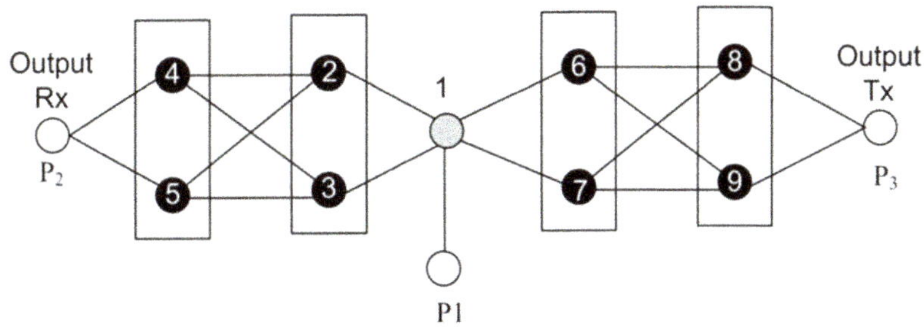

Figure 6.16 Coupling topology of the Diplexer Design IV.

0.0148, $K_{78} = -0.021$, $K_{79} = 0.0079$, $Q_E(P_1, 1) = 4.82$, $Q_E(P_2, 4) = 35.03$, $Q_E(P_2, 5) = 135.10$, $Q_E(P_3, 8) = 59.47$, and $Q_E(P_3, 9) = 86.97$.

In the Rx channel, f_{03} and f_{05} are considered as even-mode self-resonant frequencies, and f_{02} and f_{04} are considered as odd-mode self-resonant frequencies. Similarly, in the Tx channel, f_{07} and f_{09} are even-mode self-resonant frequencies, and f_{06} and f_{08} are odd-mode self-resonant frequencies. In the Rx channel, perturbation is applied in an inward direction using the side walls $\{a_5\}$ and $\{c_5\}$. So even-mode self-resonant frequencies f_{03} and f_{05} are higher than their respective odd-mode self-resonant frequencies f_{02} and f_{04} of the Rx channel and transmission zero appears above the passband of the Rx channel. In the Tx channel, perturbation is applied in an outward direction using side walls $\{b_5\}$ and $\{d_5\}$ of the Tx channel. As a result, both odd-mode self-resonant frequencies are higher than their respective even-mode self-resonant frequencies and design-specified transmission zero appears below the passband of the Tx channel. The dimensions of this diplexer are provided in Figure 6.15. Each side-wall dimension of the single-mode SIHC, which is placed at the junction of the Rx and Tx filters, is 9.27 mm that is corresponding to 9.6 GHz. Figure 6.17(a) shows the diplexer responses using the coupling matrix and EM simulation and measurement. The measured minimum insertion loss in the passband of Rx and Tx channels are 1.94 dB and 1.98 dB, respectively. Measured spurious responses and the fabricated prototype are shown in Figure 6.17(b). The spurious

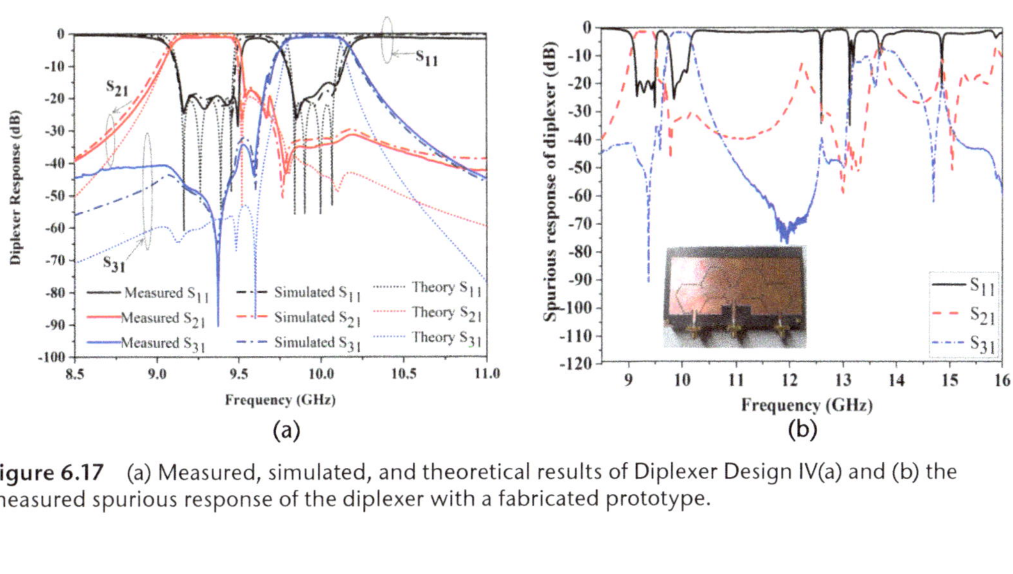

Figure 6.17 (a) Measured, simulated, and theoretical results of Diplexer Design IV(a) and (b) the measured spurious response of the diplexer with a fabricated prototype.

responses are measured from 8.5 to 16 GHz, which shows higher-order modes of SIHC associated with Rx and Tx channels.

6.1.6.2 Diplexer Design IV(b)

In this section, the diplexer is designed using a pair of SIHCs with two design-specified transmission zeros on both Rx and Tx channels and its schematic is shown in Figure 6.18. The Rx channel has five poles and the Tx channel has four poles. The design specifications are as follows. The center frequency is at 9.6 GHz, the FBW is 8.6%, the Rx and Tx channels' bandwidth are 370 MHz and 315 MHz, respectively. The return losses in both passbands of Rx and Tx channels are 22 dB. The passband of the Rx channel is from 9.153 to 9.523 GHz. The passband of the Tx channel is from 9.662 to 9.977 GHz. Transmission zeros for the Rx channel are at 9.08 GHz and 9.58 GHz on both sides of the passband. Similarly, transmission zeros for the Tx channel are at 9.5 GHz and 10.0 GHz on both sides of the passband.

The coupling topology of this diplexer is same as the previous design (Figure 6.16). Based on the design specifications and the given coupling topology, the coupling coefficients of the diplexer are calculated. De-normalized self-resonant frequencies, coupling coefficient, and external quality factors are: f_{01} = 9.55 GHz, f_{02} = 9.467 GHz, f_{03} = 9.32 GHz, f_{04} = 9.185 GHz, f_{05} = 9.363 GHz, f_{06} = 9.947 GHz,

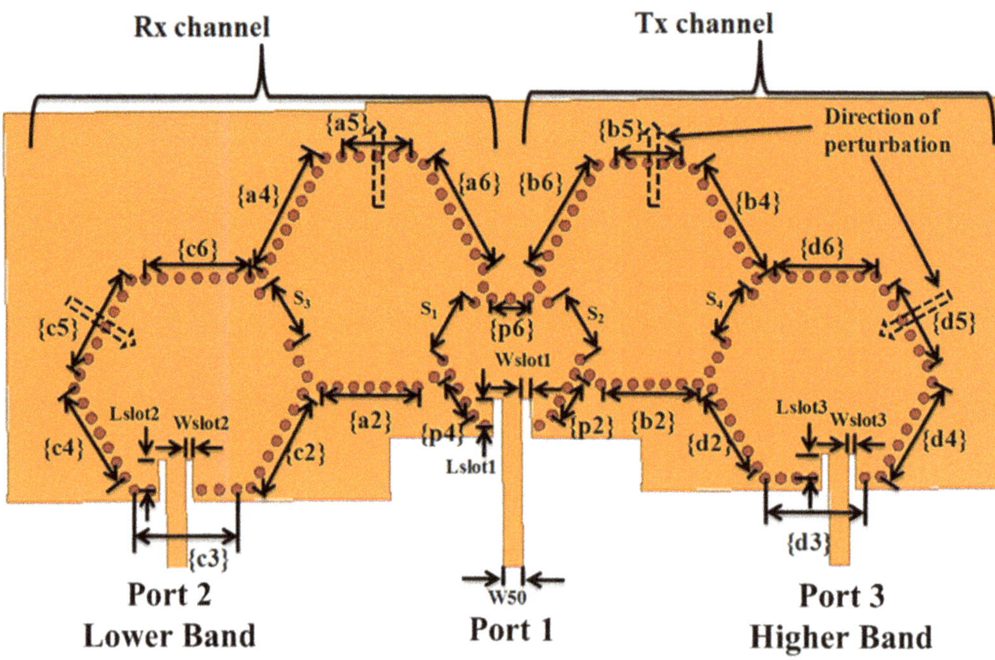

Figure 6.18 Schematic of Diplexer Design IV(a), with the dimensions as follows: L_{slot1} = 3.2 mm, W_{slot1} = 1.23 mm, $\{p_2\}$ = 5.7 mm, $\{p_4\}$ = 5.7 mm, and $\{p_6\}$ = 4.6 mm. Rx channel: $\{c_1\}$ = 11.775 mm, $\{c_2\}$ = 13.2 mm, $\{c_3\}$ = 13.025 mm, $\{c_4\}$ = 12.6 mm, $\{c_5\}$ = 13.2 mm, $\{c_6\}$ = 13.2 mm, $\{a_1\}$ = 12.575 mm, $\{a_2\}$ = 12.0 mm, $\{a_3\}$ = 11.775 mm, $\{a_4\}$ = 16.0 mm, $\{a_5\}$ = 8.8 mm, $\{a_6\}$ = 14.7 mm, S_1 = 8.075 mm, S_3 = 7.375 mm, L_{slot2} = 3.65 mm, and W_{slot2} = 1.0 mm. Tx channel: $\{d_1\}$ = 10.875 mm, $\{d_2\}$ = 11 mm, $\{d_3\}$ = 12.475 mm, $\{d_4\}$ = 13.2 mm, $\{d_5\}$ = 8 mm, $\{d_6\}$ = 12.6 mm, $\{b_1\}$ = 12.425 mm, $\{b_2\}$ = 12.0 mm, $\{b_3\}$ = 10.875 mm, $\{b_4\}$ = 15.4 mm, $\{b_5\}$ = 8.0 mm, $\{b_6\}$ = 14.7 mm, s_2 = 8.025 mm, S_4 = 6.875 mm, W_{slot3} = 0.95 mm, and L_{slot3} = 1.95 mm. (© 2017 Wiley [5].)

$f_{07} = 9.725$ GHz, $f_{08} = 9.646$ GHz, $f_{09} = 9.847$ GHz, $K_{12} = 0.0154$, $K_{13} = 0.0455$, $K_{16} = 0.0115$, $K_{17} = 0.0531$, $K_{24} = 0.0322$, $K_{25} = -0.0116$, $K_{34} = 0.0033$, $K_{35} = 0.0246$, $K_{68} = 0.0251$, $K_{69} = -0.0104$, $K_{78} = 0.0125$, $K_{79} = 0.0234$, $Q_E(P_1, 1) = 12.10$, $Q_E(P_2, 4) = 136.0$, $Q_E(P_2, 5) = 27.45$, $Q_E(P_3, 8) = 145.83$, and $Q_E(P_3, 9) = 32.11$.

In the Rx channel, perturbation is done in an outward direction using the side wall $\{a_5\}$ and in an inward direction using the side wall $\{c_5\}$, as shown in Figure 6.18. The odd-mode frequency (f_{02}) of SIHC $\{a\}$ and even-mode frequency (f_{05}) of SIHC $\{c\}$ are at a higher frequency than even-mode (f_{03}) and odd-mode (f_{04}) frequencies of their respective SIHCs. Both these conditions have made it possible to bring the transmission zeros on both sides of the passband. In the Tx channel, both design-specified transmission zeros appear on both sides of the passband using the same perturbation techniques as in the Rx section.

The responses of the diplexer are shown in Figure 6.19(a). The measured minimum insertion losses in the passbands of the Rx and Tx channels are 1.48 dB and 2.05 dB, respectively. The measured spurious response and the fabricated prototype are shown in Figure 6.19(b). This spurious response is measured from 8.5 to 16 GHz and exhibits higher-order modes of SIHC associated to the Rx and Tx channels. The isolation of all the diplexer designs (IV(a) and IV(b)) mentioned above is demonstrated in Figure 6.20.

6.1.7 Reconfigurable Diplexer Design with Tunable Perturbation

In this section, diplexer with tunable perturbation is implemented using varactor diode. The schematic and dimension of the diplexer with tunable perturbation are shown in Figure 6.21. The diplexer design specifications and its analysis were already discussed in Section 6.1.3. Here, the side walls $\{a_5\}$ and $\{b_5\}$ of Rx and Tx channels are employed to design a tunable perturbation-based diplexer. In Section 6.1.3, it is observed through simulation that the modal characteristics of a dual-mode SIHC and design-specified transmission zero are affected by varying *inset*. This tunability can be realized practically by introducing varactor diodes in both Rx and Tx channels. The Keysight ADS circuit simulator is used to analyze the tunable perturbation

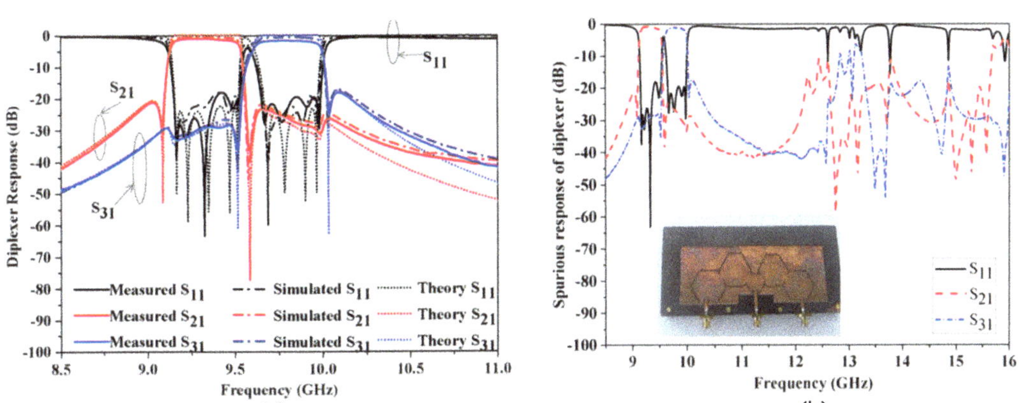

(a) (b)

Figure 6.19 (a) Measured, simulated, and theoretical results of Diplexer Design IV(b) and (b) measured spurious response of the diplexer with fabricated prototype. (© 2017 Wiley [5].)

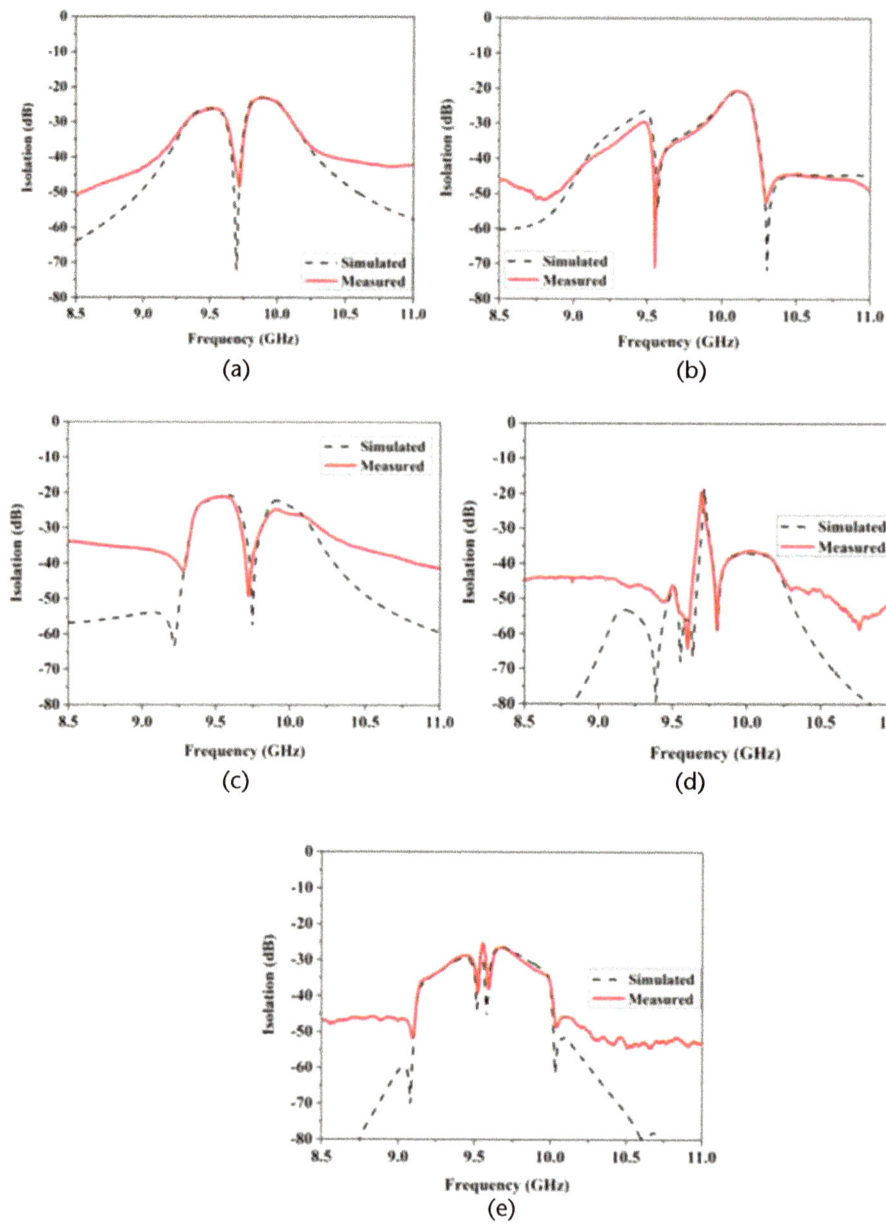

Figure 6.20 (a–e) Simulated and measured isolations between the Rx and Tx channels of Diplexer Design I, Diplexer Design IIb (Asymmetric Diplexer Design), Diplexer Design III, Diplexer Design IV(a), and Diplexer Design IV(b), respectively. (© 2017 Wiley [5].)

and its schematic is shown in Figure 6.22. Before incorporating varactor diode in design, circuit simulation is carried out based on coupling coefficients of Diplexer Design I mentioned in Section 6.1.3.

The circuit simulation and EM simulation are presented in Figure 6.23 where it is observed that both are agreeing fairly. When ground coplanar waveguides (GCPWs) are introduced through side walls $\{a_5\}$ and $\{b_5\}$ of both Rx and Tx channels, this perturbation disturbs the self-resonant frequencies of both dual-mode SIHCs. One

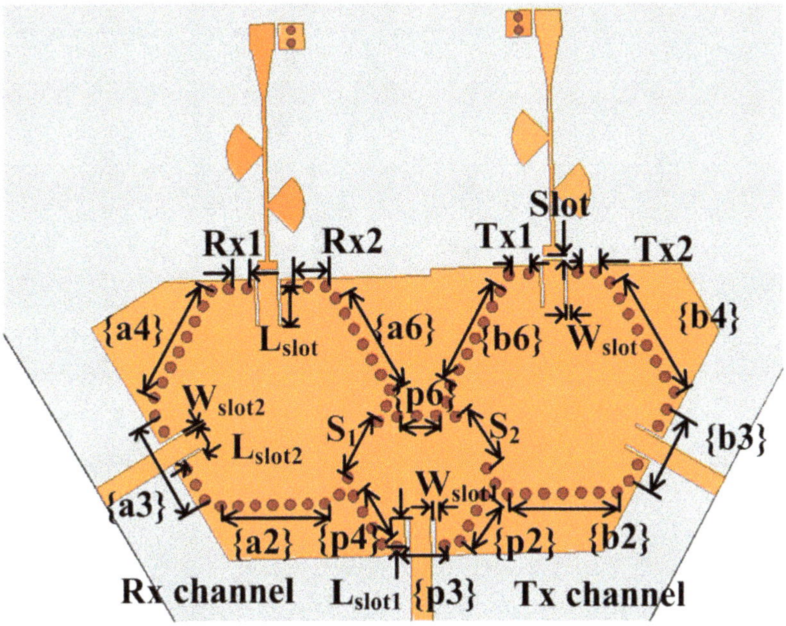

Figure 6.21 Schematic of a diplexer with tunable perturbation. The dimensions of the diplexer are provided in Figure 6.4, except Rx1 = 2.25 mm, Rx2 = 4.3 mm, Tx1 = 2.4 mm, Tx2 = 2.2 mm, L_{slot} = 4.43 mm, W_{slot} = 0.4 mm, and *Slot* = 0.6 mm.

end of the GCPW is inserted inside the dual-mode cavity, whereas another end is connected to the varactor diode. The external coupling between GCPW and concerning modes of that respective channels are determined using (6.21):

$$Q_e = \frac{f_0}{\Delta f_{+90^\circ} - \Delta f_{-90^\circ}} \tag{6.21}$$

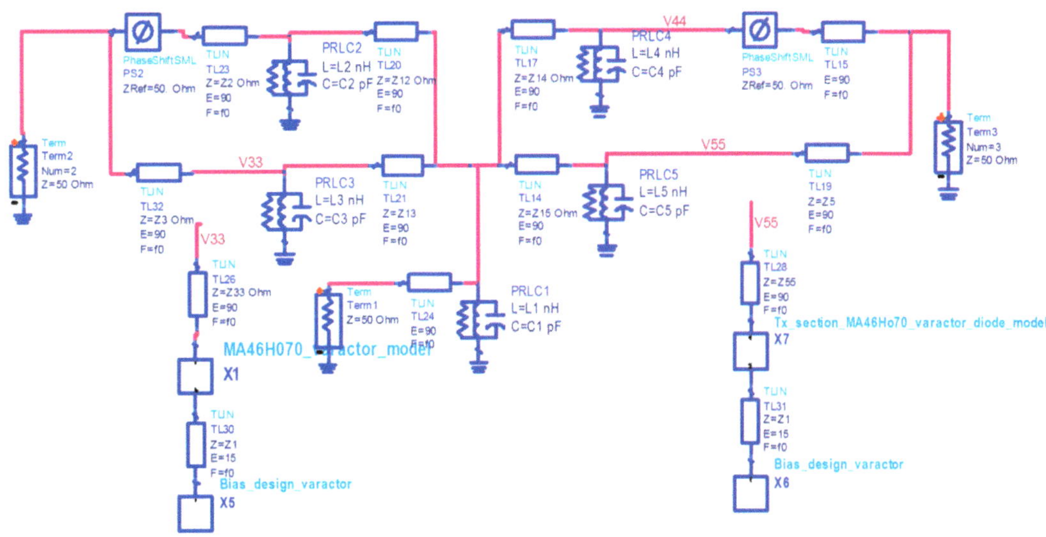

Figure 6.22 ADS circuit model of the tunable diplexer.

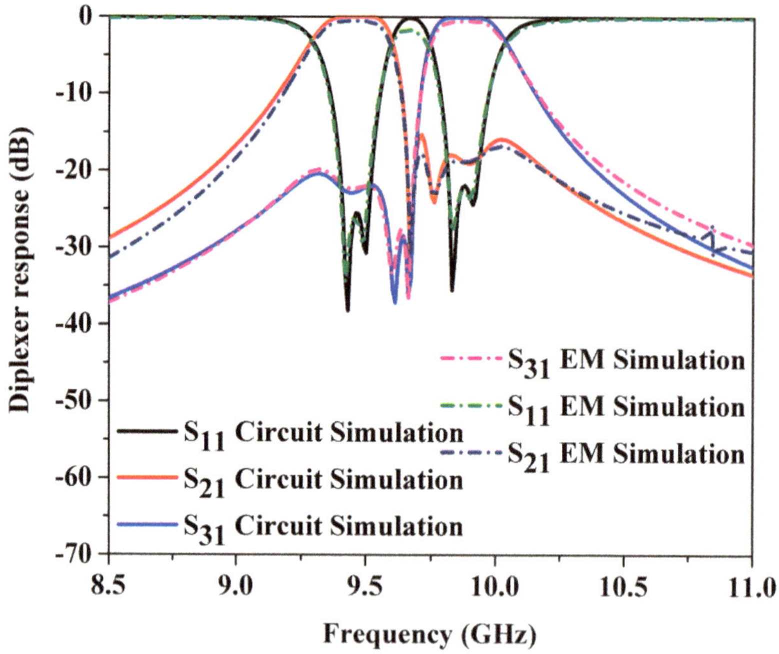

Figure 6.23 Circuit and EM-simulated responses of the diplexer shown in Figure 6.4.

where $\Delta f_{+90°}$ and $\Delta f_{-90°}$ are the frequencies at which the S_{11} phase shows +90° and −90° difference with respect to the phase of center frequency (f_0). The calculated external coupling of GCPWs are incorporated in the circuit simulator as shown in Figure 6.22. It is observed that the external quality factor between GCPW and the odd modes of both channels are extremely high, which means that hardly any coupling exists between them. Perturbations using side walls $\{a_5\}$ and $\{b_5\}$ do not affect odd-mode self-resonant frequency of that respective channel as mentioned in Section 6.1.2.1.

To design tunable perturbation, the varactor diode plays a great role. In the design, MA46H070-1056 varactor diodes are used. The circuit model of the varactor diode is provided in Figure 6.24. The characteristic of varactor diode is that junction capacitance varies from 1.1 to 0.15 pF with reverse DC bias voltage from 0 to 20V. Figure 6.25(a) exhibits the effect of varying Rx channel's reverse bias voltage on design-specified transmission zero when the Tx bias voltage is kept constant. During the Rx bias variation from 0 to 20V, the maximum 170-MHz shift of the Rx transmission zero is observed, whereas it does not affect the other Rx channel's transmission zeros, which appear because of the loading effect of the Tx channel. It is also observed that the transmission zero shifts towards higher frequency with increasing reverse bias voltage due to reduction of varactor diode's capacitance.

Similarly, Figure 6.25(b) shows the effect of the varying Tx channel's reverse bias voltage on design specified transmission zero when the Rx bias voltage is kept constant. During this bias variations, the maximum 180-MHz shift of the transmission zero is found. However, it does not affect the other transmission zeros that appear due to loading effect of the Rx channel on the Tx channel. Here, while increasing reverse bias voltage (Tx channel), shunt capacitance of the self-resonating even-mode

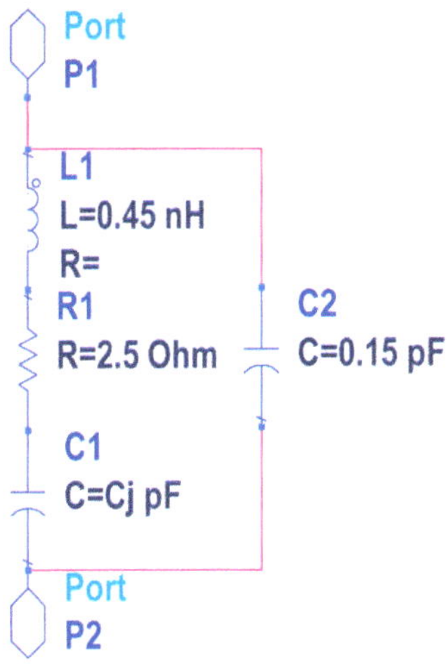

Figure 6.24 Circuit model of the varactor diode (Ma46H070-1056).

frequency is reduced. Hence, the Tx channel's design specified that transmission zero is shifting towards a lower frequency. The simulation and measured responses are shown in Figure 6.26 at two bias voltages. The fabricated prototype of the tunable diplexer is shown in Figure 6.27.

6.2 Dual-Band Diplexer Design Using a Single-Mode SIHC

Unlike the previous design, in this designed dual-band diplexer, the common junction is a SIW-based Y-junction. In both Rx and Tx channels, the power is coupled through the iris window of the Y-junction connected to Port 1. Both channels consist of dual-band bandpass filters where the Rx channel is at a lower frequency and the Tx channel is at a higher frequency. The schematic of the dual-band filter is shown in Figure 6.28. Resonators {*a*} and {*b*} are single-mode SIHCs attached to

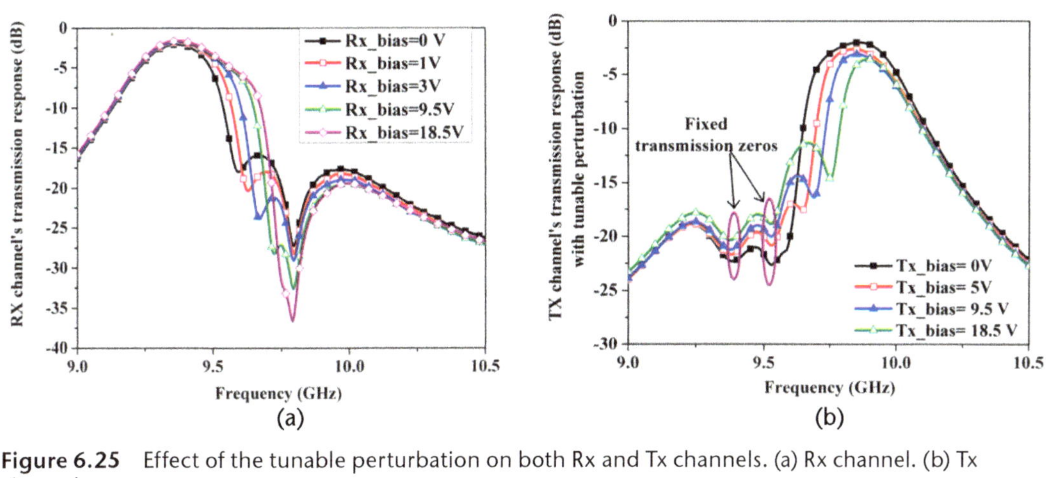

Figure 6.25 Effect of the tunable perturbation on both Rx and Tx channels. (a) Rx channel. (b) Tx channel.

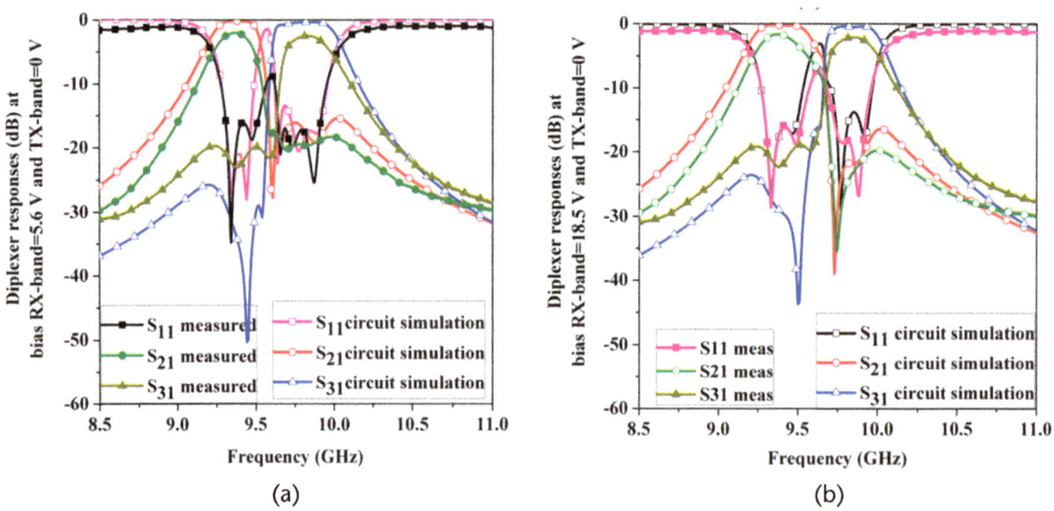

Figure 6.26 Measured and circuit-simulated responses of the design at two different bias points. (a) Rx channel bias at 5.6V and Tx at 0V. (b) Rx channel bias at 18.5V and Tx at 0V.

Figure 6.27 Fabrication prototype of the SIHC-based diplexer with tunable perturbation.

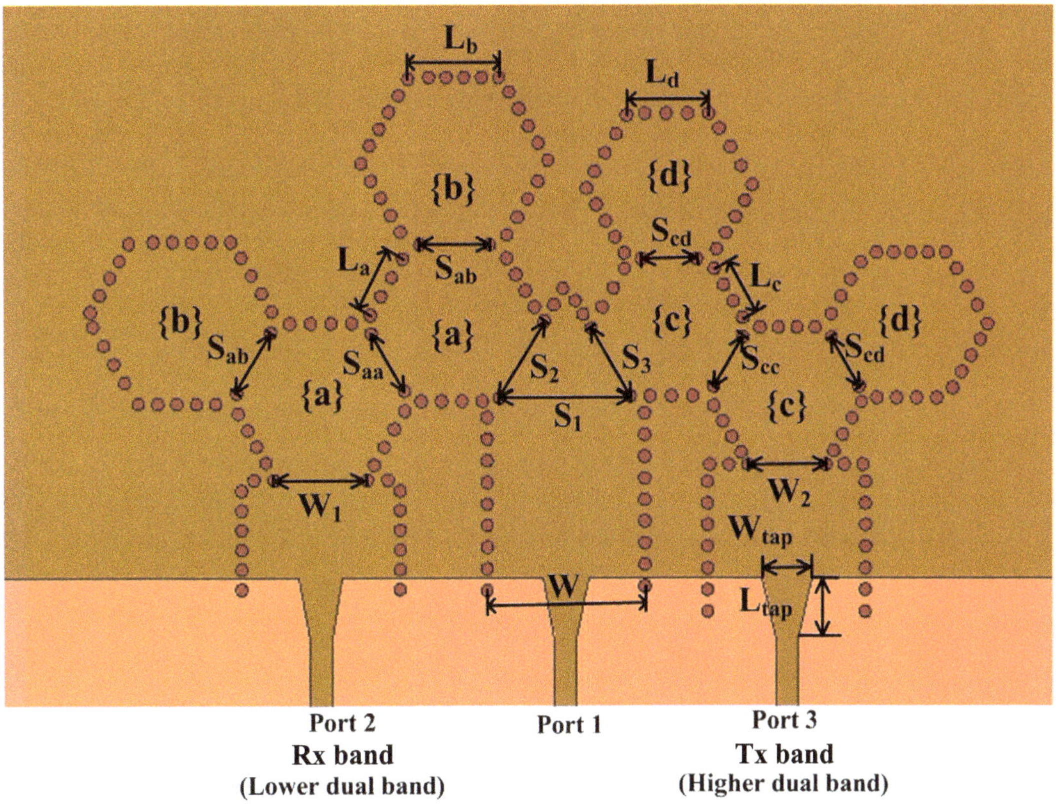

Figure 6.28 Schematic of dual-band diplexer with its dimension as follows: L_a = 10.19 mm, L_b = 10.76 mm, L_c = 9 mm, L_d = 9.5 mm, S_1 = 14.1 mm, S_2 = 9.5 mm, S_3 = 8.3 mm, S_{aa} = 7.1 mm, S_{ab} = 7.55 mm, S_{cc} = 6.1 mm, S_{cd} = 5.98 mm, W = 17 mm, W_1 = 9.975 mm, W_2 = 8.475 mm, W_{tap} = 5.4 mm, and L_{tap} = 6.6 mm.

the Rx channel. Similarly, {c} and {d} are also single-mode SIHCs attached to the Tx channel. The coupling topology of the dual-band diplexer is shown in Figure 6.29 where resonators {a}, {b}, {c}, and {d} in Figure 6.28 are represented by resonators 1, 2, 3, and 4, respectively. In this topology, two inline resonators are connected with two shunt resonators in both channels. This coupling topology provides dual-band response in both Rx and Tx channels. In this diplexer, all the resonators are single-mode hexagonal cavities. Each band comprises two poles and one transmission zero.

Design specifications of this diplexer are as follows:

- *Rx band:* Lower Rx passband ranges from 8.5 GHz (f_{L1}) to 8.57 (f_{L11}) GHz, return loss 15 dB in the passband, and transmission zero is at 8.7 GHz, which is above the passband. Higher Rx-band passband is 8.8 GHz (f_{L22}) to 8.86

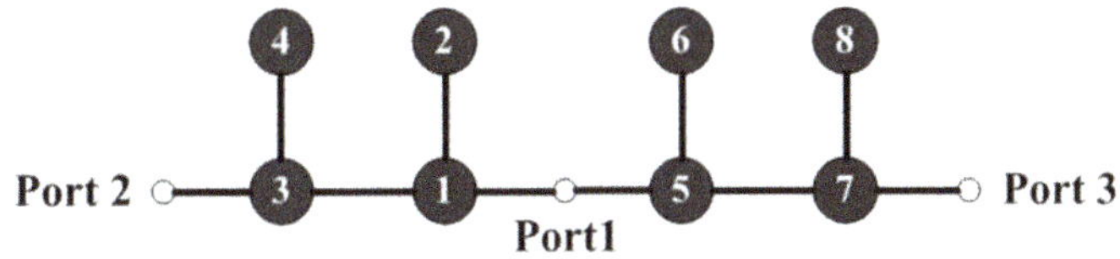

Figure 6.29 Coupling topology of the dual-band diplexer.

GHz (f_{L2}), return loss in the passband is 25 dB, and a transmission zero is at 8.7 GHz, which is below the Rx higher passband. Two transmission zeros are between the passband of Rx channel, which are improving the selectivity.

- *Tx band:* The lower Tx-passband range limited within 9.9 GHz (f_{H1}) to 10.0 GHz (f_{H11}), passband return loss is 18.0 dB, and transmission zero is at 10.10 GHz that is above the Tx lower passband. Higher Tx passband limits from 10.2 GHz (f_{H2}) to 10.25 GHz (f_{H22}). Passband return loss is 20.0 dB and the design-specified Tx higher band transmission zero is placed at 10.10 GHz. Synthesis procedure of designing dual-band diplexer is as follows.

Previously in the diplexer synthesis procedure, two frequency transformations are carried out. Here one more linear frequency transformation is required to realize the dual-band diplexer.

Step 1: Bandpass to lowpass frequency transformation as shown in Figure 6.30(a) where both Rx and Tx filter are transformed from f-to-Ω plane as given in (6.22).

$$\Omega = \frac{f_0}{\Delta f}\left(\frac{f}{f_0} - \frac{f_0}{f}\right) \tag{6.22}$$

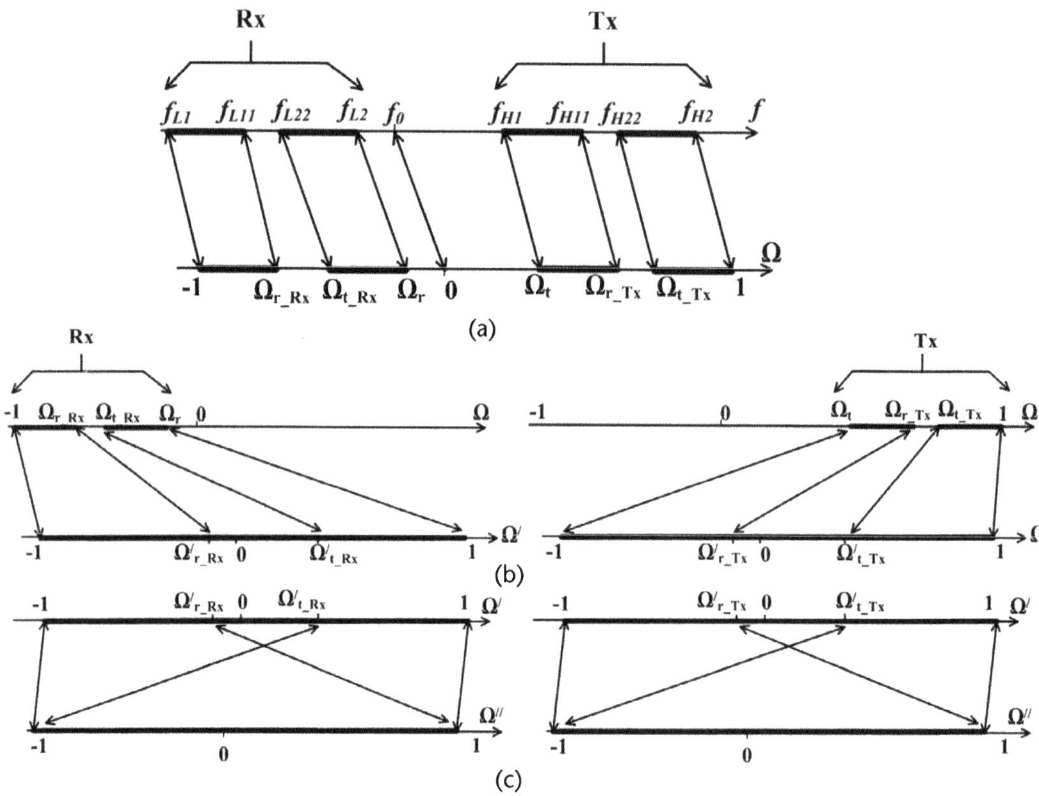

Figure 6.30 Three-stage frequency mapping for dual-band diplexer analysis. (a) Frequency mapping from the Ω domain to the *f* domain. (b) Rx channel: Frequency mapping from Ω′ to Ω domain and Ω″ to Ω′ domain. (c) Tx channel: Frequency mapping from Ω′ to Ω domain and Ω″ to Ω′ domain.

Step 2: Linearly transform Rx and Tx filter from Ω-plane to Ω′-plane where both the channels are separate bandpass filters as shown in Figure 6.30(b). This frequency transformation can be expressed as:

$$\begin{gathered}
\Omega = a_1\Omega' + b_1 \quad \text{for Rx filter} \\
\Omega = a_2\Omega' + b_2 \quad \text{for Tx filter} \\
\left.\begin{aligned} a_1 &= (\Omega_r + 1)/2 \\ b_1 &= (\Omega_r - 1)/2 \end{aligned}\right\} \text{Rx Channel} \qquad \left.\begin{aligned} a_2 &= (1 - \Omega_t)/2 \\ b_2 &= (1 + \Omega_t)/2 \end{aligned}\right\} \text{Tx Channel}
\end{gathered} \tag{6.23}$$

After calculating a_1, b_1, a_2, and b_2, other frequency passband edges of both channels are required to be calculated. Both Ω'_{r_Rx} and Ω'_{t_Rx} are passband edges of Rx channel in Ω′ domain. Similarly, Ω'_{r_Tx} and Ω'_{t_Tx} are passband edges of Tx channel in Ω′ domain. These passband edges are expressed in (6.24):

$$\begin{aligned}
\Omega'_{r_\mathrm{Rx}} &= \left(\Omega_{r_\mathrm{Rx}} - b_1\right)/a_1 \\
\Omega'_{t_\mathrm{Rx}} &= \left(\Omega_{t_\mathrm{Rx}} - b_1\right)/a_1 \\
\Omega'_{r_\mathrm{Tx}} &= \left(\Omega_{r_\mathrm{Tx}} - b_2\right)/a_2 \\
\Omega'_{t_\mathrm{Tx}} &= \left(\Omega_{t_\mathrm{Tx}} - b_2\right)/a_2
\end{aligned} \tag{6.24}$$

Step 3: Linearly transform the dual-band Rx filter to two individual Rx lower and Rx higher bands separately. Similarly, linearly transform the Tx filter to two separate Tx lower and Tx higher bands as shown in Figure 6.30(c). Here frequency transformations, in the Rx band, from Ω′ to Ω″ are expressed in (6.25):

$$\begin{gathered}
\Omega' = a_3\Omega'' + b_3 \quad \text{for Rx lower band and} \\
\Omega' = a_4\Omega'' + b_4 \quad \text{for Rx higher band} \\
\left.\begin{aligned} a_3 &= \left(\Omega'_{r_\mathrm{Rx}} + 1\right)/2 \\ b_3 &= \left(\Omega'_{r_\mathrm{Rx}} - 1\right)/2 \end{aligned}\right\} \text{Rx Channel lower band} \\
\left.\begin{aligned} a_4 &= \left(1 - \Omega'_{t_\mathrm{Rx}}\right)/2 \\ b_4 &= \left(1 + \Omega'_{t_\mathrm{Rx}}\right)/2 \end{aligned}\right\} \text{Rx Channel higher band}
\end{gathered} \tag{6.25}$$

Similarly, in the Tx dual-band filter, linear frequency transformation is carried out from the Ω′ to Ω″ domain. This frequency transformation leads to separate Tx lower-frequency and Tx higher-frequency bands. This frequency transformation on both Tx higher and lower passbands is expressed in (6.26).

For TX channel lower band: $\Omega' = a_5\Omega'' + b_5$

For TX channel higher band: $\Omega' = a_6\Omega'' + b_6$

$$\left.\begin{aligned} a_5 &= \left(\Omega'_{r\_\mathrm{Tx}} + 1\right)/2 \\ b_5 &= \left(\Omega'_{r\_Tx} - 1\right)/2 \end{aligned}\right\} \text{Tx Channel lower band}$$

$$\left.\begin{aligned} a_6 &= \left(1 - \Omega'_{t\_\mathrm{Tx}}\right)/2 \\ b_6 &= \left(1 + \Omega'_{t\_\mathrm{Tx}}\right)/2 \end{aligned}\right\} \text{Tx Channel higher band} \tag{6.26}$$

$$M = \begin{bmatrix} & P_1 & P_2 & P_3 & 1 & 2 & 3 & 4 & 5 & 6 & 7 & 8 \\ P_1 & 0 & 0 & 0 & 0.307 & 0 & 0 & 0 & 0.32 & 0 & 0 & 0 \\ P_2 & 0 & 0 & 0 & 0 & 0 & 0.307 & 0 & 0 & 0 & 0 & 0 \\ P_3 & 0 & 0 & 0 & 0 & 0 & 0 & 0 & 0 & 0 & 0.32 & 0 \\ 1 & 0.307 & 0 & 0 & 0.791 & 0.176 & 0.107 & 0 & 0 & 0 & 0 & 0 \\ 2 & 0 & 0 & 0 & 0.176 & 0.756 & 0 & 0 & 0 & 0 & 0 & 0 \\ 3 & 0 & 0.307 & 0 & 0.107 & 0 & 0.791 & 0.177 & 0 & 0 & 0 & 0 \\ 4 & 0 & 0 & 0 & 0 & 0 & 0.177 & 0.756 & 0 & 0 & 0 & 0 \\ 5 & 0.32 & 0 & 0 & 0 & 0 & 0 & 0 & -0.824 & 0.117 & 0.114 & 0 \\ 6 & 0 & 0 & 0 & 0 & 0 & 0 & 0 & 0.117 & -0.873 & 0 & 0 \\ 7 & 0 & 0 & 0.32 & 0 & 0 & 0 & 0 & 0.114 & 0 & -0.824 & 0.117 \\ 8 & 0 & 0 & 0 & 0 & 0 & 0 & 0 & 0 & 0 & 0.117 & -0.873 \end{bmatrix} \tag{6.27}$$

After carrying out frequency transformations, the polynomials related to the dual-band diplexer are derived using the diplexer synthesis procedure, which was discussed earlier in this chapter. Based on the diplexer synthesis procedure and coupling topology, the coupling matrix of the specific dual-band diplexer is obtained and provided in (6.27).

Calculated self-resonating frequencies of all the single-mode resonators shown in Figure 6.28, external Q and coupling coefficients are extracted from (6.27) and are given as: {a} → $(f_{01}) = (f_{03}) = 8.68$ GHz, {b} → $(f_{02}) = (f_{04}) = 8.71$ GHz, {c} → $(f_{05}) = (f_{07}) = 10.07$ GHz, {d} → $(f_{06}) = (f_{08}) = 10.11$ GHz. $Q_e(P_1, 1) = 57.6984$, $Q_e(P_1, 5) = 53.1056$.

Rx channel external quality factor: $Q_e(P_2, 3) = 57.6984$.

Tx channel external quality factor: $Q_e(P_3, 7) = 53.1056$.

The coupling coefficients calculated between two synchronous resonators are as follows:

$$K_{ij} = \pm\frac{f_{p2}^2 - f_{p1}^2}{f_{p2}^2 + f_{p1}^2} \tag{6.28}$$

where $k_{i,j}$ are the coupling coefficients between i and j resonators and f_{p1} and f_{p2} are the lower and higher split resonant frequencies when resonators i and j are coupled. The values of coupling coefficients are: $K_{1,2} = K_{3,4} = 0.0325$, $K_{1,3} = 0.0197$, $K_{5,6} = K_{7,8} = 0.0215$, $K_{5,7} = 0.0210$.

In this design topology, as shown in Figure 6.29, resonators 1 and 3 have same resonant frequency. The coupling between the synchronous resonators is determined by the inductive iris window (S_{aa}). When the resonant frequencies of both the resonators are different, which is called asynchronous resonator, such as resonators 1 and 2 in Figure 6.29, the coupling coefficients are determined by the inductive iris window S_{ab}, which is calculated using (6.29).

$$k = \pm\frac{1}{2}\left(\frac{f_{02}}{f_{01}} + \frac{f_{01}}{f_{02}}\right) \cdot \sqrt{\left(\frac{f_{p2}^2 - f_{p1}^2}{f_{p2}^2 + f_{p1}^2}\right)^2 - \left(\frac{f_{02}^2 - f_{01}^2}{f_{02}^2 + f_{01}^2}\right)^2} \quad (6.29)$$

where f_{01} and f_{02} are the self-resonant frequencies of 1 and 2, respectively. f_{p1} and f_{p2} are the lower and higher split resonant frequencies when both 1 and 2 resonators are coupled.

The external quality factors can be calculated by varying the iris windows attached to input and output ports. In the present design, the iris windows S_2 and S_3 attached to input Port 1 determine input external quality factor of the Rx and Tx channels, respectively.

$$Q_e = \frac{f_0}{\Delta f_{+90°} - \Delta f_{-90°}} \quad (6.30)$$

where $\Delta f_{+90°}$ and $\Delta f_{-90°}$ are the frequencies at which the S_{11} phase shows +90° and −90° difference with respect to the phase of the center frequency (f_0). Thus, theoretical couplings are realized in design. The design is fabricated using Rogers RT/Duroid 5880 with a thickness of 0.787 mm. The theoretical, simulated, and measured responses of the designed dual-band diplexers are exhibited in Figure 6.31(a, b). The measured isolation of the diplexer is found below 35 dB as shown in Figure 6.32(a, b) shows the fabricated prototype.

6.3 Triplexer Design

In this section, a compact triplexer is realized using single and dual-mode hexagonal cavities in SIW technology. Triplexer is a four-port network where one is input port and remaining are output channels of specific frequency band. The input common

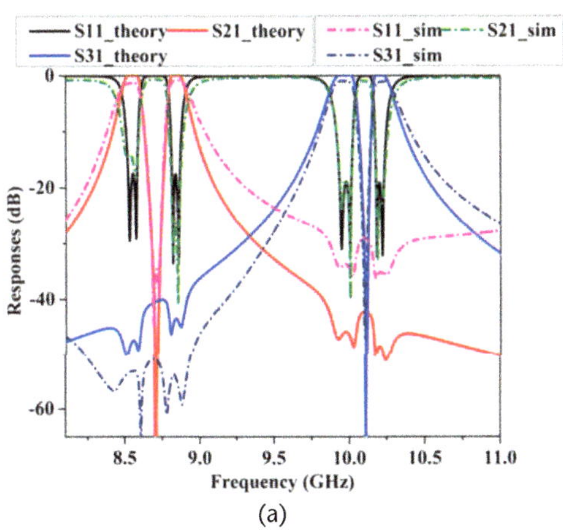

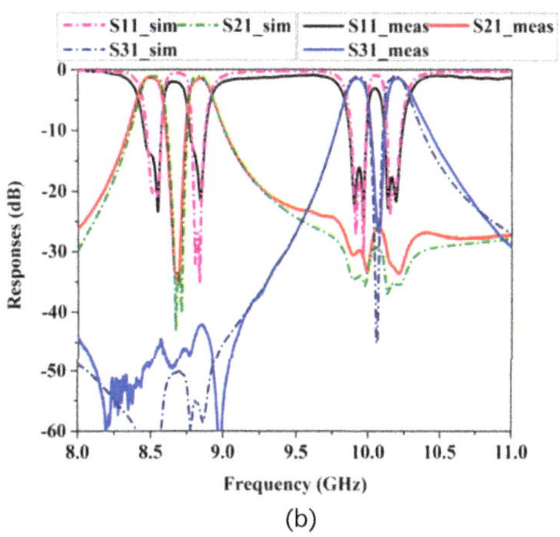

Figure 6.31 (a) Simulated and theoretical S-parameters of the dual-band diplexer. (b) Measured and simulated responses of the dual-band diplexer.

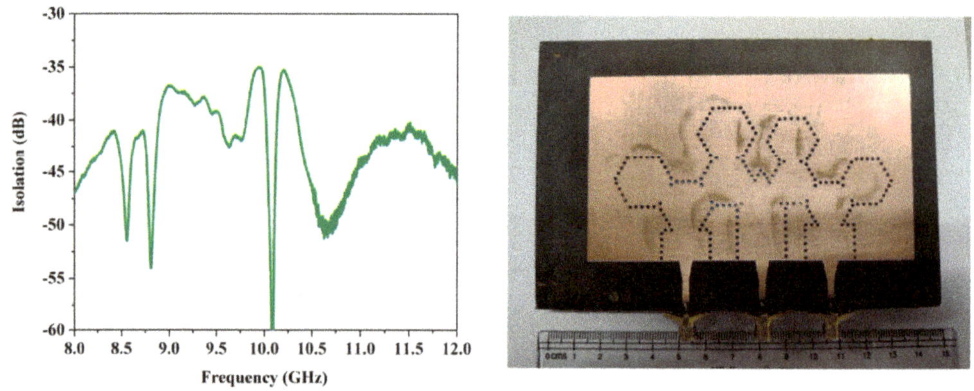

(a) (b)

Figure 6.32 (a) Measured isolation between both channels, and (b) fabricated prototype of dual-band diplexer.

junction is a single-mode hexagonal cavity of which three side walls are exploited to couple power into three different channels through iris coupling (H-coupling). The common resonator is attached to each single-mode hexagonal cavity of the three channels [8]. These single-mode resonators are connected to a single dual-mode hexagonal cavity of their respective channel. Each of the three channels has three poles: one by a single-mode resonator and reaming two by a dual-mode SIHC. However, the common resonator's dimension is chosen in such a way that it contributes an additional pole to an Rx channel and therefore it becomes fourth-order bandpass filter. The schematic of the triplexer and its dimensions are provided in Figure 6.33. The coupling topology of the design is provided in Figure 6.34. In the schematic, the common resonator is named as $\{a\}$. Besides that, all three single-mode resonators associated to Rx, Mx and Tx channels are named as $\{b\}$, $\{c\}$, and $\{d\}$, respectively. The dual-mode resonators $\{L\}$, $\{M\}$, and $\{U\}$ are attached to Rx, Mx, and Tx channels, respectively, and therefore, the Rx channel consists of $\{b\}$ and $\{L\}$, the Mx channel consists of $\{c\}$ and $\{M\}$ and, similarly, the Tx channel consists of $\{d\}$ and $\{U\}$ single and dual-mode resonators. In the coupling topology, single mode resonators $\{a\}$, $\{b\}$, $\{c\}$, and $\{d\}$ are represented by resonators 1, 2, 3, and 4, respectively; and dual modes of resonators $\{L\}$, $\{M\}$, $\{U\}$ are represented by 5, 6, 7, 8, 9, and 10, respectively.

$$M = \begin{bmatrix} & P_1 & P_2 & P_3 & P_4 & 1 & 2 & 3 & 4 & 5 & 6 & 7 & 8 & 9 & 10 \\ P_1 & 0 & 0 & 0 & 0 & 0.899 & 0 & 0 & 0 & 0 & 0 & 0 & 0 & 0 & 0 \\ P_2 & 0 & 0 & 0 & 0 & 0 & 0 & 0 & 0 & -0.408 & 0.326 & 0 & 0 & 0 & 0 \\ P_3 & 0 & 0 & 0 & 0 & 0 & 0 & 0 & 0 & 0 & 0 & -0.434 & 0.284 & 0 & 0 \\ P_4 & 0 & 0 & 0 & 0 & 0 & 0 & 0 & 0 & 0 & 0 & 0 & 0 & -0.414 & 0.307 \\ 1 & 0.899 & 0 & 0 & 0 & -0.033 & 0.390 & 0.508 & 0.506 & 0 & 0 & 0 & 0 & 0 & 0 \\ 2 & 0 & 0 & 0 & 0 & 0.390 & 0.747 & 0 & 0 & 0.156 & 0.087 & 0 & 0 & 0 & 0 \\ 3 & 0 & 0 & 0 & 0 & 0.508 & 0 & 0.152 & 0 & 0 & 0 & 0.202 & 0.157 & 0 & 0 \\ 4 & 0 & 0 & 0 & 0 & 0.506 & 0 & 0 & -0.673 & 0 & 0 & 0 & 0 & 0.184 & 0.120 \\ 5 & 0 & -0.408 & 0 & 0 & 0 & 0.156 & 0 & 0 & 0.593 & 0 & 0 & 0 & 0 & 0 \\ 6 & 0 & 0.326 & 0 & 0 & 0 & 0.087 & 0 & 0 & 0 & 1.028 & 0 & 0 & 0 & 0 \\ 7 & 0 & 0 & -0.434 & 0 & 0 & 0 & 0.202 & 0 & 0 & 0 & -0.179 & 0 & 0 & 0 \\ 8 & 0 & 0 & 0.284 & 0 & 0 & 0 & 0.157 & 0 & 0 & 0 & 0 & 0.255 & 0 & 0 \\ 9 & 0 & 0 & 0 & -0.414 & 0 & 0 & 0 & 0.184 & 0 & 0 & 0 & 0 & -0.984 & 0 \\ 10 & 0 & 0 & 0 & 0.307 & 0 & 0 & 0 & 0.120 & 0 & 0 & 0 & 0 & 0 & -0.563 \end{bmatrix} \tag{6.31}$$

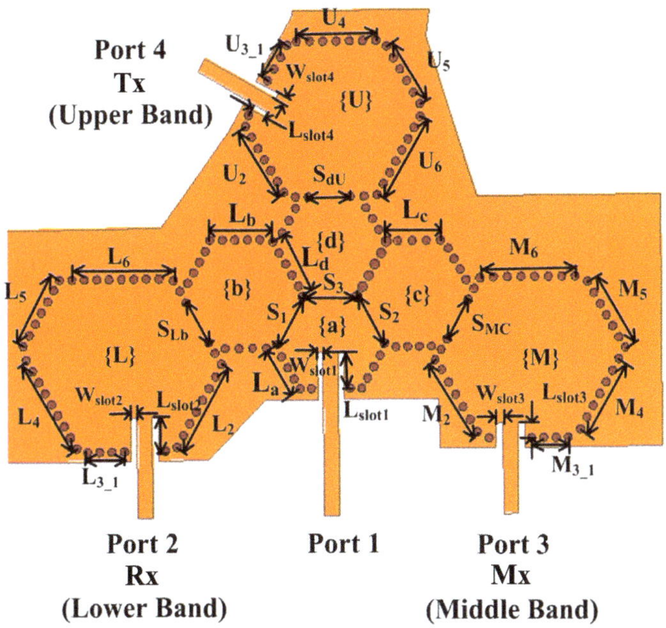

Figure 6.33 Schematic of triplexer with the dimensions: $L_a = 9.23$ mm, $L_b = 10.41$ mm, $L_c = 9.815$ mm, $L_d = 9.62$ mm, $L_{slot1} = 6.2$ mm, $W_{slot1} = 0.9$ mm, $S_1 = 8.5$ mm, $S_2 = 8.5$ mm, $S_3 = 8.55$ mm, $W_{slot2} = 1.1$ mm, $W_{slot3} = 1.2$ mm, $W_{slot4} = 0.9$ mm, $L_{slot2} = 5.25$ mm, $L_{slot3} = 2.3$ mm, $L_{slot4} = 2.5$ mm, $L_2 = 14.75$ mm, $L_{3\_1} = 6$ mm, $L_4 = 16$ mm, $L_5 = 11.8$ mm, $L_6 = 16.35$ mm, $S_{Lb} = 7.23$ mm, $M_2 = 13.1$ mm, $M_{3\_1} = 6$ mm, $M_4 = 12.6$ mm, $M_5 = 11.9$ mm, $M_6 = 15.3$ mm, $S_{MC} = 7.1$ mm, $U_2 = 12$ mm, $U_{3\_1} = 6.3$ mm, $U_4 = 12.6$ mm, $U_5 = 10.5$ mm, $U_6 = 14$ mm, and $S_{dU} = 6.675$ mm.

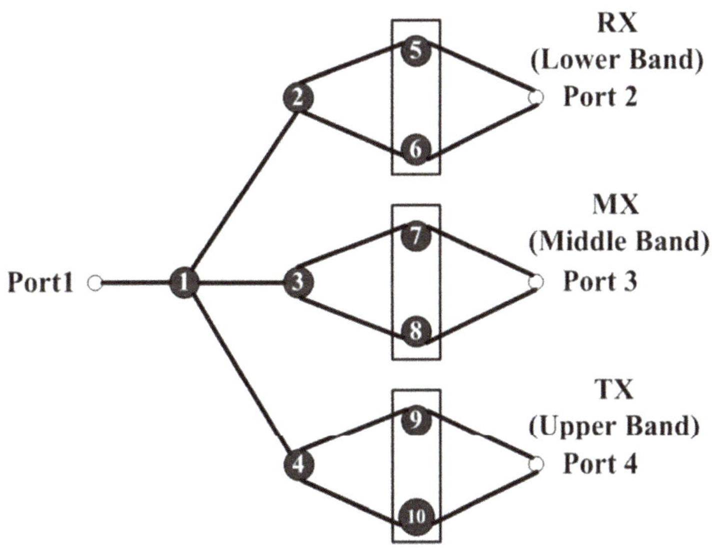

Figure 6.34 Coupling topology of the triplexer.

Perturbation technique is applied to all three dual-mode resonators in order to realize bandpass filters for their respective channel. Each channel has one design-specified transmission zero because of a dual-mode resonator. Other than the design-specified transmission zero, it has six more transmission zeros because ($N_{p\mathrm{Tx}}$ + $N_{p\mathrm{Mx}}$), ($N_{p\mathrm{Rx}}$ + $N_{p\mathrm{Tx}}$), and ($N_{p\mathrm{Rx}}$ + $N_{p\mathrm{Mx}}$) are additional transmission zeros for the Rx, Mx, and Tx channels, respectively, as mentioned in the triplexer design synthesis procedure. These transmission zeros appear because of loading effect on the respective channel by other two channels. The design specifications of the triplexer are as follows:

- *Lower band (Rx band):* f_{L1} = 8.7 GHz, f_{L2} = 9.0 GHz, return loss is 21 dB in the passband, and the transmission zero is at 8.26 GHz.
- *Middle band (Mx band):* f_{M1} = 9.27 GHz, f_{M2} = 9.48 GHz, return loss is 24 dB in the passband, and the transmission zero is at 8.88 GHz.
- *Upper band (Tx band):* f_{H1} = 9.757 GHz, f_{H2} = 10 GHz, return loss is 20 dB in the passband, and the transmission zero is at 9.38 GHz.

Based on design specifications and coupling topology, the coupling matrix of the triplexer is generated and is given in (6.31). From the coupling matrix values, external quality factors, de-normalized self-resonant frequencies of the resonators, and coupling coefficients are calculated and given next.

The calculated resonant frequencies of single mode resonator are: {a} → (f_{01}) = 9.35 GHz, {b} → (f_{02}) = 8.85 GHz, {c} → (f_{03}) = 9.23 GHz, and {d} → (f_{04}) = 9.78 GHz.

The calculated resonant frequencies of the dual-mode SIHCs are: {L}: f_{05} = 8.95 GHz and f_{06} = 8.68 GHz, {M}: f_{07} = 9.44 GHz and f_{08} = 9.16 GHz, and {U}: f_{09} = 9.99 GHz and f_{10} = 9.70 GHz.

The calculated external quality factors and coupling coefficients are: $Q_e(P_1, 1)$ = 8.88, $Q_e(P_2, 5)$ = 43, $Q_e(P_2, 6)$ = 67.5, $Q_e(P_3, 7)$ = 38.1, $Q_e(P_3, 8)$ = 89, $Q_e(P_4, 9)$ = 41.9, $Q_e(P_4, 10)$ = 76.1, $K_{1,2}$ = 0.0544, $K_{1,3}$ = 0.0708, $K_{1,4}$ = 0.0705, $K_{2,5}$ = 0.0217, $K_{2,6}$ = 0.0121, $K_{3,7}$ = 0.0282, $K_{3,8}$ = 0.0219, $K_{4,9}$ = 0.0256, and $K_{4,10}$ = 0.0167.

In all dual-mode resonators, odd-mode resonant frequency is higher than their concerning even-mode. This would happen when the perturbations applied to the side walls L_5, M_5, and U_5 are in the outward direction. These perturbations help to achieve desired self-resonant frequencies of the dual-mode SIHCs. Besides that, feeding position, coupling slot width (W_{slot}) and length (L_{slot}) provide desired external quality factors. The coupling coefficients between two asynchronous resonators, such as single-mode resonator {a} and {b} or dual-mode resonator {L} and {b}, can be determined using (6.29) [6]. Simulations and theoretical responses (derived from coupling matrix) of the triplexer are provided in Figure 6.35 and measured and simulated responses are shown in Figure 6.36. Measured and simulated isolations are depicted in Figure 6.37(a). Measured and simulated insertion losses are 1.4 dB and 0.85 dB in the Rx band, 1.35 dB and 0.8 dB in the Mx band and 1.3 dB and 0.7 dB in the Tx band. It is observed that the isolation between the channels are below 20 dB. The fabricated prototype of the triplexer is shown in Figure 6.37(b).

6.4 SIW-Based Balun Filter and Balun Diplexer Design

In modern wireless communication system, passive circuits play an important role where the bandpass filter and the balun are integral parts of the RF front-end system. The requirement of employing balanced circuits in RF front end is that it makes the system highly immune to noise, electromagnetic interference (EMI), and radiation losses. Besides that, it also helps in improving the spurious response and dynamic range of the system. These are very crucial parameters for high-speed circuit systems.

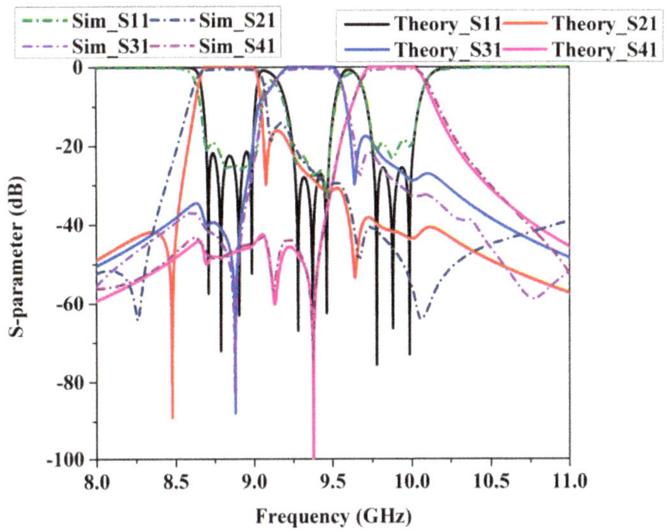

Figure 6.35 Plot of simulated and theoretical S-parameters of the triplexer.

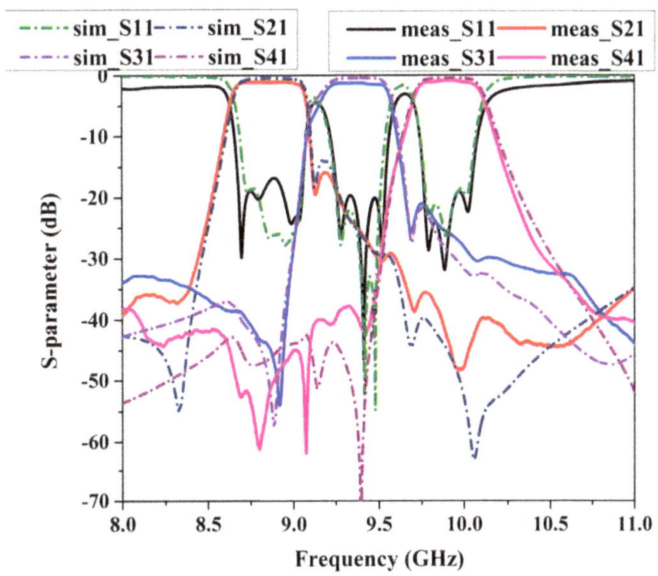

Figure 6.36 Measured and simulated S-parameters of the triplexer.

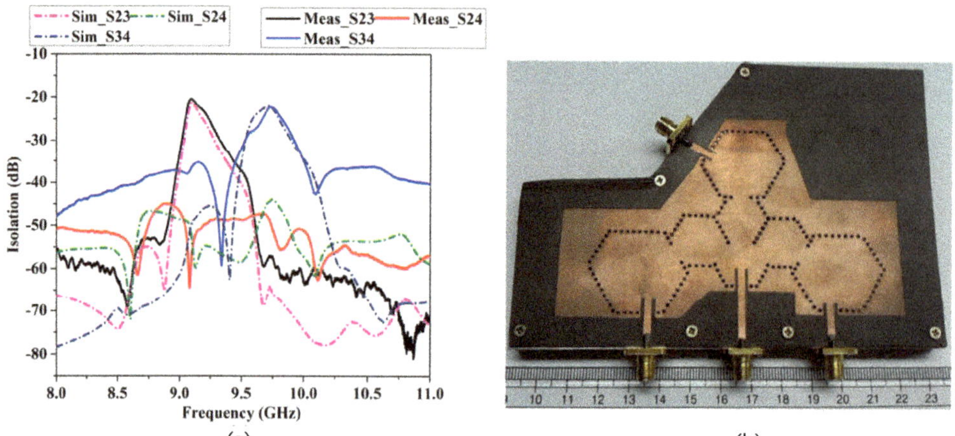

(a) (b)

Figure 6.37 (a) Plot of measured isolations between all the channels. (b) Fabricated prototype of the triplexer.

Balun is a three-port network that ensures the balance signal into the unbalance one and vice versa. One of the most popular approaches to realize balun is the conventional Marchand baluns [9] or lumped center-tap transformers [10]. However, bandpass filter passes the desired frequency band and rejects unwanted frequency. Both the balun and bandpass filter are associated components in an RF front-end system. Therefore, to make the system cost-effective and compact, it is necessary to integrate associated passive components without compromising the system's performance. A balun-BPF provides both bandpass filter and balun's characteristics simultaneously [11]. In a modern wireless communication system, integrated diplexer and balun are also widely used in the RF front-end system to bring compactness and achieve lower cost.

In this section, three different balun designs are discussed, where the first one is a fifth-order, highly selective balun bandpass filter where the input is single-ended but the output has two balanced ports. The other two designs are the balun diplexer where one is a balun diplexer with a single dual-mode SIHC at each channel and a balun diplexer with a pair of SIHCs at each channel. The dual-mode resonator's inherent characteristic is that both its modes are orthogonal. This dual-mode's orthogonality, with a proper feeding position and perturbation, leads to realize balanced output.

6.4.1 Balun Filter (Design I)

The schematic of the balun filter along with its physical dimensions are provided in Figure 6.38. The corresponding coupling topology is shown in Figure 6.39. The resonator $\{a\}$ is a single mode resonator and is designated as 1 and dual-mode resonators $\{b\}$ and $\{c\}$ are designated as 2, 3, 4, and 5 respectively in the coupling topology. In $\{b\}$, the side wall b_5 is used for perturbation in an outward direction that allows an odd-mode frequency higher than an even-mode self-resonant frequency.

This leads to bringing the transmission zero below the passband. Similarly, in the $\{c\}$ cavity, the side wall c_5 is used for perturbation in an inward direction that

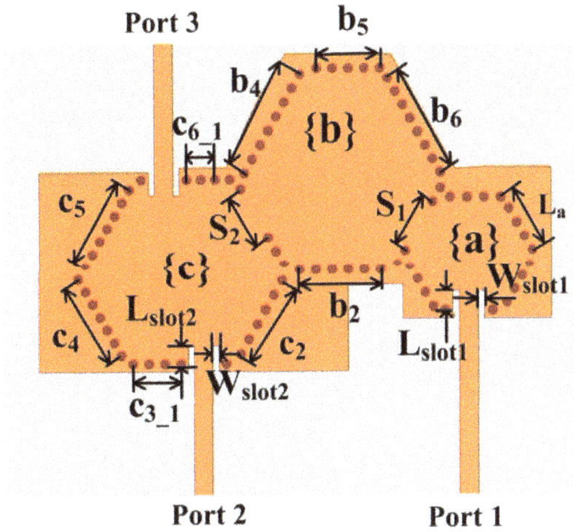

Figure 6.38 Schematic of the balun filter using dual-mode SIHC. The dimensions of the design are: $L_a = 9.35$ mm, $S_1 = 7.375$ mm, $S_2 = 7.275$ mm, $W_{slot1} = 0.9$ mm, $W_{slot2} = 0.8$ mm, $L_{slot1} = 2.55$ mm, $L_{slot2} = 2.35$ mm, $b_2 = 10.65$ mm, $b_4 = 14.7$ mm, $b_5 = 8.4$ mm, $b_6 = 14$ mm, $c_2 =$ mm, $c_{3\_1} = 6.3$ mm, $c_4 = 11.4$ mm, $c_5 = 11.7$ mm, and $c_{6\_1} = 3.8$ mm.

brings the even-mode frequency higher than the odd-mode frequency. As a consequence, the transmission zero appears above the passband. Both these filter design procedures were already discussed in Chapter 5 where the modal characteristic of the cavity with inset variations and the effect of it on transmission zeros are exhibited graphically. In the schematic in Figure 6.38, by adjusting the proper feeding position, it is possible to ensure the balanced output ports (Port 2 and Port 3) with respect to the input port (Port 1). Magnetic field distributions in the balun filter are exhibited in Figure 6.40(a, b); from this, it is evident that in the {c} cavity, both output ports are positioned to obtain an out-of-phase signal. The design specifications of this balun filter are as follows: design center frequency (f_0) = 9.70 GHz, and FBW = 4.74%. The design has two transmission zeros placed at 9.4 GHz and 10.17 GHz, and the return loss in the passband is 22.5 dB. Based on the design specifications and coupling topology, the coupling matrix of the concerning filter is provided in (6.32). From the coupling matrix, self-resonating frequencies (f_{0i}), the external quality factor ($Q_e(P_i, j)$), and coupling coefficients (K_{ij}) are calculated.

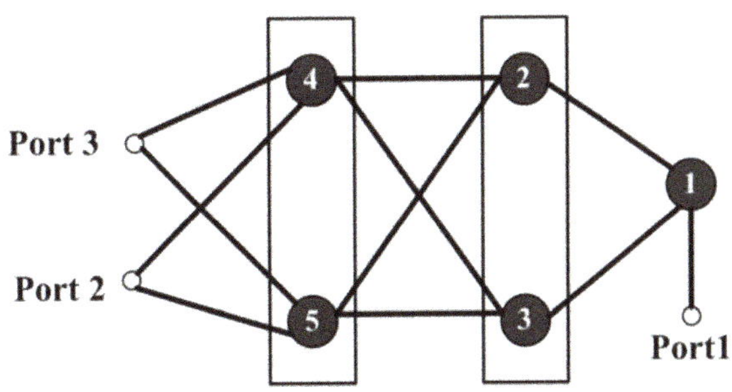

Figure 6.39 Coupling topology of the balun filter.

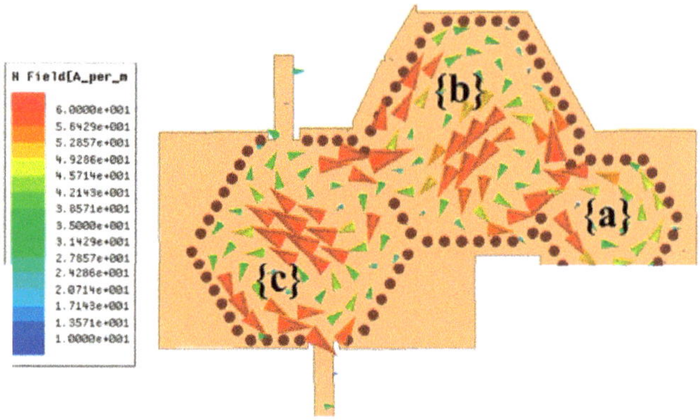

(a)

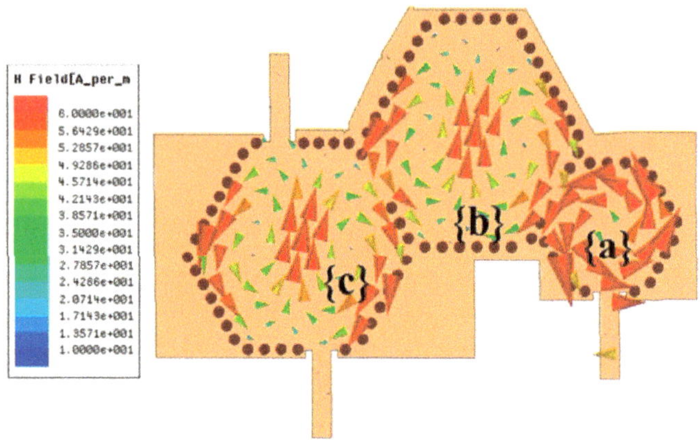

(b)

Figure 6.40 Magnetic field distribution in the balun filter for two orthogonal modes: (a) Mode 1, and (b) Mode 2.

$$M = \begin{bmatrix} & P_1 & P_2 & P_3 & 1 & 2 & 3 & 4 & 5 \\ P_1 & 0 & 0 & 0 & 1.041 & 0 & 0 & 0 & 0 \\ P_2 & 0 & 0 & 0 & 0 & 0 & 0 & -0.354 & 0.642 \\ P_3 & 0 & 0 & 0 & 0 & 0 & 0 & 0.354 & -0.642 \\ 1 & 1.041 & 0 & 0 & -0.013 & 0.454 & 0.755 & 0 & 0 \\ 2 & 0 & 0 & 0 & 0.454 & -0.928 & 0 & 0.1 & -0.391 \\ 3 & 0 & 0 & 0 & 0.755 & 0 & 0.314 & 0.1 & 0.573 \\ 4 & 0 & -0.354 & 0.354 & 0 & 0.1 & 0.1 & 1.148 & 0 \\ 5 & 0 & 0.642 & -0.642 & 0 & 0-0.391 & 0.573 & 0 & -0.371 \end{bmatrix} \quad (6.32)$$

- $f_{01} = 9.7103$ GHz, $f_{02} = 9.9231$ GHz, $f_{03} = 9.6353$ GHz, $f_{04} = 9.4468$ GHz, and $f_{05} = 9.7930$ GHz.
- $Q_e(P_1, 1) = 19.47$, $Q_e(P_2, 4) = Q_e(P_3, 4) = 168.4$, $Q_e(P_2, 5) = Q_e(P_3, 5) = 51.2$.
- $K_{1,2} = 0.0215$, $K_{1,3} = 0.0358$, $K_{2,4} = 0.0047$, $K_{2,5} = -0.0185$, $K_{3,4} = 0.0047$, and $K_{3,5} = 0.0272$.

It is observed from the coupling matrix that coupling between the output port (Port 2) with resonating nodes (4 and 5) is of the opposite polarity with respect to its counterpart.

Simulated and theoretical responses of the balun filter are provided in Figure 6.41(a). In Figure 6.41(b), the measured and simulated responses are presented. The measured results are showing good agreement with simulated and theoretical results. This filter is fabricated using Rogers RT/Duroid 5880 with relative permittivity, thickness, and loss tangent 2.2, 0.787 mm, and 0.0009, respectively. The measured passband insertion is 4.7 dB, whereas, in simulation, it is 4 dB. In Figure 6.42, measured amplitude and phase difference at the balance port are depicted. The measured amplitude imbalance in the passband is below 0.25 dB and phase difference is below 3° with the reference of 180°. The measured group delay and fabricated prototype of the balun filter are provided in Figures 6.43 and 6.44, respectively.

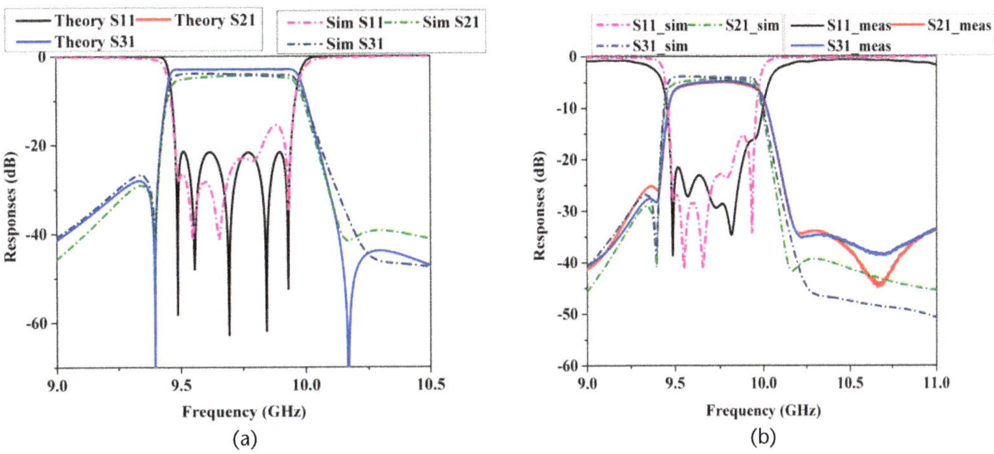

Figure 6.41 (a) Theoretical and simulated S-parameters' response of the balun filter. (b) Measured and simulated S-parameters of the balun filter.

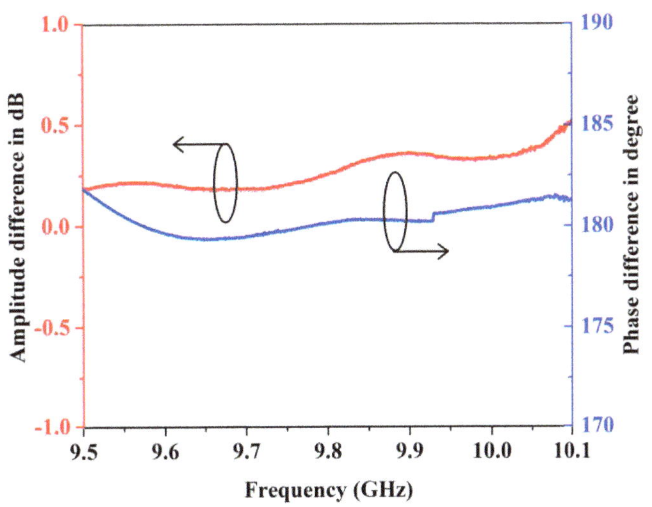

Figure 6.42 Measured amplitude and phase difference between two balanced output ports.

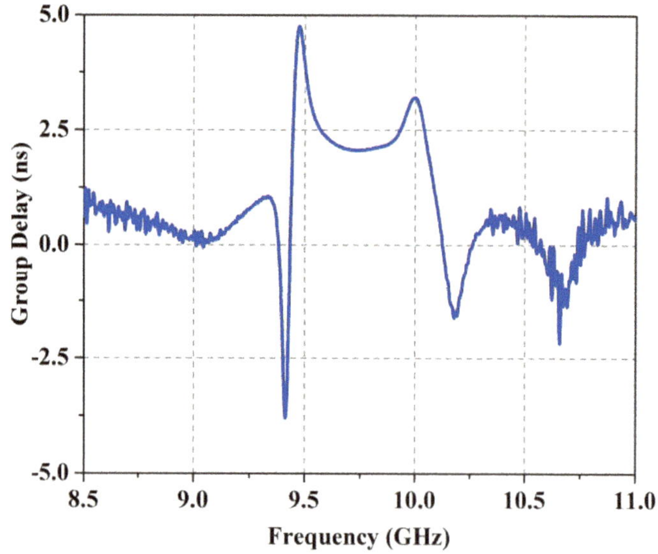

Figure 6.43 Measured group delay of the highly selective balun bandpass filter.

6.4.2 Balun Diplexer Design Using a Single Dual-Mode SIHC at Both Rx and Tx Channels (Design II)

The schematic of the balun diplexer is provided in Figure 6.45 along with its physical dimensions. The concerning coupling topology is also given in Figure 6.46. This diplexer has two channels. One is called Rx, attached to {*b*}, and another is Tx, attached to {*c*}. Each channel has a single dual-mode hexagonal cavity. As a consequence of that, both channels are acting as a two-pole filter with one transmission zero. The perturbation along the side wall of the hexagonal cavity determines the position of the transmission zero. In the design, the input common junction is a single-mode hexagonal cavity that is attached to a single-ended unbalanced Port P1,

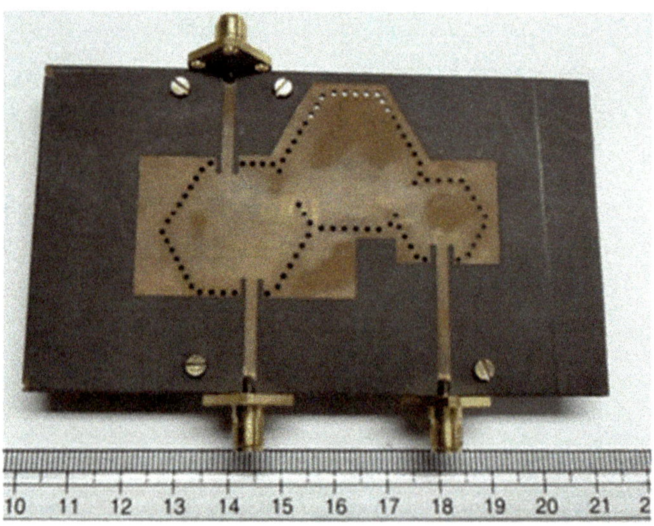

Figure 6.44 Fabricated prototype of the balun filter.

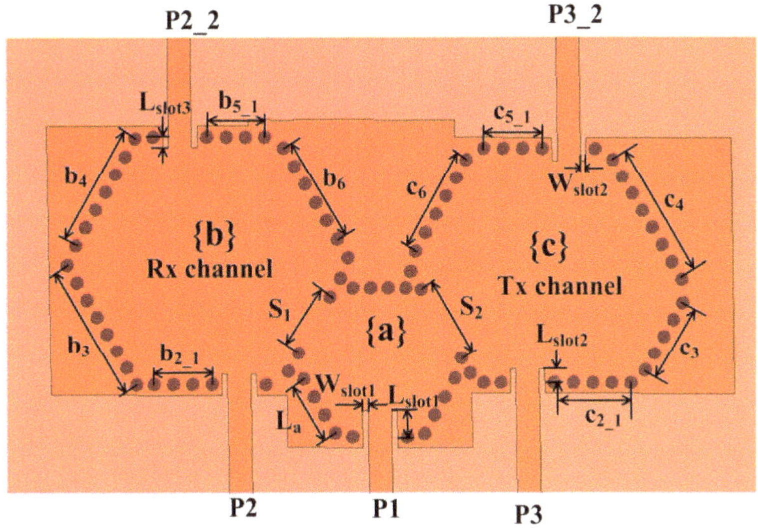

Figure 6.45 Schematic of the balun diplexer using a single dual-mode SIHC. The physical dimensions of the design are given here: L_a = 9.31 mm, S_1 = 6.575 mm, S_2 = 8.075 mm, W_{slot1} = 0.6 mm, L_{slot1} = 2.6 mm, L_{slot2} = 1.35 mm, L_{slot3} = 1.1 mm, b_{2_1} = 6.0 mm, b_3 = 14 mm, b_4 = 12.6 mm, b_{5_1} = 6.0 mm, b_6 = 12.9 mm, c_{2_1} = 8.0 mm, c_3 = 8.0 mm, c_4 = 14.0 mm, c_{5_1} = 6.0 mm, and c_6 = 10.75 mm.

whereas Rx and Tx channels are balanced. The Rx channel is attached to two balanced Ports P2 and P2_2. Similarly, the Tx channel has two balanced Ports P3 and P3_2. In Figure 6.46, resonator 1 represents single mode resonator {*a*} and 2, 3, 4, and 5 correspond to dual-mode of resonators {*b*} and {*c*} in the schematic diagram.

The design specifications of the balun diplexer are as follows. The center frequency of diplexer is at 9.6 GHz and the FBW is 5.35%.

- *Rx channel:* The passband limits 9.37 to 9.44 GHz, return loss is 25 dB, and the design specified transmission zero is at 9.58 GHz, which is above the Rx passband.
- *Tx channel:* The passband limits 9.8 to 9.89 GHz, the return loss 25 dB, and the Tx channel's design specified transmission zero is at 9.61 GHz, which is below the Tx passband.

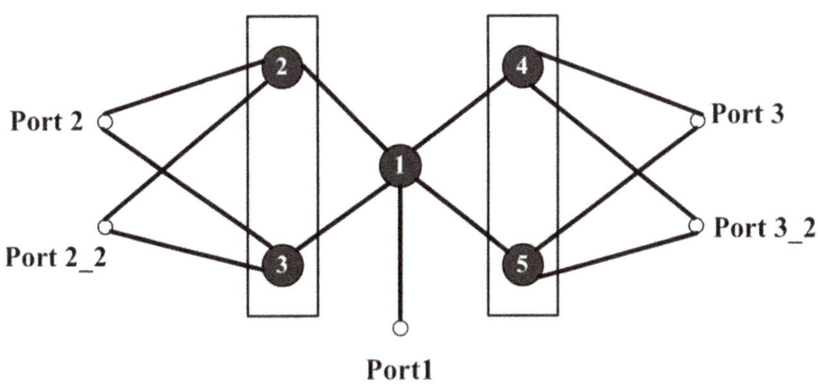

Figure 6.46 Coupling topology of Design II.

Both channels specified transmission zeros that are placed towards each other. Therefore, it improves the channel's selectivity.

Based on the design specifications, the diplexer synthesis procedure is carried out. The diplexer synthesis procedure is already discussed in great detail in this chapter. After finishing the synthesis procedure, the filter optimization technique is followed using the coupling topology of the balun diplexer. The coupling matrix is provided in (6.33). From the coupling matrix, the self-resonant frequencies (f_{0i}), external quality factor ($Q_e(P_i, {}_j)$), and coupling coefficients ($_{Kij}$) are calculated.

$$M = \begin{bmatrix} & P_1 & P_2 & P_{2\_2} & P_3 & P_{3\_2} & 1 & 2 & 3 & 4 & 5 \\ P_1 & 0 & 0 & 0 & 0 & 0 & 0.968 & 0 & 0 & 0 & 0 \\ P_2 & 0 & 0 & 0 & 0 & 0 & 0 & -0.334 & 0.136 & 0 & 0 \\ P_{2\_2} & 0 & 0 & 0 & 0 & 0 & 0 & 0.334 & 0.136 & 0 & 0 \\ P_3 & 0 & 0 & 0 & 0 & 0 & 0 & 0 & 0 & -0.384 & 0.172 \\ P_{3\_2} & 0 & 0 & 0 & 0 & 0 & 0 & 0 & 0 & 0.384 & -0.172 \\ 1 & 0.968 & 0 & 0 & 0 & 0 & -0.081 & 0.454 & 0.429 & 0.514 & 0.43 \\ 2 & 0 & -0.334 & 0 & 0.334 & 0 & 0.454 & 0.961 & 0 & 0 & 0 \\ 3 & 0 & 0.136 & 0 & -0.137 & 0 & 0.429 & 0 & 0.47 & 0 & 0 \\ 4 & 0 & 0 & -0.384 & 0 & 0.384 & 0.514 & 0 & 0 & -0.977 & 0 \\ 5 & 0 & 0 & 0.172 & 0 & -0.172 & 0.43 & 0 & 0 & 0 & -0.338 \end{bmatrix} \tag{6.33}$$

- $f_{01} = 9.64$ GHz, $f_{02} = 9.3798$ GHz, $f_{03} = 9.5035$ GHz, $f_{04} = 9.8779$ GHz, and $f_{05} = 9.7108$ GHz.
- $Q_e(P_1, 1) = 20.0$, $Q_e(P_2, 2) = Q_e(P_{2\,2}, 2) = 167.8$, $Q_e(P_2, 3) = Q_e(P_{2\,2}, 3) = 1{,}012.3$, $Q_e(P_3, 4) = Q_e(P_{3\,2}, 4) = 127.0$, and $Q_e(P_3, 5) = Q_e(P_{3\,2}, 5) = 632.9$.
- $K_{1,2} = 0.0242$, $K_{1,3} = 0.0229$, $K_{1,4} = 0.0275$, and $K_{1,5} = 0.0230$.

In the Rx channel, perturbation is applied through the side wall b_3 in an inward direction. As a consequence, the even-mode frequency is higher than the odd-mode frequency and the transmission zero appears above the passband. Similarly, in the Tx channel, perturbation is applied in an outward direction using the side wall c_3. As a result, the even-mode frequency is lower than the odd-mode frequency and the transmission zero is below the Tx passband. The modal characteristics of a dual-mode SIHC with inset variations are discussed in Chapter 5 and in Section 6.1, where the perturbation techniques applied to an SIHC-based bandpass filter and diplexer are analyzed. Using perturbation, the desired self-resonating frequencies are obtained. Besides that, by choosing a proper feeding position, coupling slot lengths and width ensure balanced output and desired external coupling coefficients. The specified coupling coefficients between two asynchronous resonators are calculated using (6.29). The calculated coupling coefficients determine the iris width S_1 and S_2. Figure 6.47(a, b) exhibit the theoretical, simulated, and measured responses of this design. The measured isolation between the output channels is depicted in Figure 6.48, and it is below 20 dB. Figure 6.49(a, b) exhibit phase and amplitude differences between balanced output port at the Rx and Tx channels, respectively.

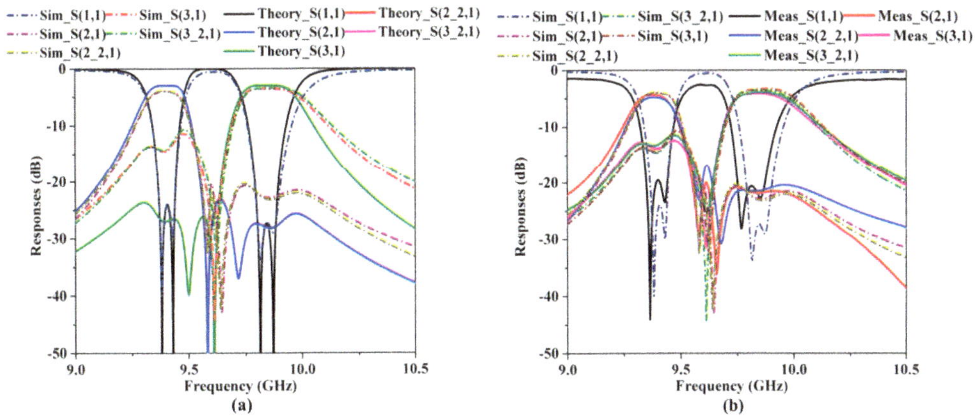

Figure 6.47 (a) Theoretical and simulated S-parameter response of the Design-II balun diplexer. (b) Measured and simulated S-parameters of the Design-II balun diplexer.

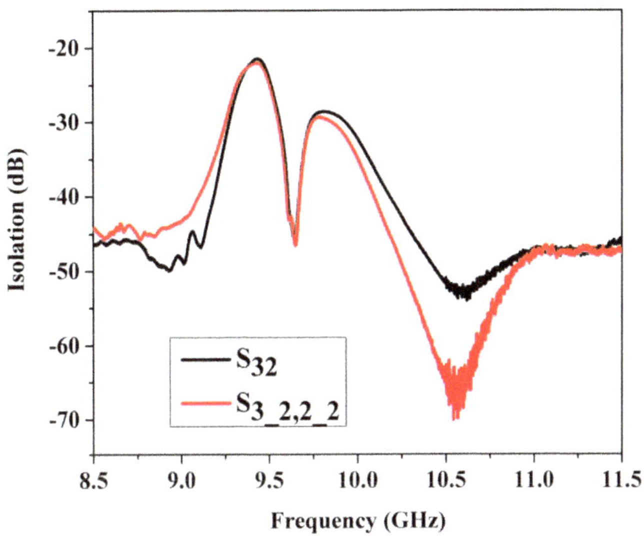

Figure 6.48 Measured isolations between the Rx and Tx channels.

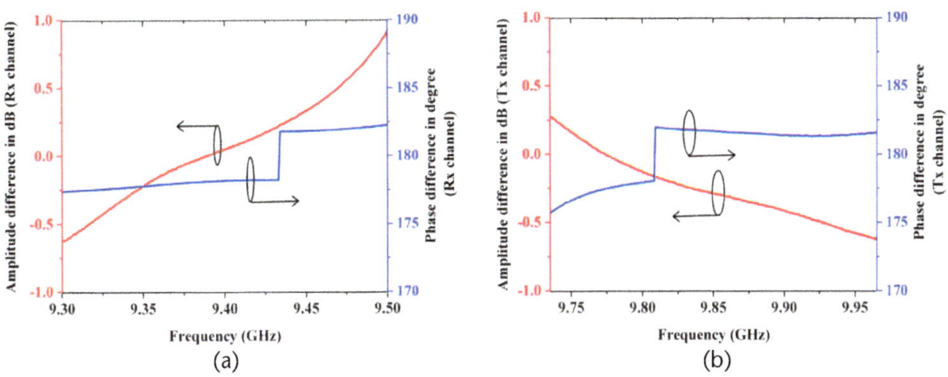

Figure 6.49 Measured amplitude and phase difference between balanced port: (a) Rx channel and (b) Tx channel.

It is observed that both Rx (P2 and P2_2) and Tx (P3 and P3_2) output ports are showing balun characteristics throughout their respective passbands. The fabricated prototype of the Design II (balun diplexer) is shown in Figure 6.50.

6.4.3 Balun Diplexer Design Using a Pair of Dual-Mode SIHC at Both Rx and Tx Channels (Design III)

The schematic of the balun diplexer is provided in Figure 6.51 along with its physical dimensions. The concerning coupling topology is also given in Figure 6.52 with similar nomenclatures of the resonators as presented in earlier designs. In this, the input common junction is a single-mode SIHC. The Rx and Tx channels have a pair of dual-mode SIHC. Therefore, both Rx and Tx channels have fourth-order bandpass filter. The balun diplexer has a single-ended input with two balanced outputs. The design specifications of the balun diplexer are provided here:

- *Design specifications:* The center frequency of the balun diplexer is at 9.50 GHz. The FBW is 8.25%.
- *Rx band:* Passband 9.1 to 9.45 GHz; return loss: 20 dB. Number of poles: 4; transmission zeros: 9.0 and 9.6 GHz.
- *Tx band:* Passband 9.65 to 9.95 GHz; return loss: 20 dB. Number of poles: 4; transmission zeros: 9.46 GHz and 10.0 GHz.

In both Rx and Tx channels, transmission zeros are kept at both sides of the passband. Therefore, it improves the selectivity of both Rx and Tx filters. Based on design specifications and coupling topology, the coupling matrix is obtained and it is provided in (6.34).

$$M = \begin{bmatrix} & P_1 & P_2 & P_{2\_2} & P_3 & P_{3\_2} & 1 & 2 & 3 & 4 & 5 & 6 & 7 & 8 & 9 \\ P_1 & 0 & 0 & 0 & 0 & 0 & 1.515 & 0 & 0 & 0 & 0 & 0 & 0 & 0 & 0 \\ P_2 & 0 & 0 & 0 & 0 & 0 & 0 & 0 & 0 & -0.355 & 0.327 & 0 & 0 & 0 & 0 \\ P_{2\_2} & 0 & 0 & 0 & 0 & 0 & 0 & 0 & 0 & 0.355 & -0.327 & 0 & 0 & 0 & 0 \\ P_3 & 0 & 0 & 0 & 0 & 0 & 0 & 0 & 0 & 0 & 0 & 0 & 0 & -0.464 & 0.05 \\ P_{3\_2} & 0 & 0 & 0 & 0 & 0 & 0 & 0 & 0 & 0 & 0 & 0 & 0 & 0.464 & -0.05 \\ 1 & 1.515 & 0 & 0 & 0 & 0 & 0.177 & 0.232 & 0.865 & 0 & 0 & 0.1 & 0.85 & 0 & 0 \\ 2 & 0 & 0 & 0 & 0 & 0 & 0.232 & 0.587 & 0 & 0.108 & -0.46 & 0 & 0 & 0 & 0 \\ 3 & 0 & 0 & 0 & 0 & 0 & 0.865 & 0 & 0.614 & 0.278 & 0.279 & 0 & 0 & 0 & 0 \\ 4 & 0 & -0.355 & 0.355 & 0 & 0 & 0 & 0.108 & 0.278 & 0.575 & 0 & 0 & 0 & 0 & 0 \\ 5 & 0 & 0.327 & -0.327 & 0 & 0 & 0 & -0.46 & 0.279 & 0 & 0.518 & 0 & 0 & 0 & 0 \\ 6 & 0 & 0 & 0 & 0 & 0 & 0.1 & 0 & 0 & 0 & 0 & -0.577 & 0 & 0.327 & -0.322 \\ 7 & 0 & 0 & 0 & 0 & 0 & 0.85 & 0 & 0 & 0 & 0 & 0 & -0.592 & 0.092 & 0.322 \\ 8 & 0 & 0 & 0 & -0.464 & 0.464 & 0 & 0 & 0 & 0 & 0 & 0.327 & 0.092 & -0.579 & 0 \\ 9 & 0 & 0 & 0 & 0.05 & -0.05 & 0 & 0 & 0 & 0 & 0 & -0.322 & 0.322 & 0 & -0.585 \end{bmatrix} \tag{6.34}$$

From the coupling matrix, the self-resonant frequencies (f_{0i}), external quality factor ($Q_e(P_i, j)$), and coupling coefficients (K_{ij}) are calculated. The common single-mode resonator's ({a}) self-resonant frequency (f_{01}) = 9.45 GHz.

The self-resonant frequencies of other resonators are provided here: f_{02} = 9.2911 GHz, f_{03} = 9.2807 GHz, f_{04} = 9.2957 GHz, f_{05} = 9.3175 GHz, f_{06} = 9.7474 GHz,

Figure 6.50 Fabricated prototype of the Design-II balun diplexer.

$f_{07} = 9.7535$ GHz, $f_{08} = 9.7482$ GHz, and $f_{09} = 9.7506$ GHz; $Q_e(P_1, 1) = 5.3$, $Q_e(P_2, 4) = Q_e(P_{2\;2}, 4) = 96.3$, $Q_e(P_2, 5) = Q_e(P_{2\;2}, 5) = 113.5$, $Q_e(P_3, 8) = Q_e(P_{3\;2}, 8) = 56.370$, $Q_e(P_3, 9) = Q_e(P_{3\;2}, 9) = 4{,}854.5$; and $K_{1,2} = 0.0191$, $K_{1,3} = 0.0713$, $K_{1,6} = 0.0082$, $K_{1,7} = 0.0700$, $K_{2,4} = 0.0089$, $K_{2,5} = -0.0379$, $K_{3,4} = 0.0229$, $K_{3,5} = 0.0230$, $K_{6,8} = 0.0269$, $K_{6,9} = -0.0274$, $K_{7,8} = 0.0076$, and $K_{7,9} = 0.0265$.

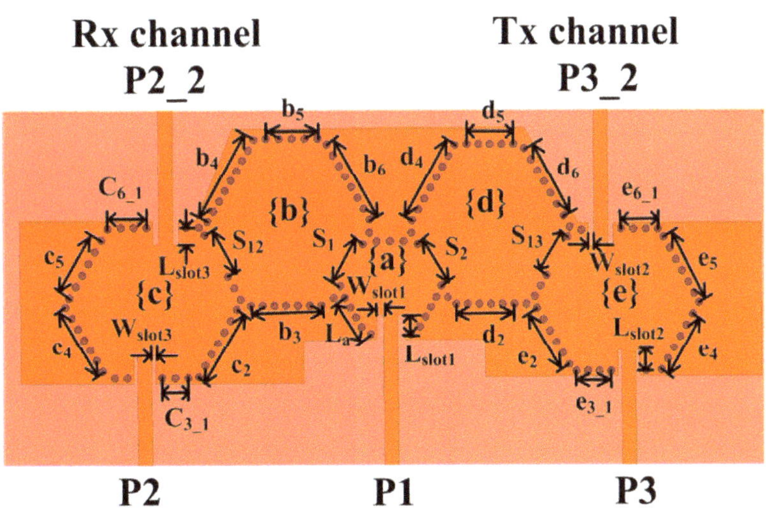

Figure 6.51 Schematic of a balun diplexer using a pair of dual-mode SIHCs. The dimensions of the design are given here: $L_a = 9.44$ mm, $S_1 = 7.9$ mm, $S_2 = 7.9$ mm, $S_{12} = 7.88$ mm, $S_{13} = 7.3$ mm, $b_3 = 12.0$ mm, $b_4 = 16.0$ mm, $b_5 = 8.8$ mm, $b_6 = 14.7$ mm, $c_2 = 13.2$ mm, $c_{3\_1} = 4.4$ mm, $c_4 = 12.6$ mm, $c_5 = 11.0$ mm, $c_{6\_1} = 6.6$ mm, $d_2 = 10.0$ mm, $d_4 = 14.7$ mm, $d_5 = 8.0$ mm, $d_6 = 13.2$ mm, $e_2 = 11.0$ mm, $e_{3\_1} = 6.0$ mm, $e_4 = 11.0$ mm, $e_5 = 12.5$ mm, $e_{6\_1} = 6.3$ mm, $L_{slot1} = 3.2$ mm, $L_{slot2} = 3.5$ mm, $L_{slot3} = 2.9$ mm, $W_{slot1} = 1.23$ mm, $W_{slot2} = 1.0$ mm, and $W_{slot3} = 0.7$ mm.

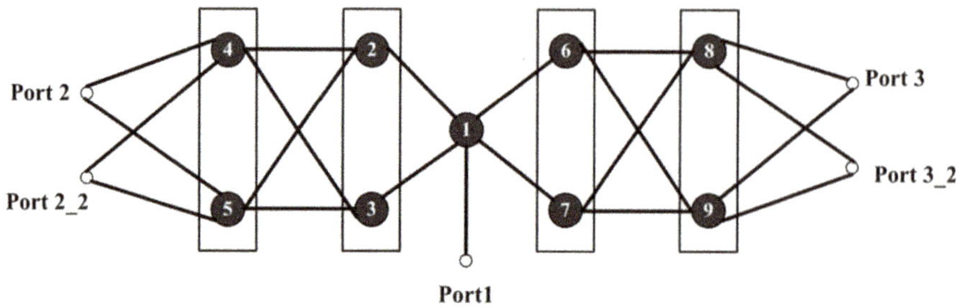

Figure 6.52 Coupling topology of Design III.

In the Rx channel, the side walls b_5 of the dual-mode cavity {b} and c_5 of {c} are used for perturbation. Similarly, in the Tx channel, the side walls d_5 of the dual-mode cavity {d}, and e_5 of {e} are used for perturbation. This helps to obtain desired self-resonant frequencies of the cavities. Proper feeding positions and coupling slot lengths and widths ensure both the output ports to be balanced. Figure 6.53(a, b) show the theoretical, simulated, and measured responses of the balun diplexer. In the Rx passband, the measured and simulated insertion losses are 4.25 dB and 3.45 dB, respectively. In the Tx passband, the measured and simulated insertion losses are 4.1 dB and 3.5 dB, respectively.

The measured and simulated isolation between both the channels are depicted in Figure 6.54 where it shows that the isolation is below 25 dB. Figure 6.55(a, b) exhibit phase and amplitude difference between the balanced output ports at the Rx and Tx channels, respectively. It is observed that both Rx (P2 and P2_2) and Tx (P3 and P3_2) output ports are showing balun characteristics throughout their respective passbands. The measured group delay of Design III is shown in Figure 6.56, which shows constant group delay in the Rx and Tx passbands. The fabricated prototype of Design III (balun diplexer) is shown in Figure 6.57.

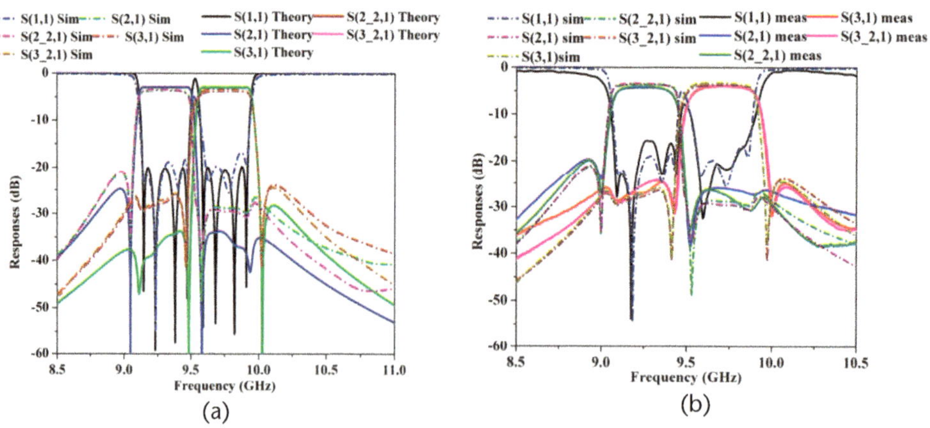

Figure 6.53 (a) Theoretical and simulated S-parameters of the Design-III balun diplexer. (b) Measured and simulated S-parameters of the Design-III balun diplexer.

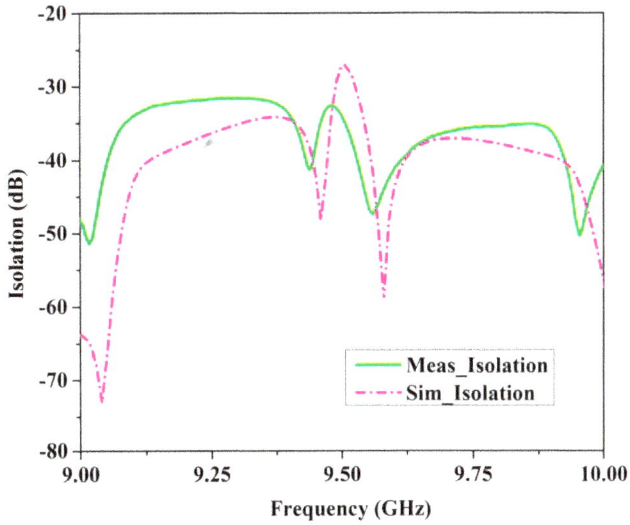

Figure 6.54 Measured isolations between the Rx and Tx channels of the Design-III balun diplexer.

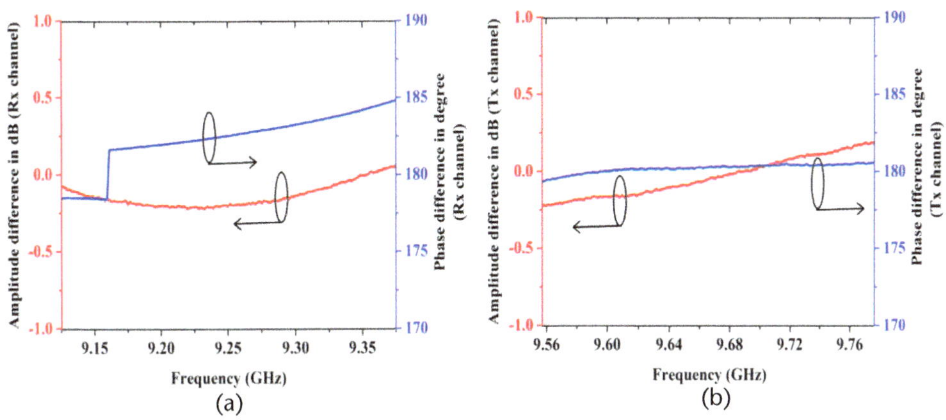

Figure 6.55 Measured amplitude and phase difference between balanced ports: (a) Rx channel, and (b) Tx channel.

6.5 SIW-Based Hybrid Branch Line Coupler

The hybrid branch line coupler plays a crucial role in implementing the mixer and balanced amplifier. It is used to achieve better return loss and spurious signal rejection. In the mixer, hybrid branch line coupler is employed to design a double-balanced mixer circuit, which removes the image IF signal. Until now, few papers have reported on the HMSIW-based coupler design. The advantage of the HMSIW over SIW for a designing coupler is that it miniaturizes the overall SIW-based coupler size and enhances the bandwidth.

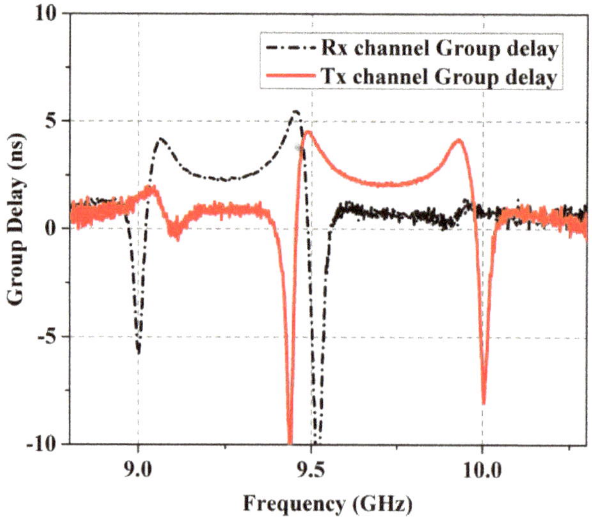

Figure 6.56 Measured group delay of the Design-III balun diplexer.

In this section, a novel HMSIW-based branch-line coupler is realized using planar PCB technology. The design principle is based on traditional hybrid branch line coupler, which is a 4-port network providing equal amplitude and quadrature or 90° phase outputs at the desired frequency band. Another port acts as an isolated port. It has 17.0% bandwidth. In band, the maximum amplitude imbalance is 1.25 dB, whereas the phase difference between outputs is 86° to 93°. The schematic of the designed half-mode branch-line coupler (HMBLC) is provided in Figure 6.58 along with its dimensions [12]. Multiple small H-plane steps are used along vertical side wall (a), which varies the width of the HMBLC along the longitudinal direction. This improves impedance matching as well as the bandwidth of the hybrid coupler.

To control the coupling and bandwidth, four vias are placed at a_{offset} and b_{offset} of the diameter d_1, as shown in the schematic. Figure 6.59 shows that, changing the positions of vias (a_{offset} and b_{offset}) along the horizontal and vertical side walls, it affects the power coupling between all four ports. Figure 6.59(a, b) plot the simulated

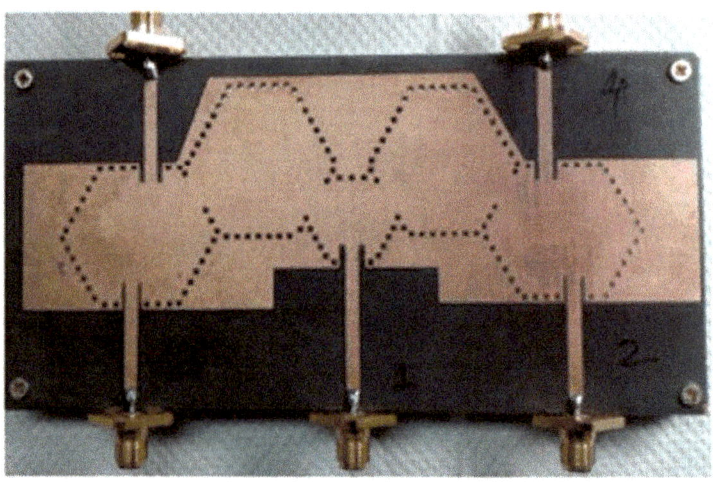

Figure 6.57 Fabricated prototype of the designed Design-III balun diplexer.

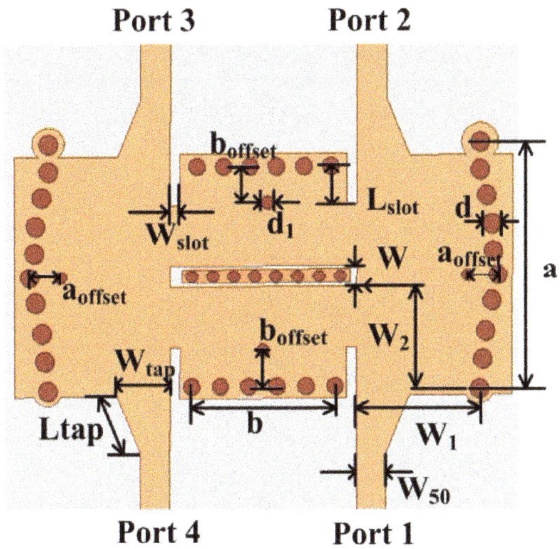

Figure 6.58 Schematic of the HMSIW-based hybrid branch line coupler. The dimensions are b = 11.5 mm, a = 20 mm, L_{tap} = 4.9 mm, W_{tap} = 4.3 mm, a_{offset} = 2.5 mm, b_{offset} = 2.9 mm, d = 1.5 mm, d_1 = 1.0 mm, W = 1.8 mm, W_1 = 9.9 mm, W_2 = 8.3 mm, W_{50} = 2.4 mm, W_{slot} = 0.8 mm, and L_{slot} = 3.1 mm. (© 2019 IEEE [12].)

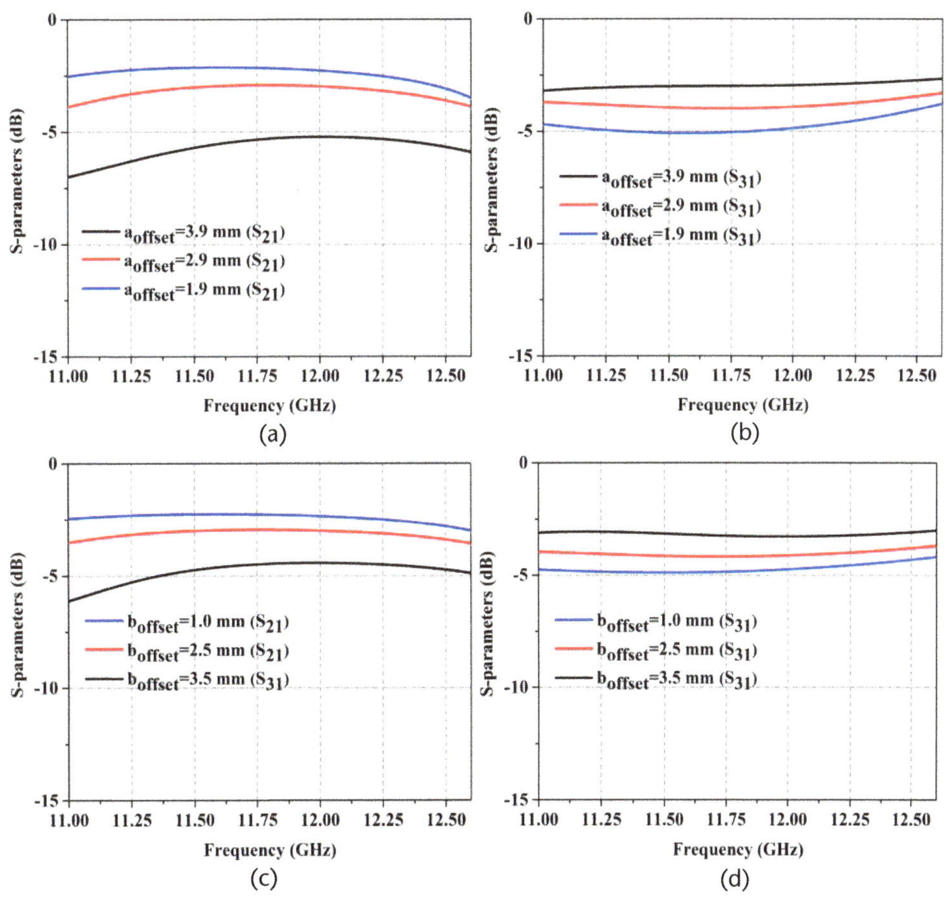

Figure 6.59 Variation of S-parameters with changes in via positions: (a) simulated S_{21} and (b) simulated S_{31} with a_{offset} variations and (c) simulated S_{21} and (d) simulated S_{31} with b_{offset} variations. (© 2019 IEEE [12].)

S_{21} and S_{31} with a_{offset} variations. Here, a_{offset} is changed from 1.9 to 3.9 mm. With increasing the length of a_{offset}, S_{21} decreases, whereas S_{31} increases. Similarly, Figure 6.59(c, d) exhibit the effect of b_{offset} variation on S_{21} and S_{31} of the HMBLC. b_{offset} is varied from 1.0 mm to 3.5 mm. With the increase of b_{offset}, S_{21} decreases, whereas S_{31} increases. By proper adjusting of a_{offset} and b_{offset}, equal power can be distributed to Port 2 and Port 3 with respect to Port 1.

Figure 6.60 shows the measured and simulated responses of the designed HMBLC. The operating bandwidth of the HMSIW extends from 10.8 to 12.8 GHz. The return loss is below 16 dB throughout the operational band. The measured maximum amplitude and phase differences at the output ports are −1.25 dB and 86° to 93°, respectively as shown in Figure 6.61. The fabricated prototype of the HMBLC is provided in Figure 6.62.

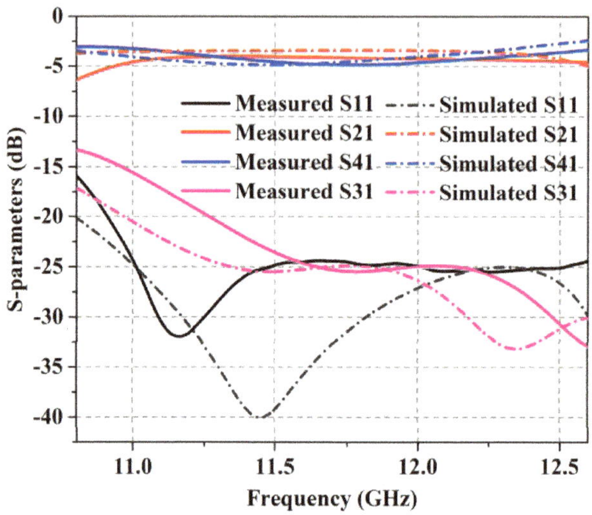

Figure 6.60 Simulated and measured responses of the half-mode hybrid BLC. (© 2019 IEEE [12].)

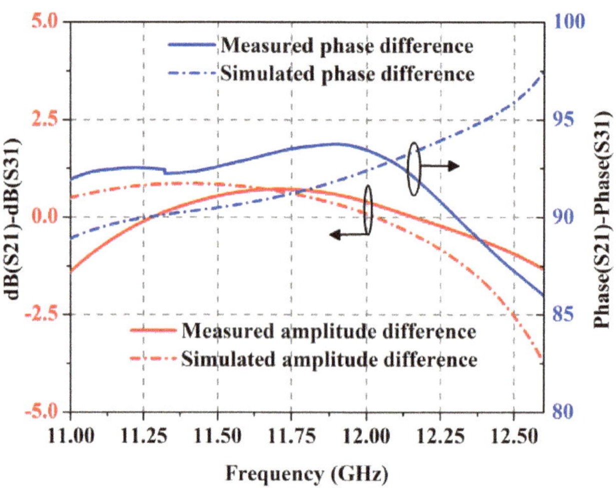

Figure 6.61 Measured and simulated phase difference and amplitude imbalance of output ports. (© 2019 IEEE [12].)

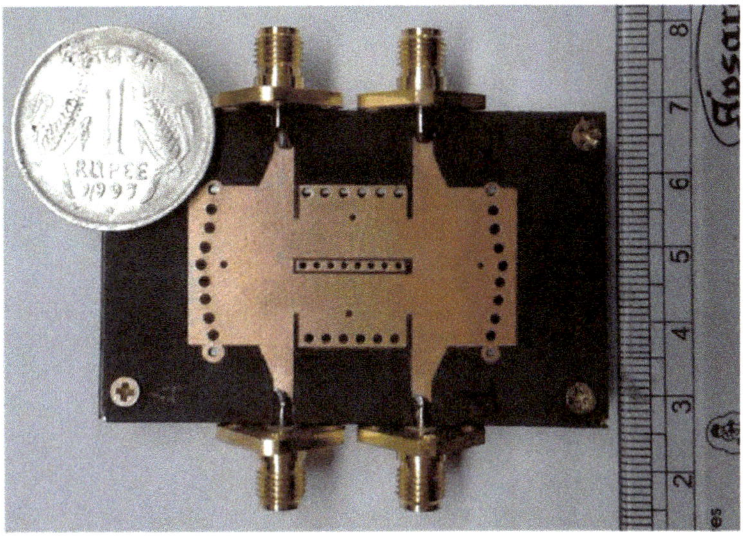

Figure 6.62 Fabricated prototype of the HMSIW-based hybrid 3-dB branch line coupler. (© 2019 IEEE [12].)

6.6 Conclusion

In this chapter, the SIW-based hexagonal cavity resonators are used to realize various SIW based passive circuit such as a diplexer, triplexer, balun filter, balun diplexer, and hybrid branch line coupler. The perturbation technique in the SIHC is utilized extensively to achieve the desired responses in each passband of the multiplexer network. First, dual-mode SIHC diplexers are designed and fabricated. Diplexers based on single or two dual-mode SIHCs in Rx and Tx channels have been discussed. Varying perturbation through the side walls of the SIHC and changing feeding positions help to realize different passband responses, improved selectivity and isolation between Rx and Tx channels without changing coupling topology that leads to design flexibility. Beside this, the input port is a single-mode resonator; hence, it also provides a greater degree of freedom to design diplexers. The designs are further extended by using a varactor diode-based tunable diplexer network. The varactor diode helps to realize real-time variation of perturbation in the side wall of the SIHC cavity, which helps to shift the transmission zero in the response of the diplexer. The design exhibits a wide range of transmission zero tuning. Similarly, a hexagonal SIW cavity-based and dual-band diplexer design is also demonstrated in this chapter.

A novel technique to realize an SIW-based planar balun filter is demonstrated in this chapter. The design uses SIW cavity-based higher-order modes to obtain out-of-phase performance of the output ports of the balun filter. The technique of the side-wall perturbation of the SIW hexagonal cavity is efficiently utilized to implement desired response in the passband. The concept is further extended to implement SIW-based diplexer balun characteristics with a balanced output of lower and upper bands. A similar perturbation technique was followed to achieve the desired response of the diplexer.

References

[1] Tang, H. J., et al., "Development of Millimeter-Wave Planar Diplexers Based on Complementary Characters of Dual-Mode Substrate Integrated Waveguide Filters with Circular and Elliptic Cavities," *IEEE Transactions on Microwave Theory and Techniques*, Vol. 55, No. 4, April 2007, pp. 776–782.

[2] Macchiarella, G., and S. Tamiazzo, "Novel Approach to the Synthesis of Microwave Diplexers," *IEEE Transactions on Microwave Theory and Techniques*, Vol. 54, No. 12, December 2006, pp. 4281–4290.

[3] Wang, R., et al., "Synthesis of Microwave Resonator Diplexers Using Linear Frequency Transformation and Optimization," *Progress in Electromagnetics Research*, Vol. 124, 2012, pp. 441–455.

[4] Chongder, P., K. V. Srivastava, and A. Biswas, "Realisation of Controllable Transmission Zeros by Perturbation Technique for Designing Dual-Mode Filter Using Substrate Integrated Hexagonal Cavity," *IET Microwaves, Antennas Propagation*, Vol. 8, No. 6, April 2014, pp. 451–457.

[5] Chongder, P., and A. Biswas, "Flexible Design Procedure for Realisation of Dual-Mode Substrate Integrated Hexagonal-Cavity-Based Diplexer," *IET Microw. Antennas Propag.*, Vol. 11, No. 14, November 2017, pp. 2083–2090.

[6] Hong, J., and M. Lancaster, *Microstrip Filters for RF/Microwave Applications*, Second Edition, New York: Wiley, 2011.

[7] Amari, S., U. Rosenberg, and J. Bornemann, "Adaptive Synthesis and Design of Resonator Filters with Source/Load-Multiresonator Coupling," *IEEE Transactions on Microwave Theory and Techniques*, Vol. 50, No. 8, August 2002, pp. 1969–1978.

[8] Macchiarella, G., and S. Tamiazzo, "Synthesis of Star-Junction Multiplexers," *IEEE Transactions on Microwave Theory and Techniques*, Vol. 58, No. 12, December 2010, pp. 3732–3741.

[9] Leong, Y. C., K. S. Ang, and C. H. Lee, "A Derivation of a Class of 3-Port Baluns from Symmetrical 4-Port Networks," *2002 IEEE MTT-S International Microwave Symposium Digest*, Vol. 2, June 2002, pp. 1165–1168.

[10] Ang, K. S., Y. C. Leong, and C. H. Lee, "Analysis and Design of Miniaturized Lumped-Distributed Impedance-Transforming Baluns," *IEEE Transactions on Microwave Theory and Techniques*, Vol. 51, No. 3, March 2003, pp. 1009–1017.

[11] Yeung, L. K., and K.-L. Wu, "An LTCC Balanced-to-Unbalanced Extracted-Pole Band-Pass Filter with Complex Load," *IEEE Transactions on Microwave Theory and Techniques*, Vol. 54, No. 4, June 2006, pp. 1512–1518.

[12] Chongder, P., and A. Biswas, "Novel Design of Compact Half-Mode Substrate Integrated Waveguide Based Hybrid Branch-Line Coupler," *2019 IEEE MTT-S International Microwave and RF Conference (IMARC)*, Mumbai, India, 2019, pp. 1–4.

CHAPTER 7

SIW Active Circuits

In Chapter 6, various SIW-based passive components designs were discussed. In this chapter, low phase noise oscillator design and its application in designing compact receiver system, namely, SOM design, will be discussed.

7.1 Low Phase Noise Oscillator

A low phase noise oscillator is an integral part of an RF front-end transceiver system. The spectral purity is one of the essential factors that judge the performance of the oscillator. The phase noise represents the spectral purity of the oscillator. An ideal oscillator's output signal is $V_{out}(t) = A_0 \cos(\omega_0 t + \Phi_0)$. However, noise generated by active and passive components modulates the signal produced by the oscillator. The noisy oscillator output signal is expressed as:

$$V_{out}(t) = \left(A_0 + a(t)\right)\cos\left(\omega_0 t + \Phi(t)\right) \tag{7.1}$$

Here A_0 and ω_0 are oscillator's actual amplitude and frequency, whereas $a(t)$ and $\Phi(t)$ are random amplitude and phase noise, respectively. In general, thermal noise, shot noise, and flicker ($1/f$) noise sources in an oscillator contribute to amplitude and phase noise that results in random side-band noise in an oscillator. Amplitude noise can easily be controlled by adopting amplitude-restoring mechanism of oscillator. The amplitude limiter circuit can be used to remove amplitude variations caused by random noise. However, it is very difficult to remove phase noise from an oscillator response. Phase noise plays an important role in spectral purity performance of the oscillator. In absence of phase noise, the ideal oscillator output is a sine wave signal, which is a delta function in frequency spectrum as shown in Figure 7.1(a). However, in a practical oscillator, presence of phase noise spreads the frequency spectrum in the vicinity of the carrier frequency as shown in Figure 7.1(b). Phase noise is a relative quantity, which is measured as the noise power (P_n) in one side band contained in a specified bandwidth (B) at an offset frequency ($\Delta\omega$) relative to the carrier output power (P_s), as shown in Figure 7.1(b).

$$\text{Phase Noise } L(\Delta\omega) = 10\log_{10}\frac{\text{Power Spectral Density at } \left(\omega_0 + \Delta\omega\right)}{P_{\text{carrier},\omega_0}} \tag{7.2}$$

When the bandwidth is 1 Hz, the ratio is expressed as dBc/Hz.

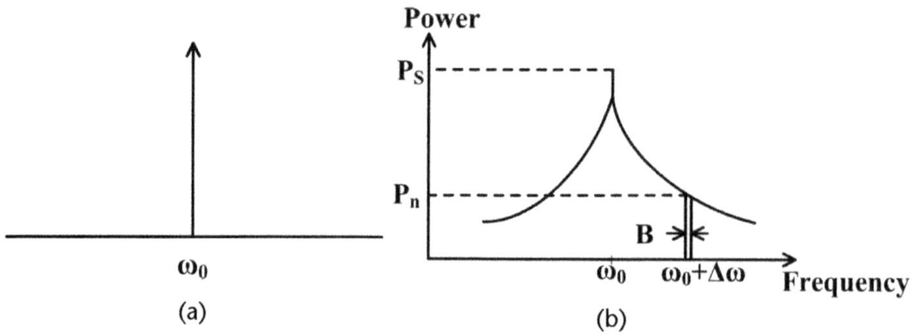

Figure 7.1 (a) Ideal signal (absence of phase noise). (b) Practical signal (with presence of phase noise).

7.1.1 Leeson's Phase Noise Model

In [1], the author described a general linear time invariant (LTI) model for phase noise in the oscillator. The model is implemented based on an amplifier with a feedback resonator as shown in Figure 7.2. When additive noise is injected into the oscillator system, it is important to understand its effect on the phase noise of the oscillator. The amplifier noise factor and gain are F and G, respectively. The amplifier noise figure is expressed in terms of the signal-to-noise ratio (S/N) as mentioned here:

$$\begin{aligned} F &= \frac{(S/N)_{\text{input}}}{(S/N)_{\text{output}}} = \frac{N_{\text{out}}}{N_{\text{in}}G} = \frac{N_{\text{out}}}{GKTB} \quad \text{where } N_{\text{in}} = KTB \\ N_{\text{out}} &= FGKTB \\ N_{\text{out(in)}} &= FKTB \end{aligned} \tag{7.3}$$

The input thermal-noise power N_{in} = kTB, where K is a Boltzmann's constant, which is 1.374×10^{-23}J/K. T is the resistor temperature in Kelvin. P_{avs} is the oscillator's output power at the input of the amplifier and the thermal noise power at the input of the amplifier is denoted by $N_{out(in)}$. The model corresponding to phase perturbation analysis of the amplifier is shown in Figure 7.3(a), and its vector

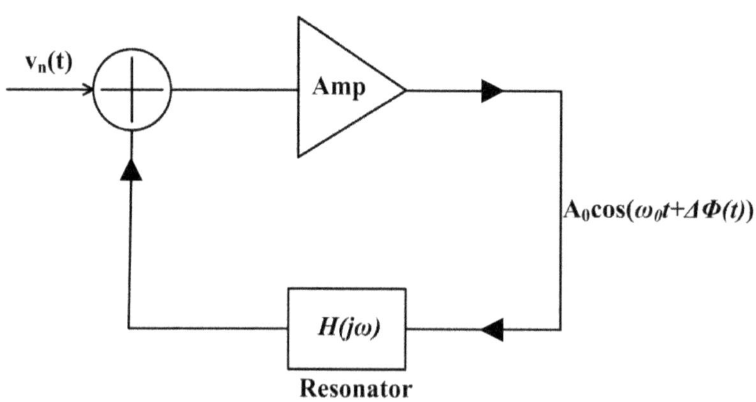

Figure 7.2 LTI model of a parallel-feedback oscillator.

representation is shown in Figure 7.3(b) where phasor form of pure sinusoidal signal rotates at ω_0 and associated additive noise rotates at $\Delta\omega$ for $i = 1$. A similar phasor diagram can be drawn for associated additive noise rotates at $-\omega_0$ by drawing the same circle but in reverse direction for $i = 2$.

The thermal noise voltage $v_{n,rms}$ at the input of the amplifier can be expressed in terms of thermal noise power at the input of the amplifier input with resistance R. Similarly, the signal voltage $v_{s,rms}$ can be expressed in terms of the oscillator's output power with the same amplifiers.

$$N_{\text{out(in)}} = \frac{v_{n,rms}^2}{R} \qquad v_{n,rms} = \sqrt{FKTBR} \tag{7.4}$$

$$P_{avs} = \frac{v_{s,rms}^2}{R} \tag{7.5}$$

$$v_{s,rms} = \sqrt{P_{avs}R} \tag{7.6}$$

The phasor diagram Figure 7.3(b) shows the phase perturbation $\Delta\theta$ generated by thermal noise v_n. The phase perturbation is maximum when v_n is perpendicular to v_s [2]. By setting $B = 1$ Hz, the input phase-noise for oscillating signal at ω_0 and additive noise at $\omega_0 \pm \Delta\omega$ are expressed as:

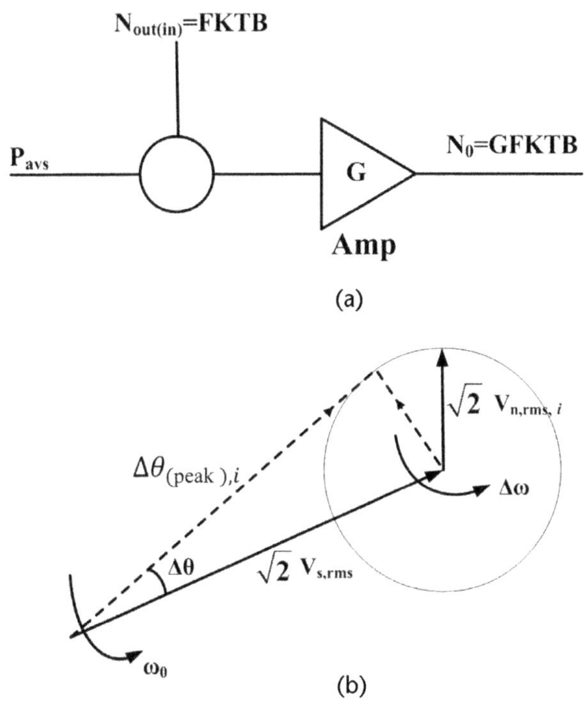

Figure 7.3 (a) Model of additive noise perturbation analysis of the amplifier. (b) Vector representation of the additive noise perturbation model (a).

$$\Delta\theta_{(\text{peak}),1} = \tan^{-1}\left(\frac{v_{n,rms,1}}{v_{s,rms}}\right) \approx \frac{|v_n|}{|v_s|} = \sqrt{\frac{FKT}{P_{avs}}}$$
$$\Delta\theta_{(rms),1} = \frac{1}{\sqrt{2}}\sqrt{\frac{FKT}{P_{avs}}} = \Delta\theta_{(rms),2} \quad (7.7)$$

$$\Delta\theta_{(rms)} = \sqrt{\Delta\theta^2_{(rms),1} + \Delta\theta^2_{(rms),2}} = \sqrt{\frac{FKT}{P_{avs}}} \quad (7.8)$$

$$S_{\Delta\theta}(\Delta\omega) = \left(\Delta\theta_{(rms)}\right)^2 = \frac{FKT}{P_{avs}} \quad (7.9)$$

For a practical transistor, the noise floor appears after some offset frequency $\Delta\omega$. Before the noise floor, there is modulation frequency close to the carrier. Figure 7.4(b) shows a flicker or $(\Delta\omega)^{-1}$ component. This flicker noise is empirically calculated. It can be seen that at the $\Delta\omega_{1/f}$ (flicker noise corner frequency), white noise and $(\Delta\omega)^{-1}$ noise are equal. The phase noise model consists of a noise-free amplifier and a phase modulator at the input, as shown in Figure 7.4.

The purity of the amplifier signal is affected by flicker noise at the frequencies adjacent to the carrier. Hence, the double-sideband phase noise spectral density in presence of flicker noise is modified as:

$$S_{\Delta\theta}(\Delta\omega) = \frac{FKT}{P_{avs}}\left(1 + \frac{\Delta\omega_{1/f}}{\Delta\omega}\right) \quad (7.10)$$

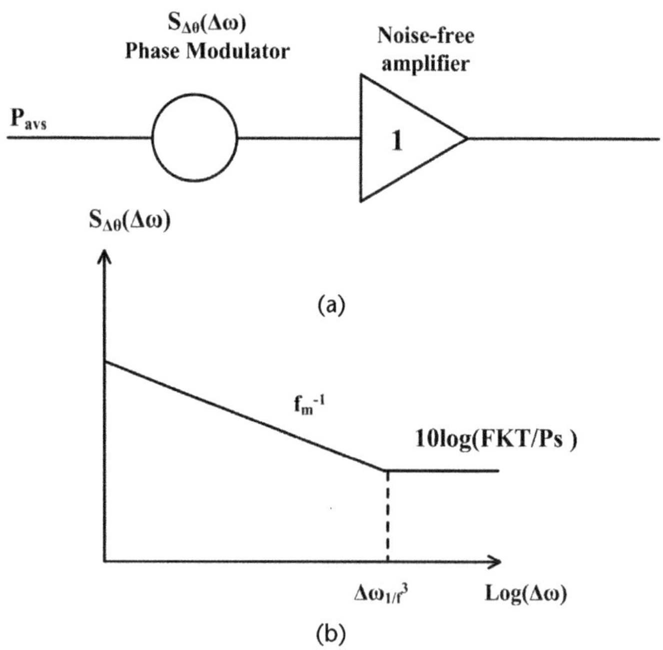

Figure 7.4 (a) Model of a noise-free amplifier with an injected phase noise and phase modulator. (b) Plot of noise spectral density with offset frequencuy.

The oscillator may be modeled as an amplifier with feedback as shown in Figure 7.5.

Based on the signal transmission theory, the phase-modulated carrier passing through bandpass filter is identical to the modulating signal passing through an equivalent lowpass filter of the oscillator model as mentioned in [3]. Figure 7.5 shows the lowpass equivalent of oscillator model to analyze the phase noise of the oscillator.

The equivalent lowpass transfer function of the single resonator is given below:

$$\hat{H}(j\Delta\omega) = \frac{1}{1 + j\left(2Q_L \Delta\omega/\omega_0\right)} \quad \text{where} \quad \frac{\omega_0}{2Q_L} = \frac{2\pi B}{2} \tag{7.11}$$

where ω_0 is the resonant frequency of the single resonator, B is bandwidth of the resonator, and Q_L is the loaded quality factor of the resonator.

The closed-loop response of the phase feedback loop as shown in Figure 7.5 is provided below:

$$\frac{\Delta\phi}{\Delta\theta} = 1 + \frac{\omega_0}{j2Q_L\Delta\omega} \tag{7.12}$$

Hence, the power spectral density of output phase noise is

$$S_{\Delta\phi}(\Delta\omega) = \left[1 + \left(\frac{\omega_0}{2Q_L\Delta\omega}\right)^2\right] S_{\Delta\theta}(\Delta\omega) \tag{7.13}$$

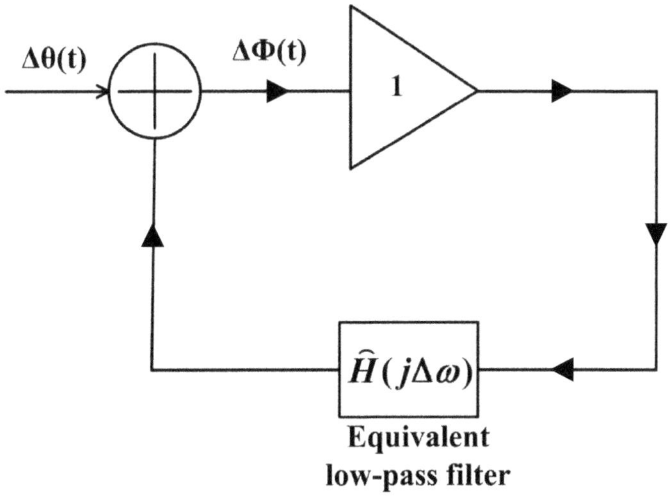

Figure 7.5 Lowpass equivalent of the oscillator's model used for phase-noise analysis.

The single sideband (SSB) power spectral density of phase noise is expressed as

$$L(\Delta\omega) = \frac{S_{\Delta\phi}(\Delta\omega)}{2} = \frac{FKT}{2P_{avs}}\left[1+\left(\frac{\omega_0}{2Q_L\Delta\omega}\right)^2\right]\left(1+\frac{\Delta\omega_{1/f}}{\Delta\omega}\right)$$
$$L(\Delta\omega) = \frac{FKT}{2P_{avs}}\left[\frac{1}{\Delta\omega^3}\frac{\omega_0^2\Delta\omega_{1/f}}{4Q_L^2} + \frac{1}{\Delta\omega^2}\left(\frac{\omega_0}{2Q_L}\right)^2 + \left(1+\frac{\Delta\omega_{1/f}}{\Delta\omega}\right)\right] \tag{7.14}$$

A typical phase noise plot is shown in Figure 7.6 where four distinct regions are visible. The initial slope is $1/f^3$, which is close to the carrier frequency. The second and third slopes are $1/f^2$ and $1/f$, respectively. The last region is horizontal, which corresponds to the thermal noise floor.

7.1.2 Low Phase Noise Oscillator Design Using the SIHC Filter

In this section, to realize low-phase noise oscillators, three different highly selective SIHC-based bandpass filters are employed in the feedback loop of the oscillator as a frequency stabilized element. These bandpass filters were already discussed in Chapter 5. Three low phase noise oscillators are designed namely, Design I, Design II, and Design III.

- *Design I:* A single triple-mode SIHC-based bandpass filter is employed as a frequency stabilized element in the feedback path of the oscillator. Using this highly selective bandpass filter, low phase noise oscillator is designed for Ku-band application.
- *Design II:* A single dual-mode SIHC-based bandpass filter is employed as a frequency stabilized element in the feedback path of the oscillator. The filter has two poles with two transmission zeros on both sides of the passband.

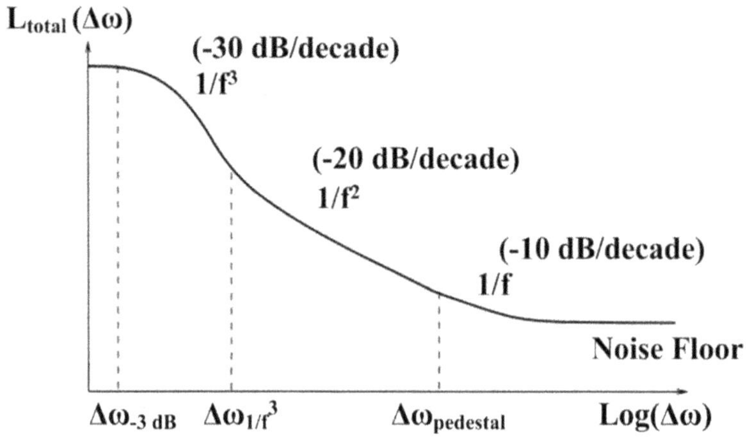

Figure 7.6 Phase noise plot using Leeson's phase noise model.

Using this highly selective bandpass filter, the low phase noise oscillator is designed for an X-band application.

- *Design III:* A pair of dual-mode SIHC-based bandpass filters is employed as a frequency stabilized element in the feedback path of the oscillator. The filter has four poles with two transmission zeros on both sides of the passband. With the help of this highly selective bandpass filter, the low phase noise oscillator is designed for the X-band application. The unloaded quality factor of a dual-mode SIHC resonator is around 500, which is better than a single-mode resonator. The external quality factors are different for all three BPFs; hence, the loaded quality factors are also different. The loaded quality factor (Q_L) can be expressed in terms of external quality factor (Q_E) and unloaded quality factor (Q_U) as (7.15):

$$\frac{1}{Q_L} = \frac{1}{Q_E} + \frac{1}{Q_U} \tag{7.15}$$

The dual-mode filters are realized using the generalized Chebyshev synthesis method, which provides the freedom to choose arbitrary transmission zeros. The existence of transmission zero ensures sharper roll-offs and better selectivity than normal Chebyshev filters that helps to design low phase noise oscillator. The inherent characteristic of a dual-mode resonator is to provide one transmission zero near the passband. Since the transmission zero is placed close to the passband edge, a large group-delay peak appears at the edge of the passband. Also, the loaded Q and group delay are directly proportional as given in (7.16), thus ensuring better phase noise performance of the oscillator.

$$\begin{aligned} Q_L &\approx \frac{\omega_0}{2}\left|\frac{d\phi}{d\omega}\right|_{\omega=\omega_0} \\ \tau_d &= \left|\frac{d\phi}{d\omega}\right|_{\omega=\omega_0} \end{aligned} \tag{7.16}$$

where ω_0 is the oscillation frequency, $\varphi(\omega)$ is the phase response of the bandpass filter, and τ_d is the group delay. Besides a loaded quality factor and group delay, insertion loss in the passband also plays crucial role in determining the phase noise performance of the oscillator. Generally, quality factor and insertion loss are both parameters that are dependent on each other as given in (7.17)

$$IL\ (\text{dB}) = 10\log\left(\frac{1}{1 - Q_L/Q_U}\right)^2 \tag{7.17}$$

Hence, the quality factor improves with weak coupling of the bandpass filter at the cost of insertion loss. However, the degradation of insertion loss of the filter reduces the oscillator's output power which affects the phase noise performance of the oscillator as the oscillator output power is inversely proportional to phase noise.

7.1.2.1 Design I

The schematic of the low phase noise oscillator is provided in Figure 7.7 where the feedback triple-mode filter is employed as a frequency stabilization element. The dimensions of the triple-mode SIHC filter were provided in Chapter 5. The higher band of the filter is exploited to realize a feedback oscillator. The substrate used to design a feedback oscillator is Rogers RT/Duroid 5880 ε_r where 2.2 = thickness = 0.787 mm, and loss tangent = 0.0009. The schematic layout of the triple-mode SIHC based feedback oscillator is shown in Figure 7.7(b) and the response of its triple mode filter in terms of S-parameters and group delay are shown in Figure 7.7(a). The group-delay peak frequency is considered as an oscillation frequency, which is 13.6 GHz. The device used here is pHEMT NE3210S01. To design an amplifier at 13.6 GHz using a common source configuration, the simultaneous conjugate matching technique is adopted. The triple-mode resonator is placed in the feedback path of the amplifier and it is connected with amplifier using a microstrip line of an electrical length of 360° or multiple of 360° at 13.6 GHz. The output signal is collected from the drain of the pHEMT device. The measured output spectrum of the oscillator is shown in Figure 7.8 with a span of 1 MHz, where the measured output power is around 1.2 dBm. The measured SSB phase noises of the oscillator is shown in Figure 7.9. The measured phase noise of the oscillator are −101.5 dBc/Hz and −126.4 dBc/Hz at 100-kHz and 1-MHz offset frequencies, respectively. The figure of merit (FOM) of the oscillator can be calculated by

$$FOM = L(\Delta f) - 20\log\left(\frac{f_0}{\Delta f}\right) + 10\log\left(\frac{P_{\mathrm{DC}}}{1\ \mathrm{mW}}\right) \tag{7.18}$$

where $L(\Delta f)$ is the phase noise at the offset frequency Δf, f_0 is the oscillation frequency, and PDC is the DC power consumption (milliwatts). The DC power consumption of the oscillator is 34 mW. The calculated FOMs of the oscillator at 1 MHz offset is −193.4 dBc/Hz. The fabricated prototype of the triple-mode SIHC filter-based feedback oscillator is shown in Figure 7.10.

7.1.2.2 Design II

Another design of the oscillator is presented here by employing dual-mode SIHC-based BPF with improved Q and selectivity as a frequency stabilization element. The details analysis of the filter was described in Chapter 5. It has two poles and two transmission zeros on both sides of the passband. The transmission zeros close to the passband improve the selectivity of the filter. The schematic of the dual-mode, filter-based feedback oscillator and the frequency response of the filter are provided in Figure 7.11(a, b), respectively. The electrical length of the feedback path is 360° or multiple of 360° at the oscillation frequency of 10 GHz. The output spectrum of the oscillator is shown in Figure 7.12(a) where the second harmonic is around 33 dB below the fundamental oscillation frequency. Figure 7.12(b) shows the output spectrum of the oscillator with a 3-MHz span around the fundamental oscillation frequency. The output oscillation power is 3.36 dBm. The measured single sideband phase noise response of the oscillator is shown in Figure 7.13. The measured SSB

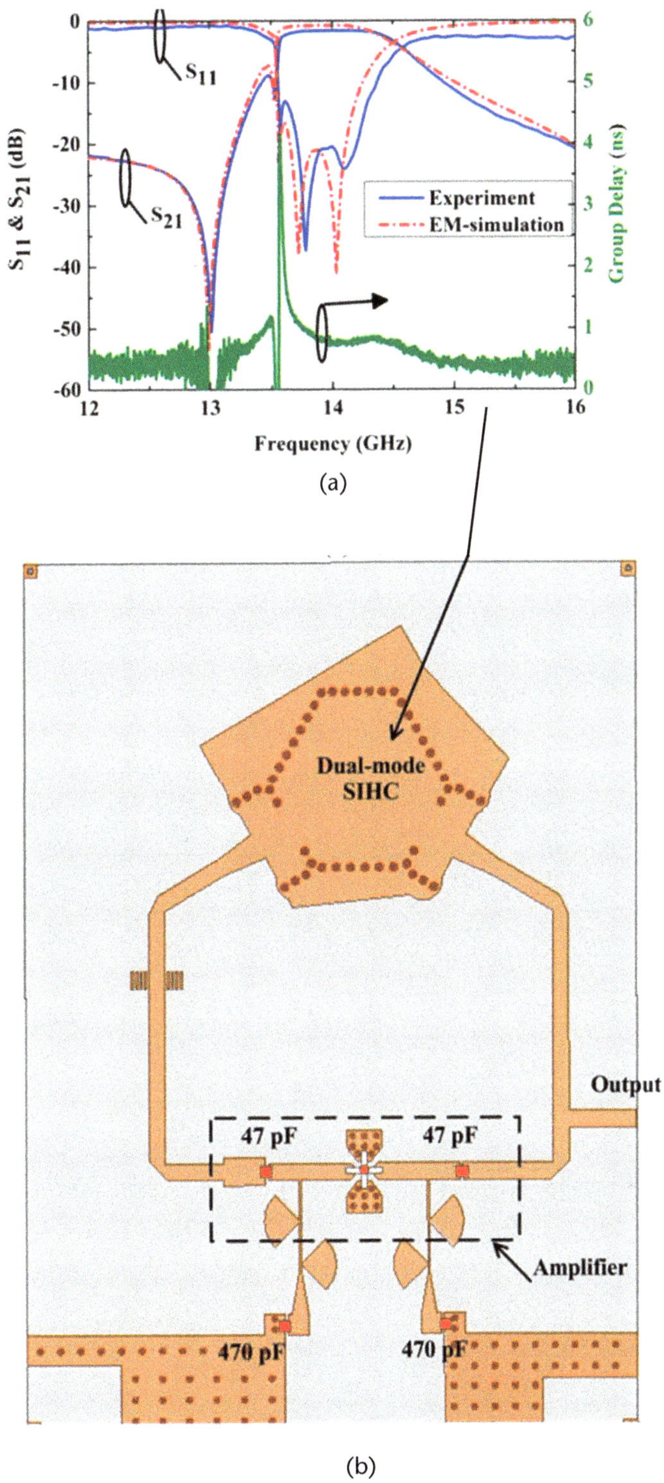

Figure 7.7 (a) Frequency stabilized filter's measured and simulated S-parameters and measured group delay response. (b) Schematic of the triple-mode SIHC filter-based feedback oscillator.

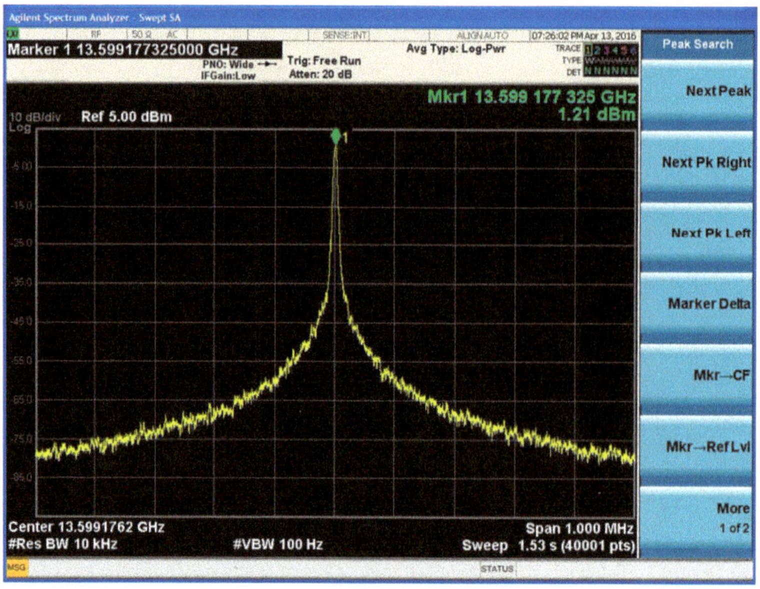

Figure 7.8 Measured output power of the oscillator.

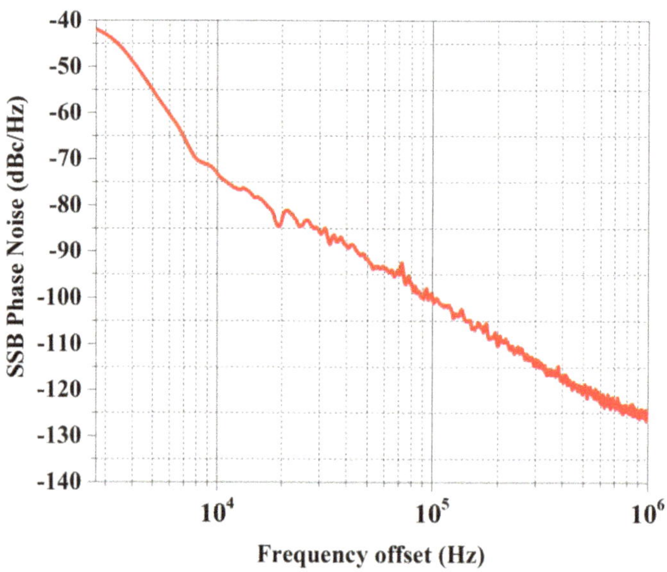

Figure 7.9 Measured phase noise of the oscillator.

phase noises of the oscillator are 110 dBc/Hz and 134 dBc/Hz at 100-kHz and 1-MHz offset frequencies respectively. The FOM of the oscillator is −199 dBc/Hz at 1 MHz. The fabricated prototype of the designed feedback oscillator is shown in Figure 7.14.

7.1.2.3 Design III

Fourth-order, highly selective, dual-mode, SIHC-based bandpass filter is employed as a frequency stabilization element in the feedback oscillator. The measured and

Figure 7.10 Fabricated prototype of pHEMT oscillator employing a single triple-mode SIHC BPF with a transmission zero above the passband.

simulated S-parameters and measured group delay of the dual-mode filter are provided in Figure 7.15(a).

The filter has two transmission zeros close to both edges of the passband. This leads to enhancement of the group-delay peak. Figure 7.15(b) shows the schematic of the feedback oscillator. The dimension of the fourth-order, dual-mode filter was

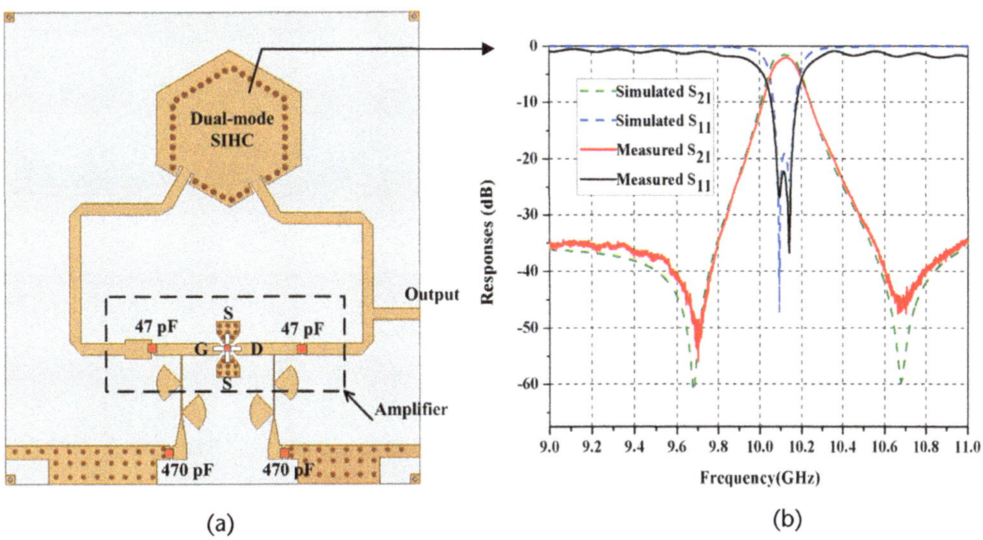

Figure 7.11 (a) Schematic of the Design-II low phase noise oscillator. The filter dimensions are provided in Chapter 6. (b) Frequency-stabilized, dual-mode, SIHC-based filter's measured and simulated S-parameters.

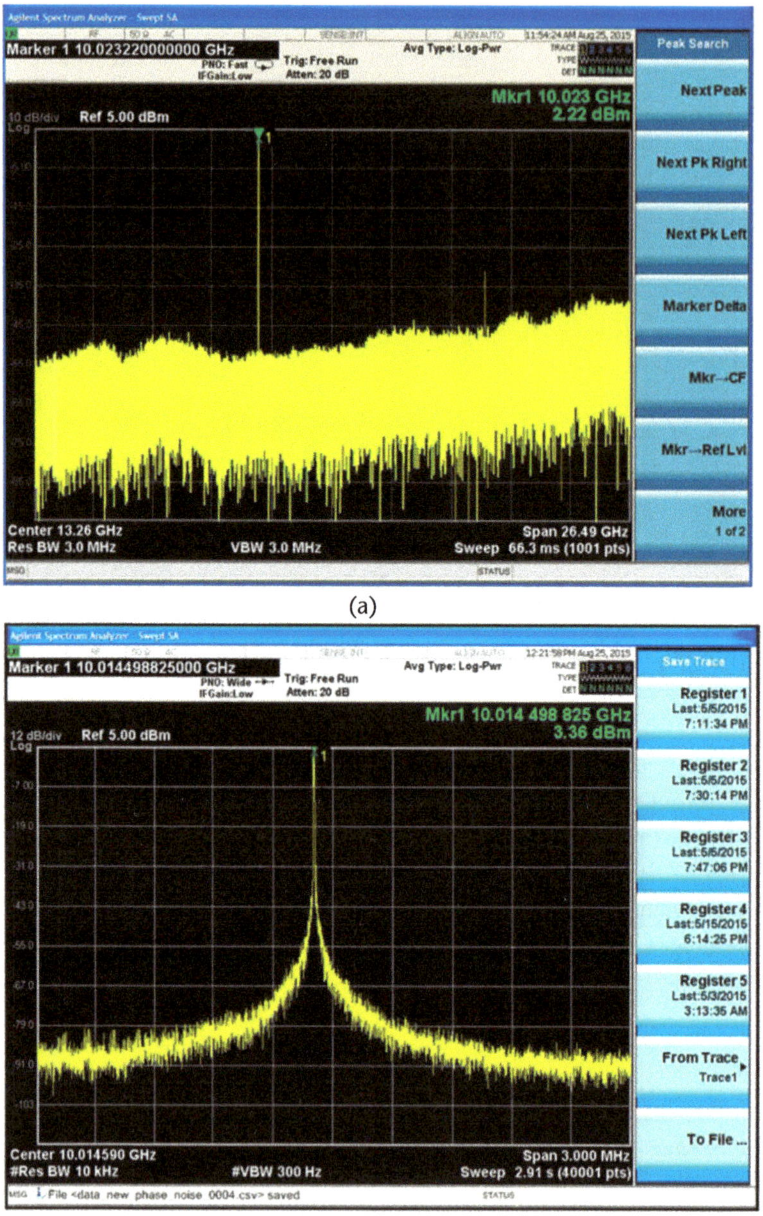

(a)

(b)

Figure 7.12 Measured output spectrum of the oscillator employing a dual-mode SIHC with transmission zeros on both sides of the passband. (a) Full-band output spectrum. (b) Measured output spectrum for a 3-MHz span around a fundamental oscillation frequency.

provided in Chapter 5. The higher group-delay peak corresponds to better phase noise performance. This technique is adopted to realize the feedback oscillator. The full-band output spectrum of the oscillator is provided in Figure 7.16(a) where it is observed that the second-harmonic frequency of the oscillator output power is around 32 dB below the output power of the fundamental oscillation frequency. Figure 7.16(b) shows the output power of the fundamental oscillation frequency

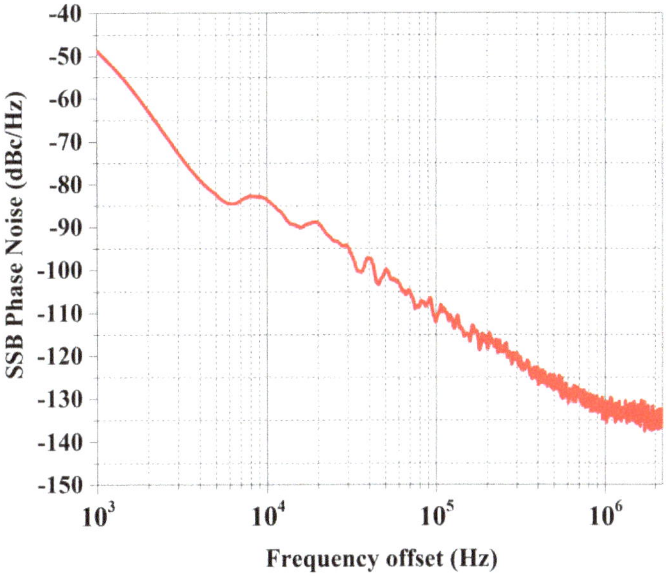

Figure 7.13 Measured phase noise of a pHEMT oscillator employing a single dual-mode SIHC with two transmission zeros on both sides of the passband.

Figure 7.14 Fabricated prototype of the pHEMT oscillator employing a single dual-mode SIHC with two transmission zeros on both sides of the passband.

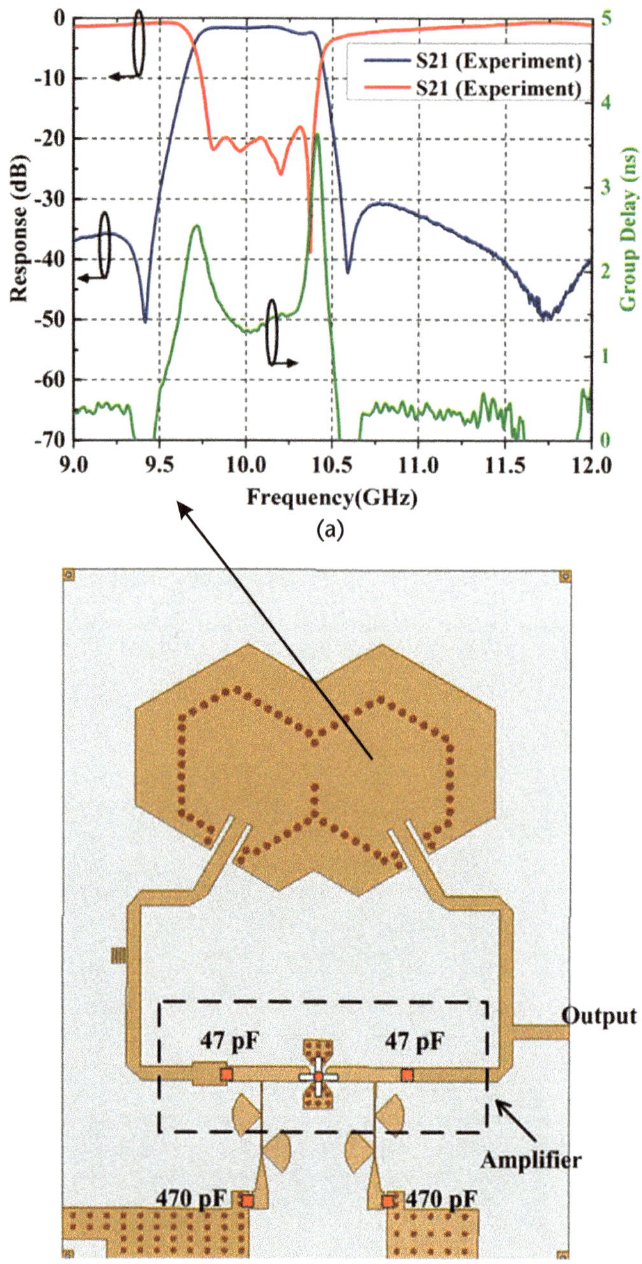

(a)

(b)

Figure 7.15 (a) Performance of the dual-mode SIHC filter. (b) Schematic of oscillator employing a pair of dual-mode SIHC with two transmission zeros on both sides of the passband.

within 1-MHz span. The oscillation frequency is at 9.878 GHz with output power of 8.51 dBm. The measured single side-band (SSB) phase noise of the oscillator is shown in Figure 7.17.

The measured SSB phase noise of the oscillator are −94.5 dBc/Hz and −123.5 dBc/Hz at 100-kHz and 1-MHz offset frequency, respectively. The FOM of the oscillator is −187.6 dBc/Hz at 1 MHz offset. Figure 7.18(a) exhibits the output power spectrum of the fundamental oscillation frequency while changing the drain bias voltage (V_{ds}) of the pHEMT device. Figure 7.18(b) shows the frequency shift of the fundamental oscillation frequency with gate and drain bias voltages. The fundamental oscillation frequency shifts towards a lower frequency from 9.8885 GHz to 9.879 GHz while increasing the gate bias voltage (V_{ds}) from −0.7V to 0V, keeping V_{ds} constant (2.4V). Similarly, the fundamental oscillation frequency moves

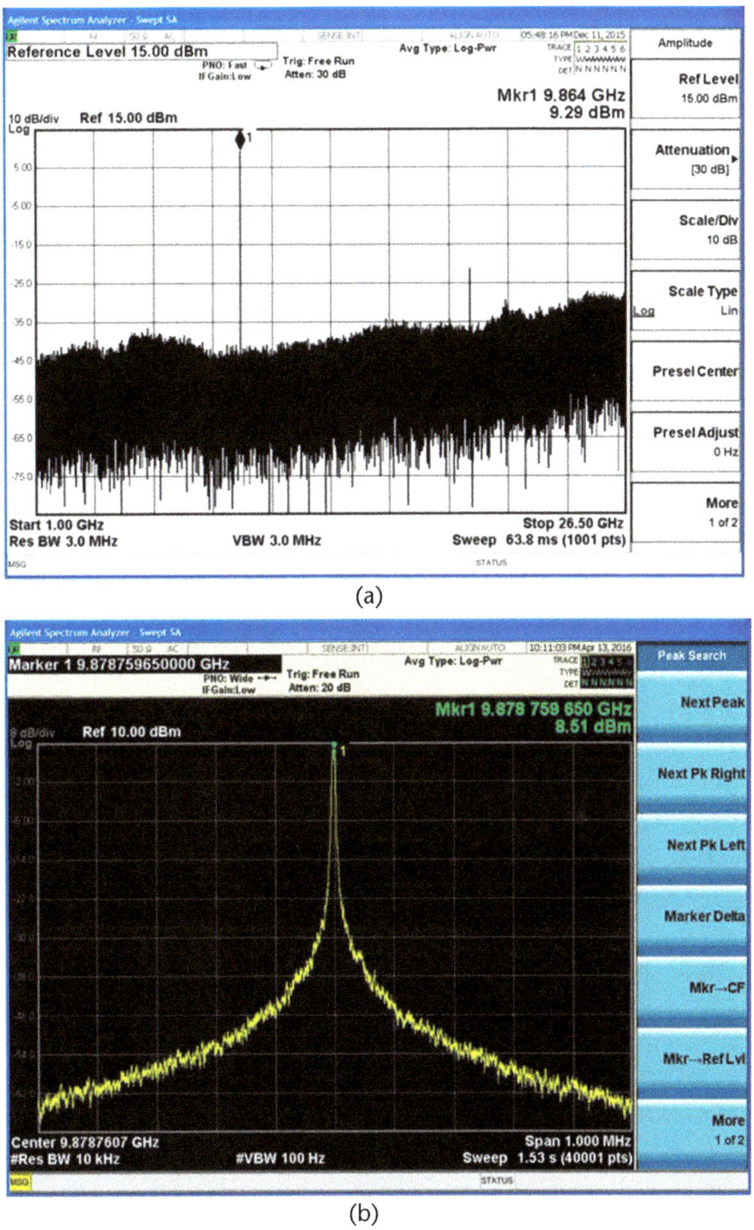

Figure 7.16 Measured output spectrum of the oscillator employing a pair of dual-mode SIHCs with transmission zeros on both sides of the passband. (a) Full-band output spectrum. (b) Measured output spectrum for a 1-MHz span.

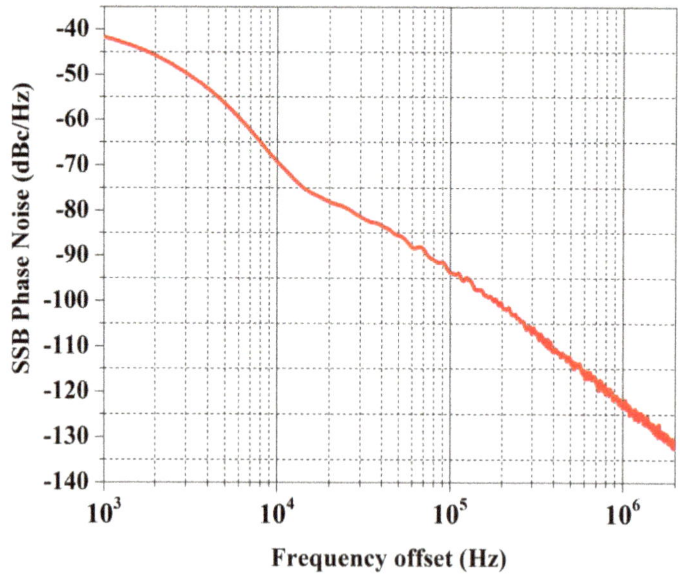

Figure 7.17 Measured phase noise of the pHEMT oscillator employing a pair of dual-mode SIHCs with two transmission zeros on both sides of passband.

towards a lower frequency from 9.881 GHz to 9.879 GHz, with the increase of a drain bias voltage (V_{ds}) from 1.15V to 2.4V while keeping V_{ds} constant (0V).

It is observed that oscillation frequency decreases with the increase of bias voltage in both cases. Device intrinsic gate-source capacitance (CGS) is more sensitive to the changes in V_{gs}, whereas device intrinsic drain-source capacitance (CDS) is less sensitive to V_{ds}. As a result, a 10-MHz shift in oscillation frequency is observed while changing V_{gs}, whereas a 2-MHz shift occurs with the change in V_{ds}. The fabricated prototype of the designed feedback oscillator is shown in Figure 7.18.

7.1.2.4 Comparison Between All Three Feedback Oscillators Design

Comparisons between all three feedback oscillators are shown in Table 7.1. The oscillation frequencies of Design-I, Design-II, and Design-III oscillators are at 13.6 GHz, 10 GHz, and 9.879 GHz, respectively. The SSB phase noise performances of all three oscillators are exhibited in Figure 7.20. It is observed that the phase noise performance of Design-II oscillator is better than other two oscillators because the loaded quality factor of Design II is higher than Design-I and Design-III oscillators. However, the Design-III oscillator has a higher oscillating output power as compared to other two designed oscillators. In both Design-II and Design-III oscillators, the first two higher-order orthogonal modes of the SIHC are exploited to design frequency stabilized BPF, whereas, in the Design-I oscillator, the next three higher-order orthogonal modes located at Ku-band are exploited to realize frequency-stabilized BPF in the feedback loop. All three oscillators are realized using pHEMT NE3210S01 device. The comparison of all three oscillators is carried out keeping DC bias voltage (V_{ds}) and current (I_{ds}) at 2V and 17 mA, respectively.

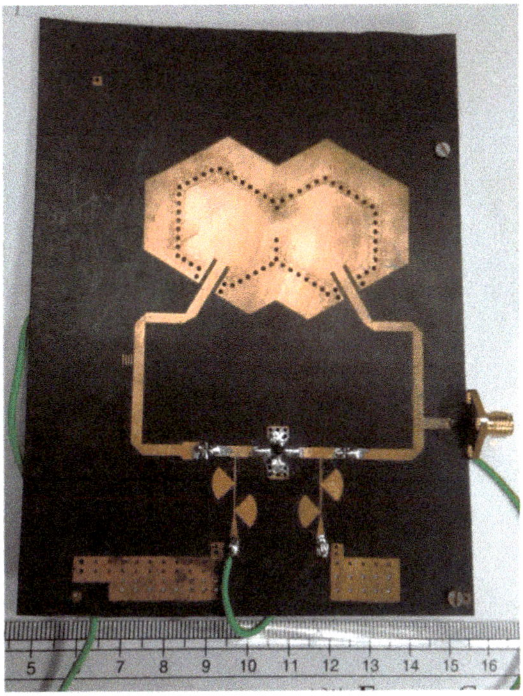

Figure 7.18 Fabricated prototype of the pHEMT oscillator employing a pair of dual-mode SIHCs with two transmission zeros on both sides of the passband.

Table 7.1 Performance Comparison Between All Three Feedback Oscillators

Design	*Frequency (GHz)*	P_{DC} *(mW)*	*Phase Noise (dBc/Hz) at 100 KHz*	*Phase Noise (dBc/Hz) at 1 MHz*	*FOM (dBc/Hz) at 1 MHz*
Design-I	13.6	34	−101.5	−126.4	−193.4
Design-II	10.0	34	−110	−134	−199.0
Design-III	9.879	34	−94.5	−123.5	−187.6

7.2 Self-Oscillating Mixer

The conventional RF front-end system consists of a local oscillator and a mixer circuit. Both of them are separate modules. However, keeping up with the advancement of wireless communication systems, the need of the hour is to miniaturize the RF front-end system without compromising the efficiency and the performance of the system. This can be made possible by combining the oscillator and the mixer circuit in a single module, namely, self-oscillating mixer (SOM), where a single device acts as an oscillator and as a mixer simultaneously. The mixing of RF and a local oscillator signal produces an intermediate frequency (IF) signal, which appears because of the nonlinearity of the active device. The SOM is introduced for the first time in [5] where MESFET is used to realize SOM for the X-band application with a 2-dB conversion gain. In [6], Tajima demonstrated a SOM design using a two-port, frequency-stabilized GaAs FET-based oscillator.

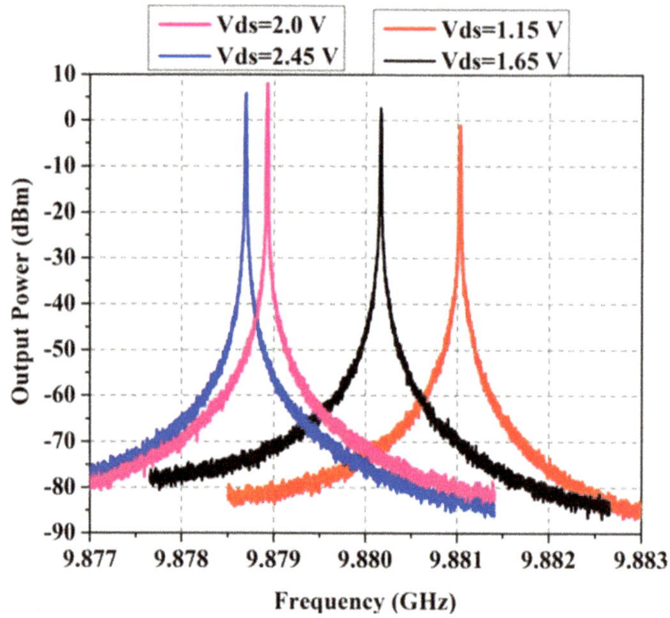

(a)

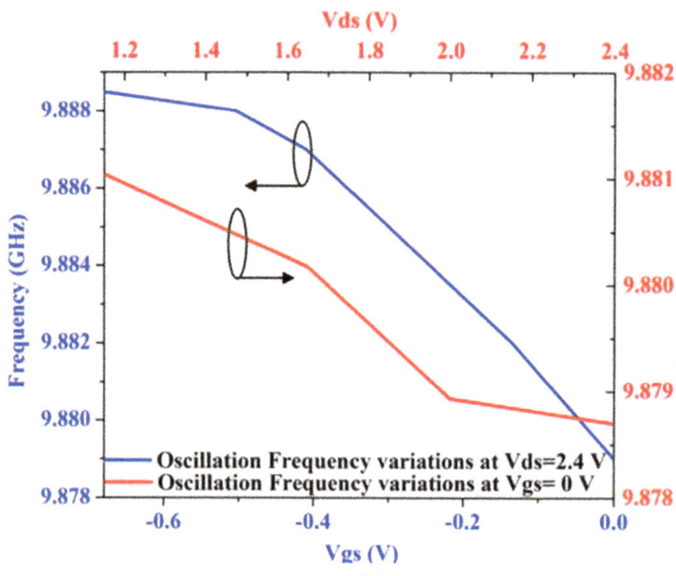

(b)

Figure 7.19 (a) Measured output spectrum of the feedback oscillator with drain bias (V_{ds}) variations. (b) Measured oscillation frequency variations with the changes of gate (V_{gs}) and drain (V_{ds}) DC bias voltage.

7.2.1 Two-Port Filtering Antenna Design Using the SIHC

The schematic of an SIHC-based two-port filtering antenna along with its dimensions is shown in Figure 7.21(a). The design consists of a single dual-mode resonator {*b*} with another single-mode resonator {*a*} and a radiating rectangular slot antenna is placed at the bottom side of resonator {a}. Both microstrip feed lines are directly

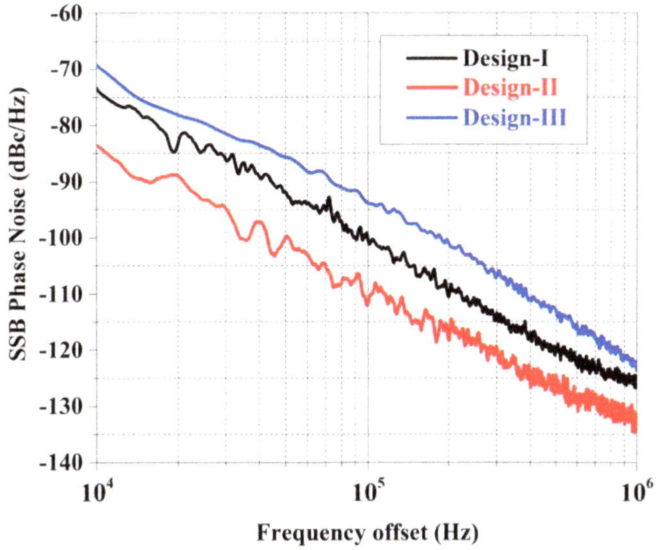

Figure 7.20 Measured phase noise of all three low phase noise oscillators.

inset into the dual-mode resonator {*b*} which acts as input and output ports. The first higher-order mode of SIHC {*a*} and the first two higher-order orthogonal modes of {*b*} are employed to design the filtering antenna. By choosing proper dimensions and position of the slot, it acts as a SIHC-backed rectangular slot antenna. The slot length (L_{slot}) is chosen half-wavelength ($\lambda/2$) at the frequency of the operation. The two-port filtering antenna is realized on Rogers RT/Duroid 5880 substrate with thickness of 0.787 mm and relative dielectric constant of 2.2. Measured and simulated S-parameters of the two-port filtering antenna are shown in Figure 7.21(b). It is observed that Port 2 of the filtering antenna has better impedance matching over the operating frequency band of the antenna compared to Port 1. Hence, the 2-port antenna works in such a way that when power is fed through Port 2, the slot antenna radiates. However, negligible radiation occurs when it is fed through Port 1.

It is also observed from Figure 7.21(b) that because of radiation effect, the insertion loss of this two-port network is much higher and –10-dB impedance bandwidth is achieved from 9.7 GHz to 10.2 GHz when Port 2 is used for excitation. Figure 7.21(b) also gives information that the two-port network behaves like a filter from 9.7 GHz to 9.94 GHz, whereas from 10 GHz to 10.2 GHz, it starts acting like a filtering antenna. This becomes clearer in Figure 7.22 where the measured and simulated gain and S-parameters' plots of the two-port filtering antenna are shown. In Figure 7.22(a), measured and simulated S-parameters and measure and simulated gain (Port 1 excited and Part 2 matched) are shown together. The maximum measured gain at Port 1 is 1.19 dBi. Similarly, in Figure 7.22(b), measured and simulated gains (Port 2 excited and Port 1 matched) and measured and simulated S-parameters are plotted together. The maximum measured gain at Port 2 is 4.7 dBi. It is observed that antenna gain follows the filtering characteristic, thereby improving filtering antenna's selectivity.

The normalized unidirectional radiation patterns of the two-port filtering antenna are shown in Figure 7.23. Since the rectangular slot is placed at the back

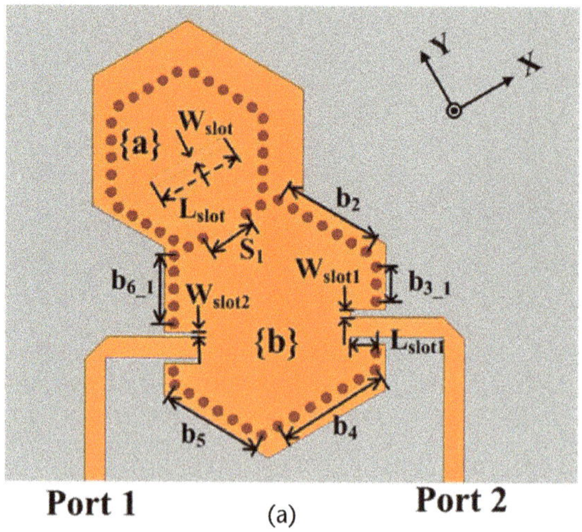

(a)

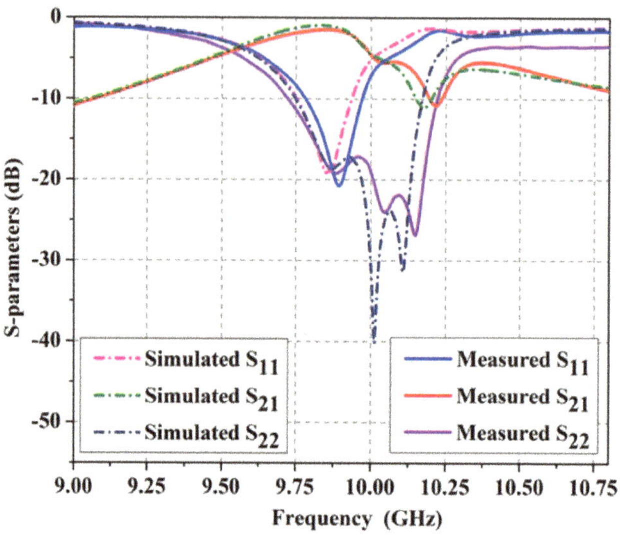

(b)

Figure 7.21 (a) Schematic of the 2-port filtering antenna. (b) Measured and simulated S-parameters of the filtering antenna. The dimensions of the 2-port SIHC based filtering antenna are: L_a = 11.0 mm, L_{slot} = 10 mm, W_{slot} = 1 mm, L_{slot1} = 3 mm, W_{slot1} = 0.8 mm, W_{slot2} = 0.6 mm, S_1 = 5.925 mm, b_2 = 12 mm, b_{31} = 4 mm, b_4 = 13.2 mm, b_5 = 11.9 mm, and b_{61} = 7.9 mm.

side of the cavity, the radiation occurs at the $-Z$ direction. The E-plane and H-plane co- and cross-polarized radiation patterns when fed by Port 2 are exhibited in Figure 7.23(a, b), respectively. The measured cross-polar level in ($-Z$) direction is below 18 dB and front-to-back ratio (FTBR) is lower than 20 dB at 10.15 GHz. Similarly, Figure 7.23(c, d) exhibit measured and simulated E-plane and H-plane co and cross-radiation patterns, respectively, when fed by Port 1 with Port 2 matched terminated. The measured cross-polar level is below 12 dB and the FTBR is lower than 8 dB at 10.15 GHz. From the radiation pattern plot, it is observed that the filtering antenna exhibits better radiation pattern when it is fed at Port 2. The fabricated prototype of the two-port filtering antenna is shown in Figure 7.24(a, b) where Figure 7.24(a)

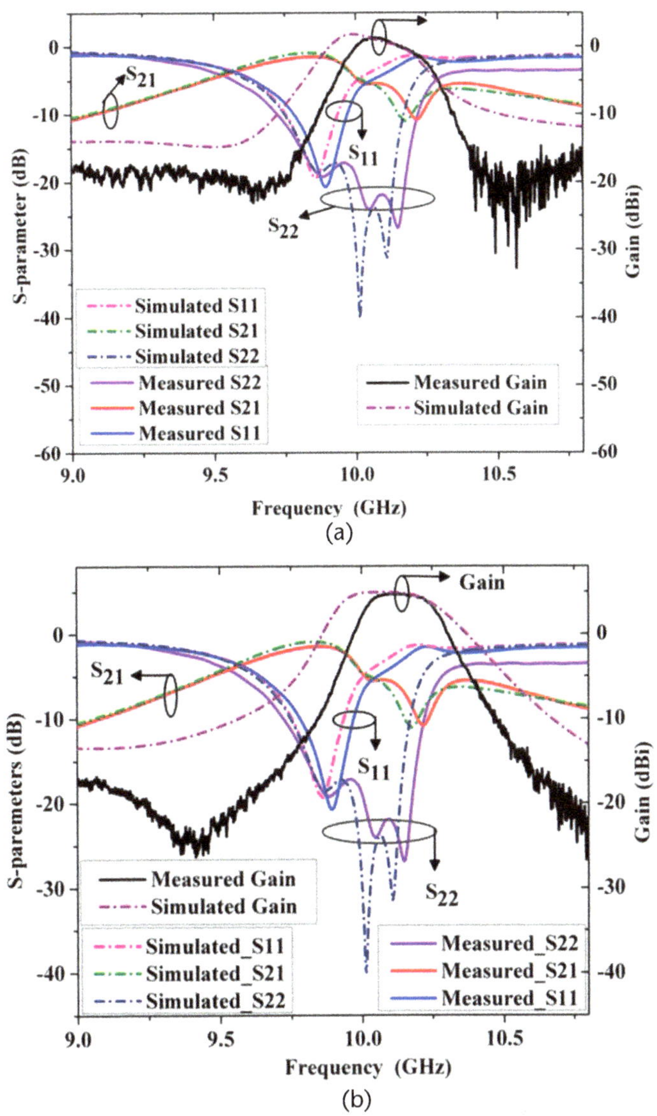

Figure 7.22 (a) Measured and simulated S-parameters the two-port filtering antenna when Port 1 is excited Port 2 is matched. (b) Measured and simulated S-parameters of the two-port filtering antenna when Port 2 is excited Port 1 is matched.

shows the top view of the filtering antenna and Figure 7.24(b) shows the bottom view of the filtering antenna.

7.2.2 SOM Design Using a Two-Port Filtering Antenna

Figure 7.25 shows the schematic of the SOM design employing a two-port filtering antenna. The designed two-port filtering antenna is connected in the feedback loop with an active device. The pHEMT NE3210S01 device is used here to design SOM. The desired oscillation frequency is kept at 9.75 GHz. Since the two-port filtering antenna's filtering characteristic is observed from 9.7 to 9.94 GHz, the two-port

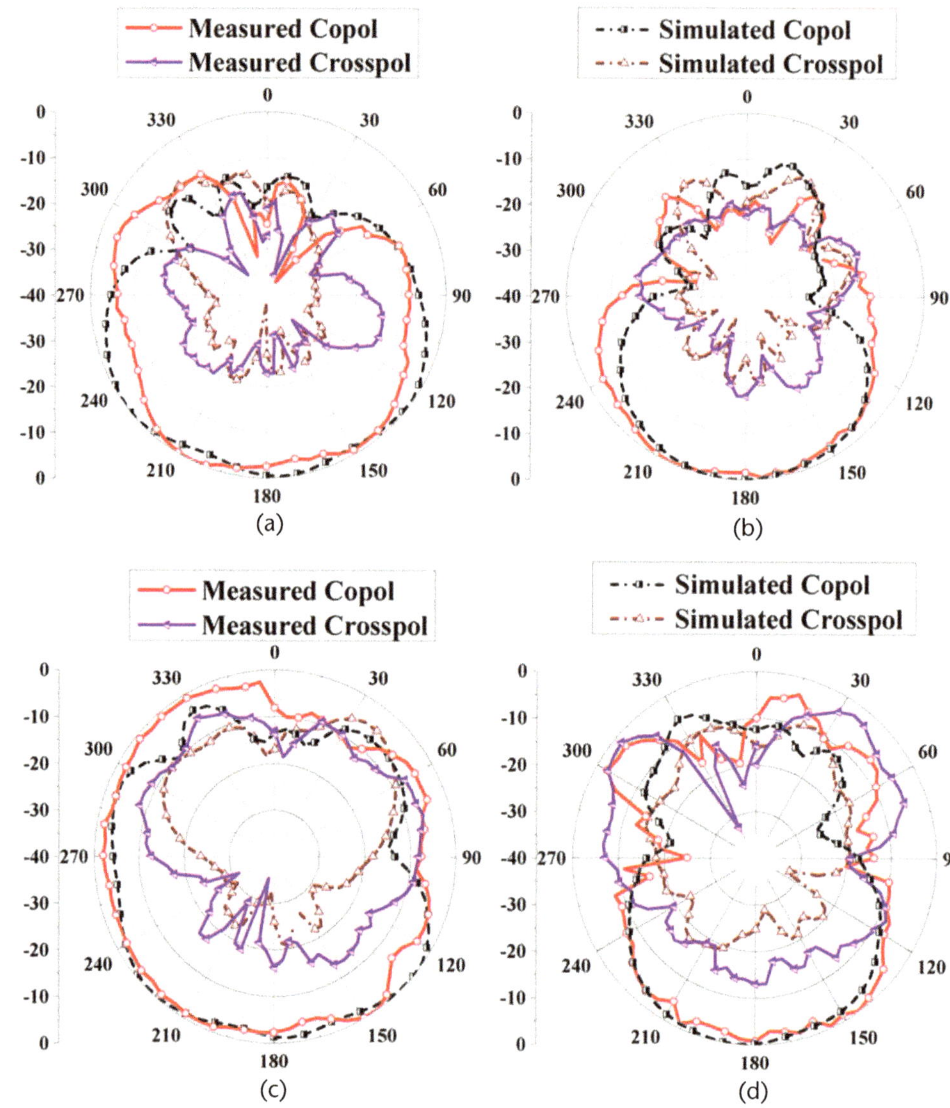

Figure 7.23 Measured radiation pattern of the two-port filtering antenna at 10.15 GHz. (a) E-plane radiation pattern when Port 2 is excited. (b) H-plane radiation pattern when Port 2 is excited. (c) E-plane radiation pattern shen Port 1 is excited. (d) H-plane radiation pattern when Port 1 is excited.

filtering antenna acts as a frequency-stabilized filter at 9.75 GHz in the feedback loop. To implement the feedback oscillator at 9.75 GHz, the pHEMT-based amplifier is connected with the 2-port filtering antenna using microstrip line of electrical length of 360° or multiple of 360°. The output signal is collected from source of the pHEMT device. The gate terminal of the pHEMT device is connected with Port 2 of the filtering antenna since Port 2 has better impedance matching which is realized using distrubuted and lumped elements. The IF output is taken from the source using distributed and lumped lowpass filter. The measured output power spectrum is shown in Figure 7.26(a) where it shows that the LO frequencies and RF are at 9.752 GHz and 10.15 GHz, respectively, and the corresponding IF is at 398 MHz.

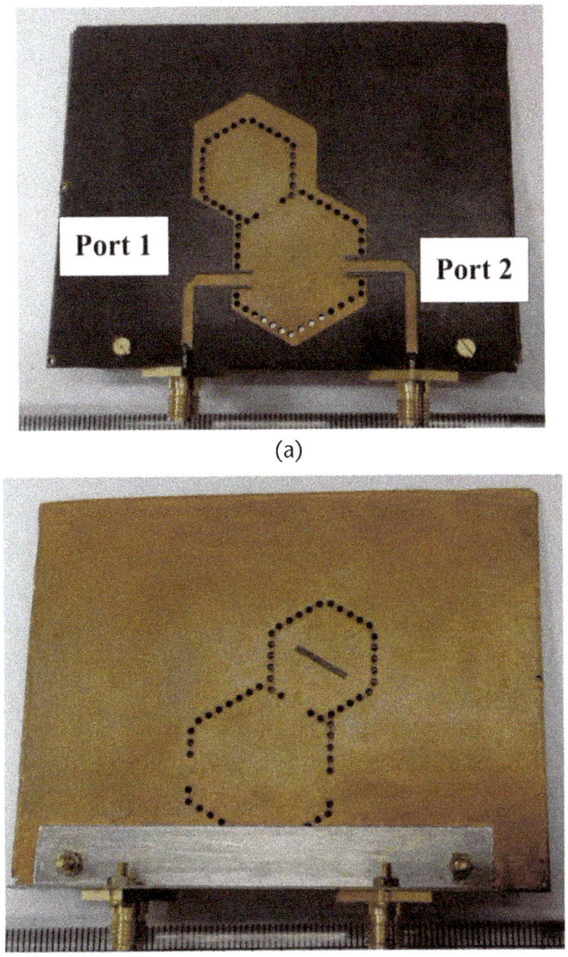

Figure 7.24 (a) Top view of the fabricated 2-port filtering antenna. (b) Bottom view of the fabricated 2-port filtering antenna.

Figure 7.26(b) shows the output spectrum of the desired IF signal when RF is at 10.15 GHz. The measured SSB phase noise of the IF signal is shown in Figure 7.27. The measured phase noises of the IF signal are −90.3 dBc/Hz and −110.2 dBc/Hz at 100-kHz and 1-MHz offset, respectively. The DC bias condition of the pHEMT are as follows: V_{ds} and I_{ds} are 2V and 10 mA, respectively. The calculated FOM at 1 MHz offset is −177.32 dBc/Hz.

Figure 7.28 shows measured conversion gain and IF power variation with RF, respectively. The conversion gain of SOM is calculated by following the procedure provided in [6]. The distance between transmitting and receiving antenna is kept 30.48 cm. The receiving antenna is a two-port filtering antenna that has a 4.7-dBi antenna gain. The transmitting antenna is a double-ridged horn antenna with 12.3-dBi gain. Based on the above-mentioned data, the calculated conversion gain plot is shown in Figure 7.28. Similarly, the IF output power variation with changes in RFs is also depicted in Figure 7.28 with a fixed RF power of 0 dBm. The fabricated prototype of the SOM is shown in Figure 7.29.

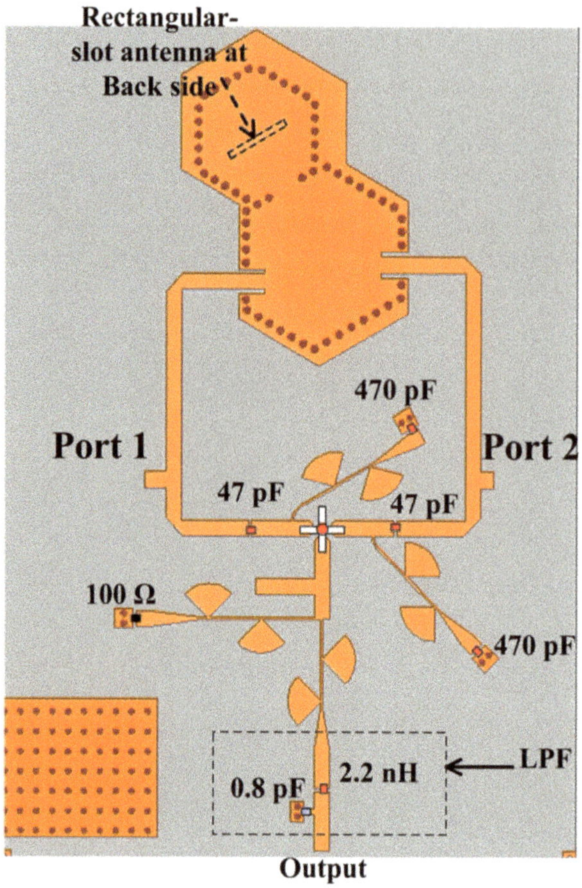

Figure 7.25 Schematic of the SOM using a two-port filtering antenna.

7.2.3 Dual-Band SOM

Figure 7.30 shows a schematic of a dual-band SOM using a quasi-Yagi dual-band filtering antenna. The dimensions related to a dual-band filtering antenna are provided in Chapter 8. In the design, a fourth-order, dual-mode, SIHC-based bandpass filter is employed as a frequency-stabilizing element in the feedback loop. The schematic of the bandpass filter and the dimensions of its design parameters are provided in Figure 7.31(a). The simulated S-parameters of the filter are shown in Figure 7.31(b). The design specifications of the filter are as follows: center frequency: 8.2 GHz; FBW: 5%; and transmission zeros: 7.78 GHz and 8.56 GHz.

Dual-band SOM is realized using the Rogers RT/Duroid 5880 substrate with a thickness of 0.787 mm and a relative dielectric constant of 2.2. The oscillation frequency of the dual-band SOM is kept at 8.1 GHz. The amplifier is designed at 8.1 GHz using the pHEMT device (NE3210S01). The amplifier is connected with a frequency-stabilized filter in the feedback loop using a microstrip transmission line of an electrical length of 360° or a multiple of 360°. The output IF signal is collected from drain terminal of the pHEMT device using microstrip based low-pass filter (LPF). A dual-band filtering antenna is designed using two quasi-Yagi

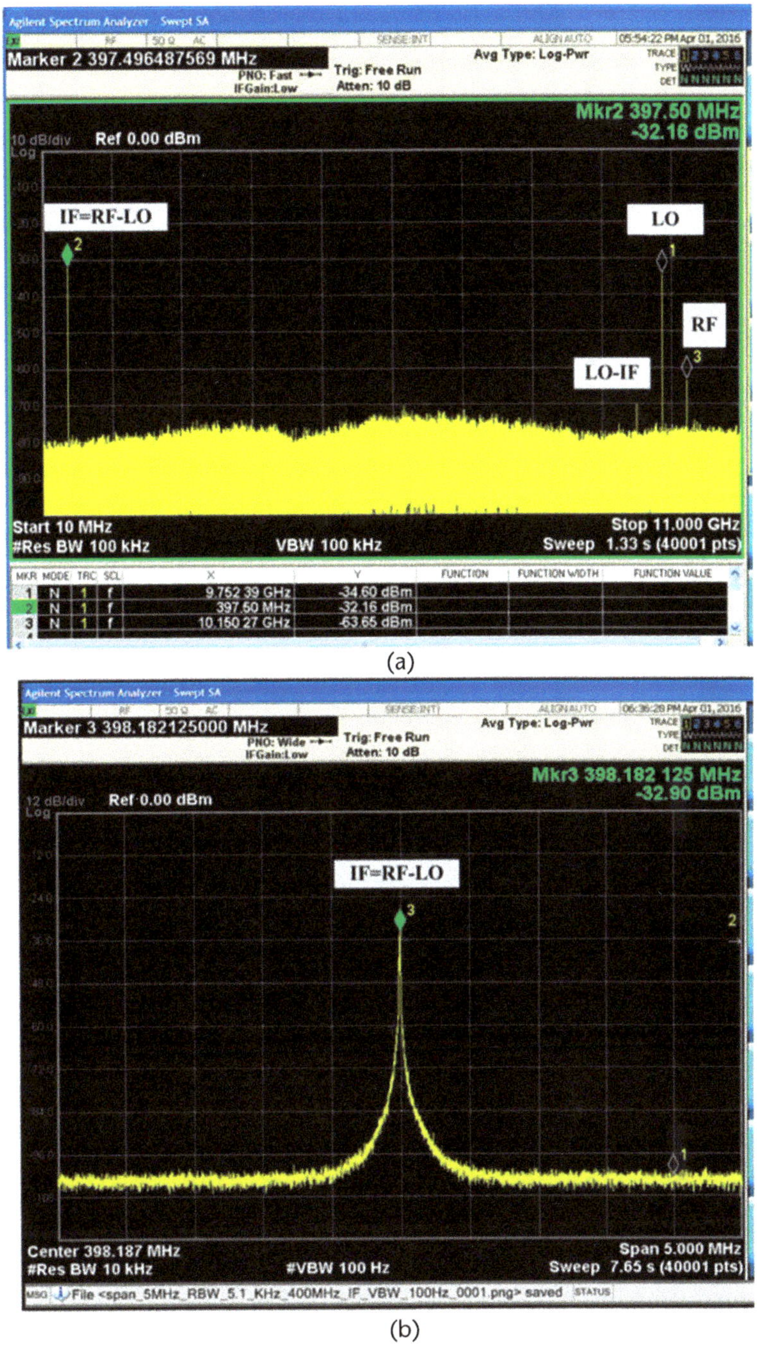

Figure 7.26 (a) Output spectrum of the SOM. (b) Spectrum of the desired IF signal when the RF signal is at 10.15 GHz.

antennas integrated with an SIHC-based balun diplexer. The dual-band filtering antenna is connected at the gate end of the pHEMT device. The center frequencies of lower and higher bands of the dual-band filtering antenna are at 9.46 GHz and 9.95 GHz with 177-MHz (1.87%) and 220-MHz (2.21%) impedance bandwidth (−10-dB S_{11}), respectively.

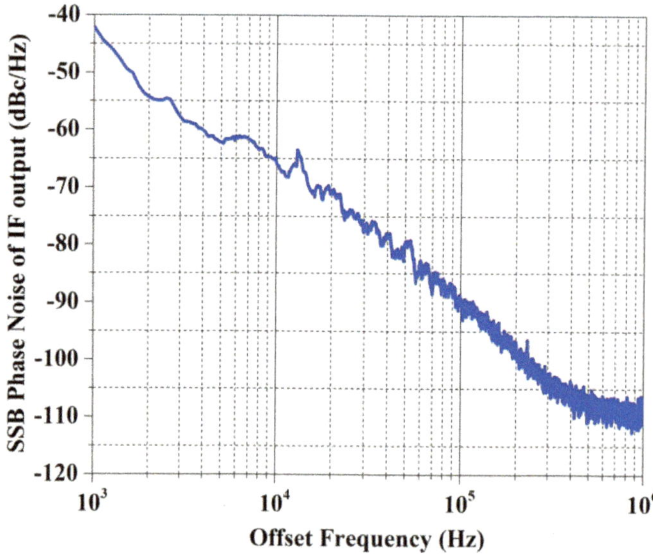

Figure 7.27 Measured SSB phase noise of the IF signal.

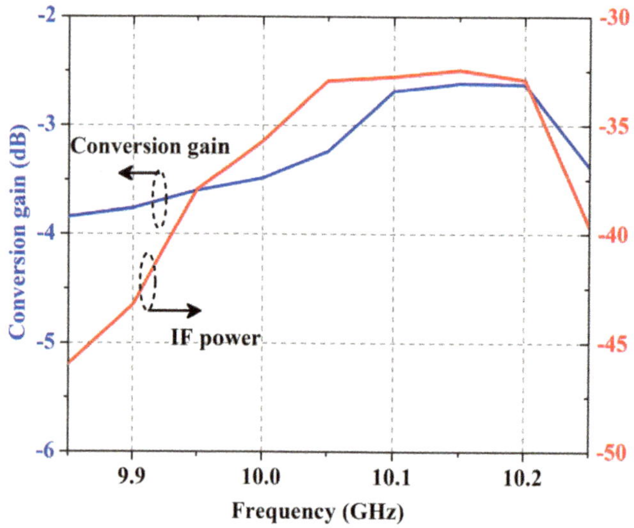

Figure 7.28 Measured IF output power and conversion gain of the SOM with RF variations when RF power is fixed at 0 dBm.

In the dual-band SOM design, lower and higher frequency bands of the dual-band filtering antenna are denoted as lower RF (f_{RFL}) and higher RF (f_{RFH}), respectively. The output spectrum of the dual-band SOM is shown in Figure 7.32. The self-oscillation frequency of the dual-band SOM is at 8.115 GHz. The lower and higher RFs are kept at 9.38 GHz (f_{RFL}) and 9.95 GHz (f_{RFH}), respectively. The desired IF signals are generated by mixing the fundamental oscillation signal with RF signals. The corresponding lower and higher IF signals are at 1.28 GHz (f_{IFL})

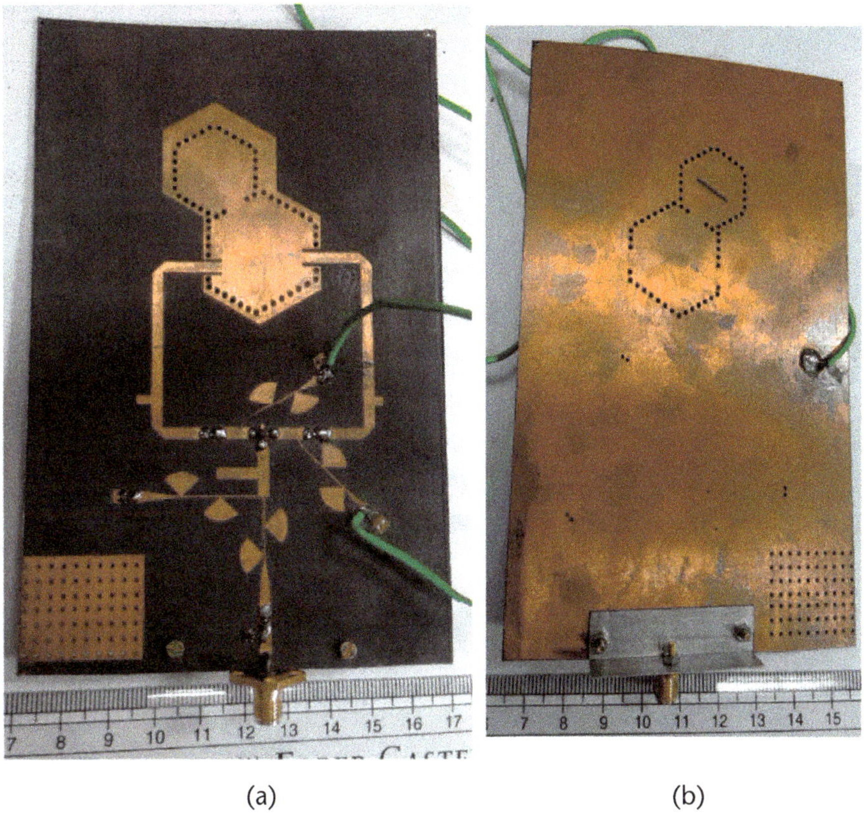

Figure 7.29 (a) Top view of the fabricated prototype of the SOM. (b) Bottom view of the fabricated prototype of the SOM.

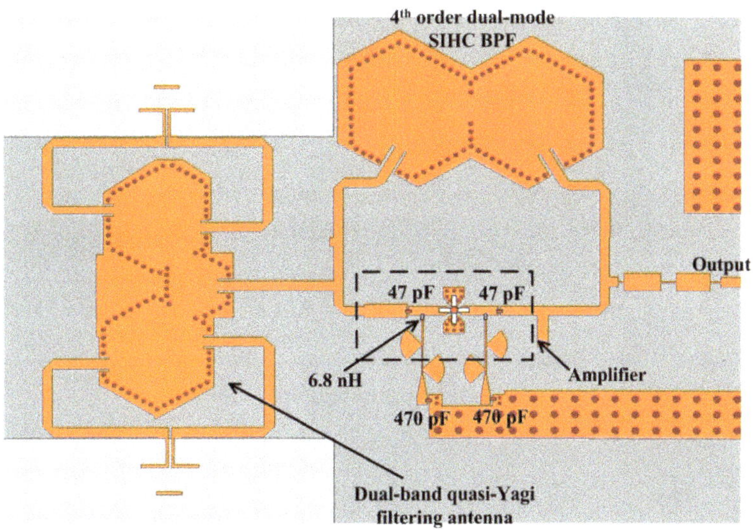

Figure 7.30 Schematic of the dual band SOM using quasi-Yagi dual-band filtering antenna.

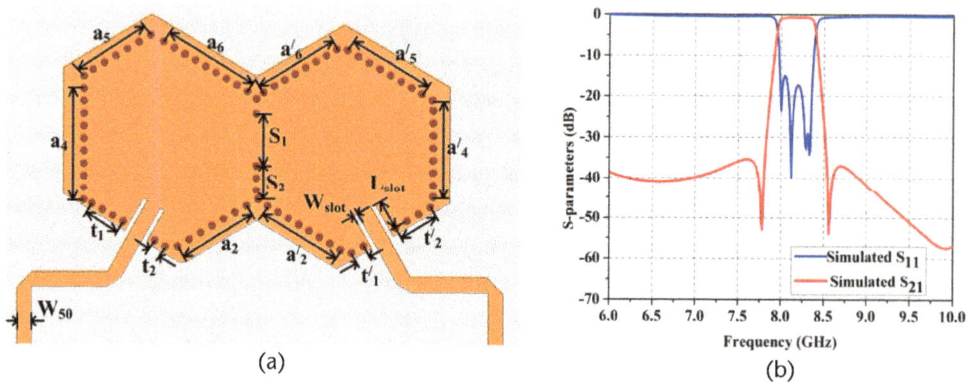

Figure 7.31 (a) Schematic of the fourth-order, dual-mode SIHC BPF employed as a frequency stabilization element in the feedback loop of the dual-band SOM. (b) Simulated S-parameters of the fourth-order, dual-mode, SIHC BPF. The dimensions related to the design parameter of dual-mode SIHC BPF are as follows: $t_1 = 6$ mm, $t_2 = 2$ mm, $a_2 = 13.2$ mm, $a_4 = 18$ mm, $a_5 = 12.9$ mm, $a_6 = 16$ mm, $S_1 = 8.325$ mm, $S_2 = 5.0$ mm, $W_{slot} = 0.8$ mm, $L_{slot} = 5.1$ mm, $a'_2 = 14$ mm, $a'_4 = 15.7$ mm, $a'_5 = 14$ mm, $a'_6 = 14$ mm, $t'_1 = 2.2$ mm, and $t'_2 = 6.6$ mm.

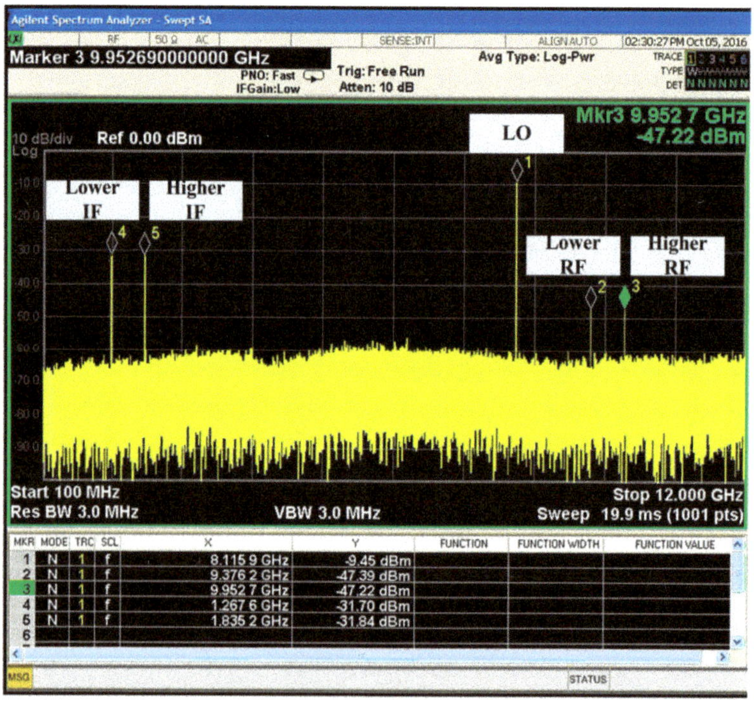

Figure 7.32 Output spectrum of the dual-band SOM.

and 1.83 GHz (f_{IFH}) with −32.6-dBm and −32-dBm output powers, respectively. The DC bias of the dual-band SOM is $V_{ds} = 2\text{V}$ and $I_{ds} = 10$ mA.

The RF signal is received by the dual-band filtering antenna from two identical, double-ridged, horn antennas with 12.3-dBi antenna gain. Both the identical transmitting antennas are kept 30.48 cm away from both quasi-Yagi antennas.

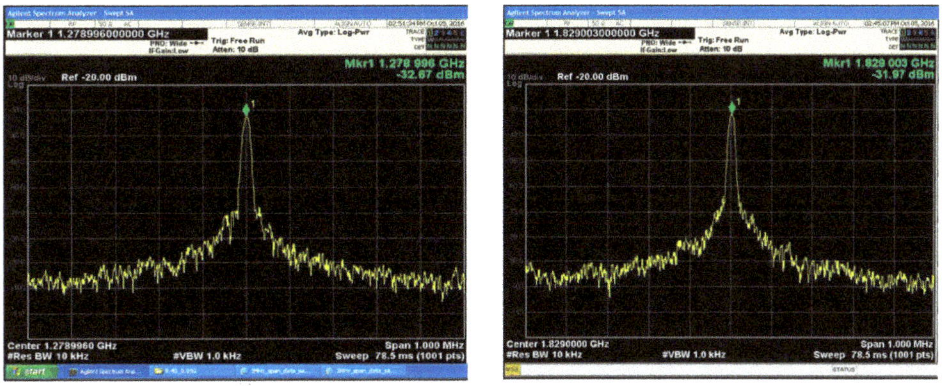

(a) (b)

Figure 7.33 Output spectrum of IF signal. (a) Output spectrum of lower-band IF signal (f_{IFL}) for the lower-band RF at 9.38 GHz (f_{RFL}). (b) Output spectrum of the higher-band IF signal (f_{IFH}) for higher-band RF at 9.95 GHz (f_{RFH}).

The lower-band and higher-band gains of the dual-band filtering antenna are 5.2 dBi and 4.98 dBi, respectively. The measured dual-band filtering antenna's gain, S-parameters, and radiation patterns are given in Chapter 8. The desired output spectrum of lower and higher IF signals are shown in Figure 7.33(a, b), respectively, while keeping both lower and higher RFs at 9.38 GHz and 9.95 GHz, respectively.

Figure 7.34 shows the measured SSB phase noise of IF signals. The measured lower IF signal (f_{IFL}) phase noises are −100 dBc/Hz and −111.6 dBc/Hz at 100-kHz and 1-MHz offsets, respectively, and measured higher IF signal (f_{IFH}) phase noises are −101.7 dBc/Hz and −112.4 dBc/Hz at 100-kHz and 1-MHz offsets,

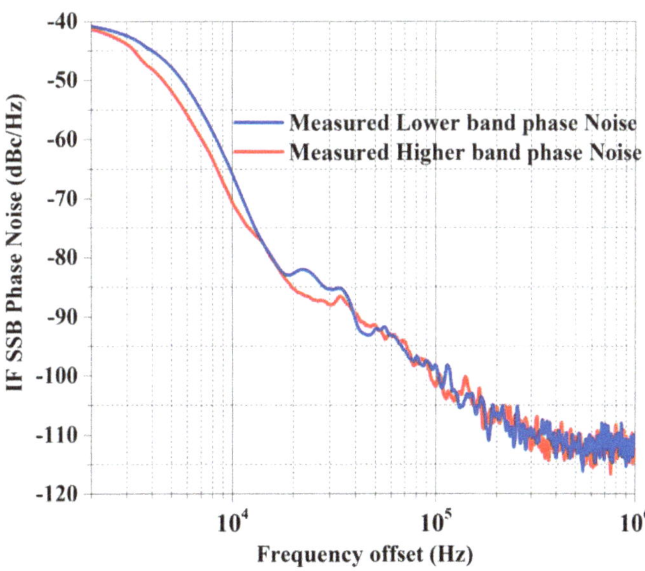

Figure 7.34 Measured SSB phase noise of lower and upper bands of the dual-band SOM.

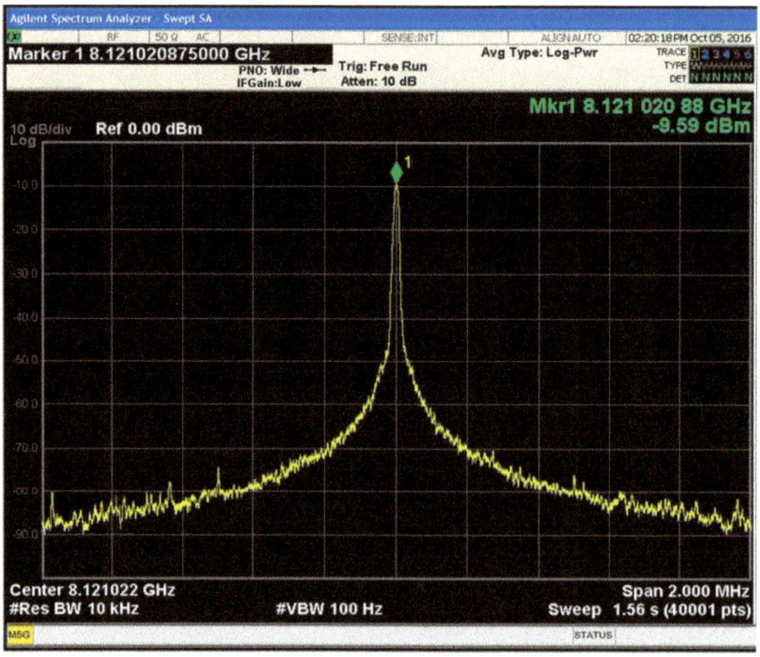

Figure 7.35 Output spectrum of fundamental self-oscillation frequency of the dual-band SOM.

respectively. The calculated FOMs based on measured phase noise of both lower (f_{IFL}) and higher (f_{IFH}) IF signal are −169 dBc/Hz and −174 dBc/Hz at a 100-kHz offset, respectively. The output power spectrum of fundamental oscillating signal is shown in Figure 7.35 within a 2-MHz span. Measured lower (f_{IFL}) and higher (f_{IFH}) IF output power variations with the changes of lower and higher-band RF are exhibited in Figure 7.36(a, b). In both cases, the IF output powers are measured using RF transmitting power from a double-ridged, horn antenna at 0 dBm.

Figure 7.37(a, b) show the measured conversion gain of lower (f_{IFL}) and upper (f_{IFH}) band IF signals of the dual-band SOM, respectively. The lower IF band (f_{IFL}) conversion gain varies from −3.5 dB to −4 dB with the variation of RF lower-band frequency (f_{RFL}) from 9.25 GHz to 9.45 GHz. The higher IF band (f_{IFH}) conversion gain varies from −3.2 dB to −2.5 dB with the variation of RF higher-band frequency (f_{RFH}) from 9.7 GHz to 10.05 GHz. The fabricated prototype of the dual-band SOM is shown in Figure 7.38(a, b).

7.3 Conclusions

In this chapter, several SIW-based active microwave circuits are discussed. The SIHC-based high Q resonators are utilized to implement low phase noise oscillators where highly selective bandpass filters act as frequency stabilization element that helps to achieve SSB phase noise below −130 dBc/Hz at an 1-MHz offset frequency. Also, various SIW-based self-oscillating mixer designs are explored. In this chapter, one compact two-port SIW-based filtering antenna design was presented for designing

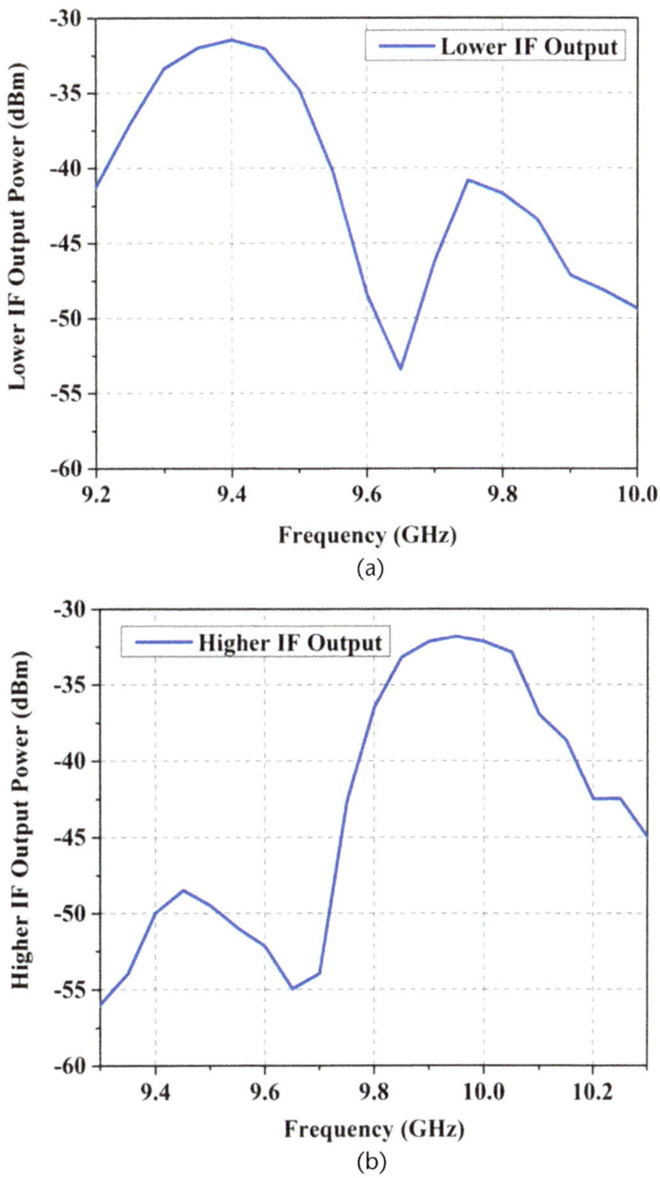

Figure 7.36 (a) Lower band IF (f_{IFL}) power variations with lower-band RF (f_{RFL}). (b) Higher band IF (f_{IFH}) power variations with higher-band RF (f_{RFH}).

SOM, where the filtering antenna acts as a radiating as well as a frequency stabilization element, thereby improving phase noise and also providing compactness. Another SOM design is presented where SIW based high Q filter network is integrated with quasi-Yagi diplexing antenna. This helps in providing SOM with low profile portable dual band receiver network design. All the designs are fabricated and measured using standard horn antenna. The integrated circuits can be good candidate for low profile portable wireless applications. All the designs are fabricated and measured using the standard horn antenna. The integrated circuits can be good candidates for low-profile portable wireless applications.

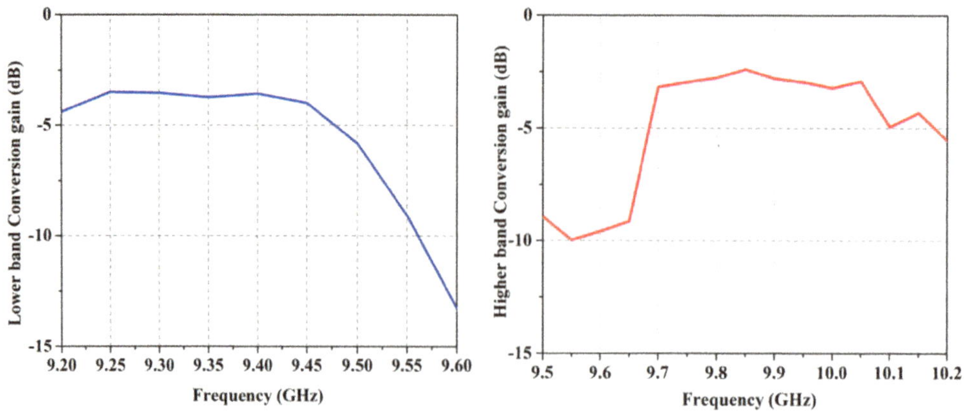

(a) (b)

Figure 7.37 (a) Lower-band conversion gain of the dual-band SOM. (b) Higher-band conversion gain of the dual-band SOM.

(a) (b)

Figure 7.38 (a) Top view of dual-band SOM. (b) Bottom view of dual-band SOM.

References

[1] Leeson, D., "A Simple Model of Feedback Oscillator Noise Spectrum," *Proceedings of the IEEE*, Vol. 54, No. 2, February 1966, pp. 329–330.

[2] Gonzalez, G., *Foundations of Oscillator Circuit Design*, Norwood, MA: Artech House, 2006.

[3] Sauvage, G., "Phase Noise in Oscillators: A Mathematical Analysis of Leesons Model," *IEEE Transactions on Instrumentation and Measurement*, Vol. 26, No. 4, December 1977, pp. 329–330.

[4] Lin, J., and T. Itoh, "Active Integrated Antennas," *IEEE Transactions on Microwave Theory and Techniques*, Vol. 42, No. 12, December 1994, pp. 2186–2194.

[5] Tajima, Y., "GaAs FET Applications for Injection-Locked Oscillators and Self-Oscillating Mixers," *IEEE Transactions on Microwave Theory and Techniques*, Vol. 27, No. 7, July 1979, pp. 629–632.

[6] Stephan, K. D., and T. Itoh, "Recent Efforts on Planar Components for Active Quasi-Optical Applications," *IEEE MTT-S International Microwave Symposium Digest*, Vol. 3, May 1990, pp. 1205–1208.

CHAPTER 8

The SIW System and Its Application

In Chapter 7, high-Q SIW filter-based low phase noise oscillator design and its application in designing a compact receiver system, namely, SOM, were discussed. In this chapter, SIW filtering antenna, SIW active integrated antenna (AIA), and SIW transceiver architecture will be discussed.

8.1 SIW Filtering Antenna and Its Application in AIA

In modern wireless communication systems, both the antenna and the filter are essential parts of RF front-end systems. The antenna is used for transmitting and receiving signals, whereas the filter passes the desired frequency band and rejects the undesired one. In general, both are kept in a cascaded fashion in RF front-end systems. However, this cascaded network may create difficulties due to a mismatch between the antenna and the filter. Since both are working in the same frequency band, any impedance mismatch at the edge of the antenna passband degrades the performance of the system. This problem can be overcome by integrating the antenna with the filter and cosimulating both of them together. This integration enhances the system's performance and reduces the circuit area, which are the key challenges in modern wireless communication systems.

The AIA consists of an active device and a radiating element integrated on the same substrate. With rapidly growing modern communication systems, AIAs have received much attention because of their compactness and low cost. There are many practical applications of AIAs, such as power combining, beam steering/switching, retro-directive arrays, and high-efficiency power amplifier designs. At the millimeter-wave frequency, higher transmission line loss, limited source power, and reduced antenna efficiency make AIA an attractive candidate.

In this chapter, four different applications of filtering antennas are implemented using SIW technology.

1. A third-order highly selective SIHC bandpass filter is exploited to realize a filtering antenna. The filtering antenna comprises a dual-mode and a single-mode hexagonal cavity. The single-mode SIHC is backed by a rectangular slot, which acts as a radiating element. The application of this filtering antenna as an AIA used as a transmitting section is also demonstrated.
2. Two diplexing antennas are presented in this chapter. One of the designs is realized using a dual-mode SIHC at the Rx and Tx channels and a single-mode SIHC-backed bow-tie slot antenna shared with both channels. The

other design comprises a pair of dual-mode SIHCs at the Rx and Tx channels and a single-mode SIHC-backed bow-tie slot antenna shared with the Rx and Tx channels. Both designs have good isolation between the Rx and Tx channels.

3. A triplexing antenna is presented here, which consists of three channels, namely, Rx, Mx, and Tx. Each channel has one dual-mode resonator and a single-mode resonator. The common junction of the triplexer is a single-mode SIHC resonator with bow-tie slot antenna.
4. To obtain a relatively larger bandwidth, a quasi-Yagi antenna is employed in the SIHC-based balun filter design. Its application in the AIA design as a transmitting section is also demonstrated here.

8.1.1 Third-Order Filtering Antenna Using SIHC (Design I)

Design I consists of a single dual-mode SIHC with another single-mode SIHC-backed radiating rectangular slot resonator. First, a third-order bandpass filter is realized using the following design specifications: center frequency at 8.87 GHz and FBW of 2.6% with the design-specified transmission zero at 8.65 GHz.

The coupling topology of the filter is shown in Figure 8.1. Based on the design specifications and coupling topology, the coupling matrix is given in (8.1).

$$M = \begin{bmatrix} & P_1 & P_2 & 1 & 2 & 3 \\ P_1 & 0 & 0 & 0.817 & 0.578 & 0 \\ P_2 & 0 & 0 & 0 & 0 & 1 \\ 1 & 0.817 & 0 & -0.702 & 0 & -0.874 \\ 2 & 0.578 & 0 & 0 & 1.082 & 0.412 \\ 3 & 0 & 1 & -0.874 & 0.412 & -0.108 \end{bmatrix} \tag{8.1}$$

From the coupling matrix, self-resonant frequencies of the resonators, external quality factor, and coupling between resonators are calculated as follows: f_{01}

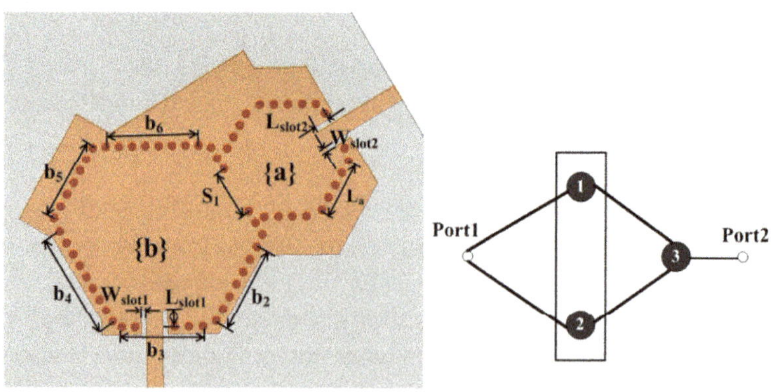

(a) (b)

Figure 8.1 (a) Schematic of the third-order SIHC filter along with dimensions: a =10.29 mm; L_{slot1} = 2.7 mm; W_{slot1} = 0.8 mm; b_6 = 14 mm; b_3 = 12.275 mm; b_2 = 12.6 mm; b_4 = 15.6 mm; b_5 = 12 mm; S_1 = 7.325 mm; and L_a = 10.29 mm. (b) Coupling topology.

= 8.9554 GHz, f_{02} = 8.7507 GHz, f_{03} = 8.8867 GHz, Q (P_1, 1) = 57.8, Q (P_2, 2) = 115.5, Q (P_2, 3) = 38.51, $k_{1,3}$ = 0.0227, and $k_{2,3}$ = 0.0107.

Figure 8.1 illustrates the schematic and dimensions of the filter. Here, L_a is the length of each side wall of the single-mode resonator {a}. To obtain the design-specified transmission zero below the passband, perturbation is applied in the outward direction using the side wall b_5. As a consequence, the even-mode frequency is lower than the corresponding odd-mode frequency of dual-mode resonator {b}. The simulated and theoretical responses of the filter are shown in Figure 8.2.

Figure 8.3 shows the schematic of an identical single-mode SIHC resonator {a} shown in Figure 8.1 with a rectangular slot but it is excited by a SIW aperture having same aperture dimension (S1). When {a} and coupling width S_1 are the same as those provided in the filter, without introducing slot, the electric fields of the fundamental and first higher-order modes of the SIHC are exhibited in Figure 8.4(a, b), respectively. As shown in Figure 8.4(c), the fundamental mode is at 9.1 GHz, and the first higher-order mode appears at 13.96 GHz. Here, the first higher-order mode is exploited to design a filtering antenna. When a rectangular slot is etched at the bottom of the SIHC, by choosing the proper dimensions and position of the slot, it acts as an SIHC-backed rectangular slot antenna. The slot length (L_{slot}) is chosen as a half-wavelength ($\lambda/2$) at the frequency of operation, and width (W_{slot}) is chosen very small. The dimensions related to the rectangular slot are provided in Figure 8.3. Without changing the dimensions of resonator {a}, the cavity-backed slot is placed in the middle, at the bottom of the cavity wall. From Figure 8.5(c), it is observed that the fundamental mode shifts from 9.1 to 9.33 GHz and the first higher-order mode shifts from 13.96 GHz to 10.05 GHz due to cavity perturbation by the slot. The electric field variations with the respective modes are exhibited in Figure 8.5(a, b). To tune the first higher oder mode in presence of slot at 8.95 GHz, the dimension of each side wall of the SIHC is increased from 10.29 to 12.21 mm, as shown in Figure 8.6.

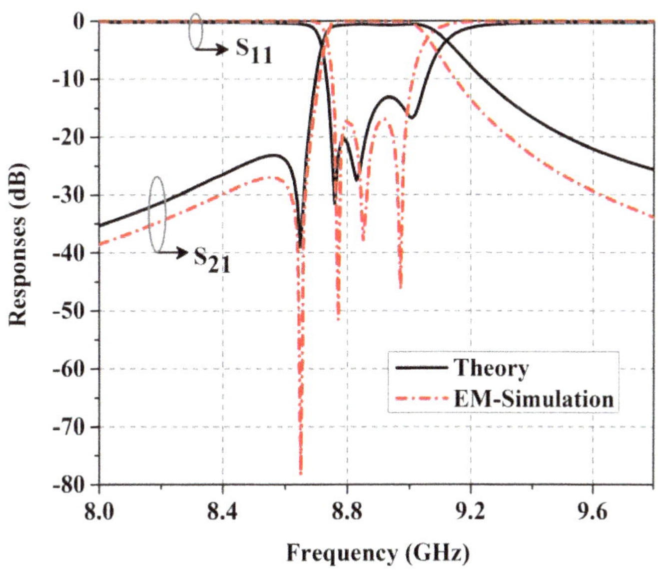

Figure 8.2 Simulated and theoretical (coupling matrix) filter responses.

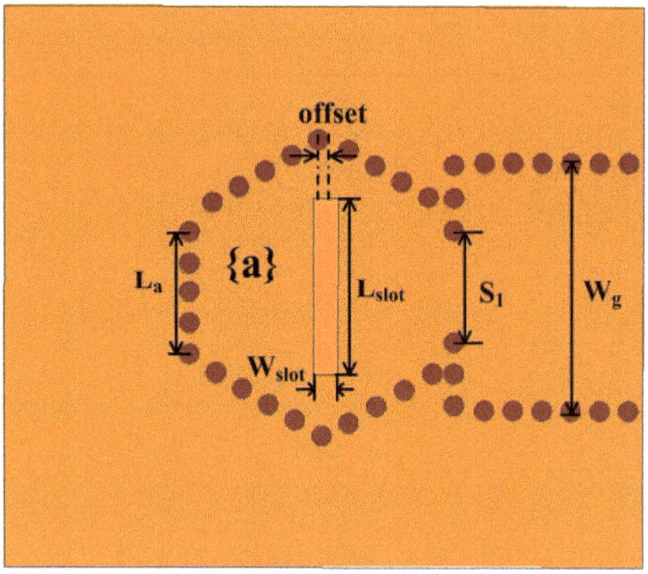

Figure 8.3 Schematic of the single-mode SIHC ({a})-backed rectangular slot antenna coupled with the SIW feeding network. Dimensions related to the schematic are provided along with the schematic. L_{slot} = 11.45 mm; W_{slot} = 1.6 mm; *offset* = –0.1 mm; S_1 = 7.325 mm; W_g = 15 mm; and L_a = 10.29 mm.

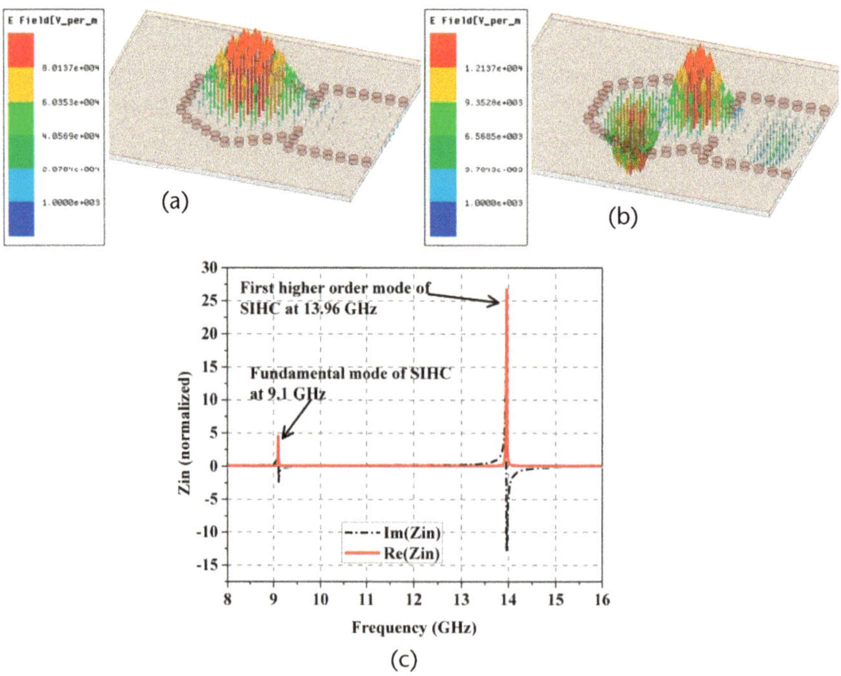

Figure 8.4 Electric field vector and normalized input impedance plot of a single mode SIHC without slot having dimension of each side-wall (L_a) 10.29 mm: (a) Electric field vector plot for fundamental mode at 9.1 GHz. (b) Electric field vector plot for the first higher-order mode at 13.96 GHz. (c) Plot of a simulated normalized input impedance (Z_{in}) of the single-mode SIHC with L_a = 10.29 mm.

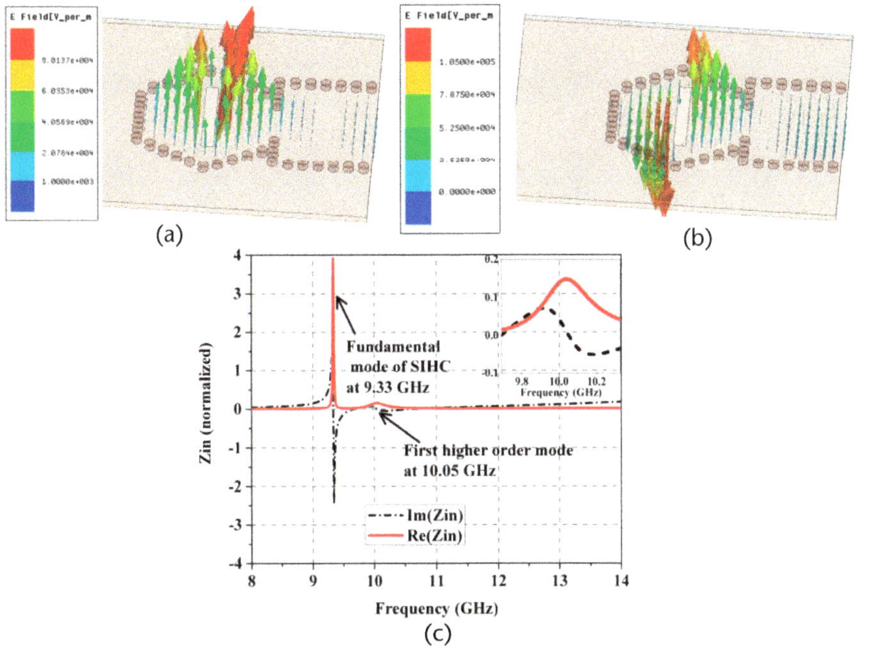

Figure 8.5 Electric field vector plot of the single-mode SIHC with a cavity-backed rectangular slot. The dimension of each side wall of SIHC (L_a) is 10.29 mm and the antenna dimension is given in Figure 8.3. (a) Electric field vector plot for the fundamental mode at 9.33 GHz. (b) Electric field vector plot for the first higher-order mode at 10.05 GHz. (c) Plot of a simulated normalized input impedance (Z_{in}) of the single-mode SIHC backed with a rectangular slot antenna with $L_a = 10.29$ mm.

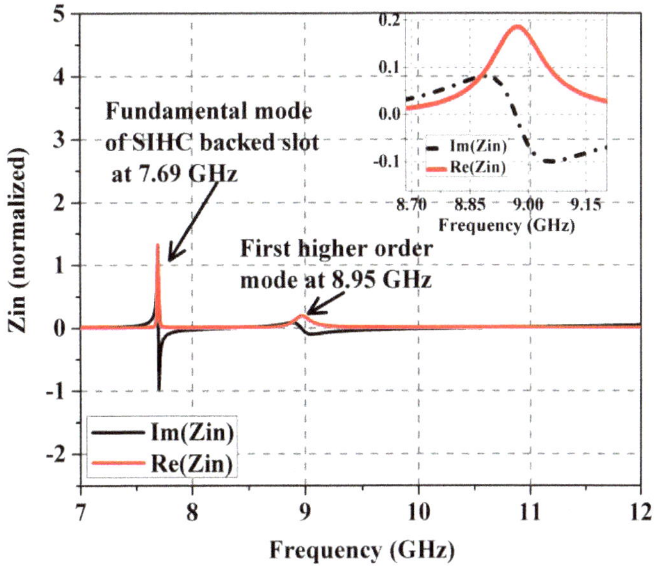

Figure 8.6 Plot of a normalized input impedance (Z_{in}) of the single-mode SIHC-backed rectangular slot antenna. Each side-wall dimension (L_a) for the SIHC is 12.21 mm and the antenna dimension is given in Figure 8.3.

The schematic of the filtering antenna and its dimensions are shown in Figure 8.7(a). In comparison with the filter shown in Figure 8.1, only the dimension of the single-mode resonator {*a*} is increased while keeping all the remaining dimensions the same for realizing filter antenna. The corresponding coupling topology is shown in Figure 8.7(b) where resonator 3 is an SIHC-backed rectangular slot antenna. The electric field distribution in the filtering antenna is shown in Figure 8.8.

It is observed that the filtering antenna is highly sensitive to various physical parameters (L_{slot}, W_{slot}, *offset*) of the radiating slots as shown in Figure 8.9. The resonant frequency of the cavity-backed slot antenna is inversely proportional to the slot length (L_{slot}). The dependence of the cavity-backed slot antenna's resonant frequency on L_{slot} is shown in Figure 8.9(a). The resonant frequency variations due to offset positioning (*offset*) of the slot from the center line of the cavity are shown in Figure 8.9(b).

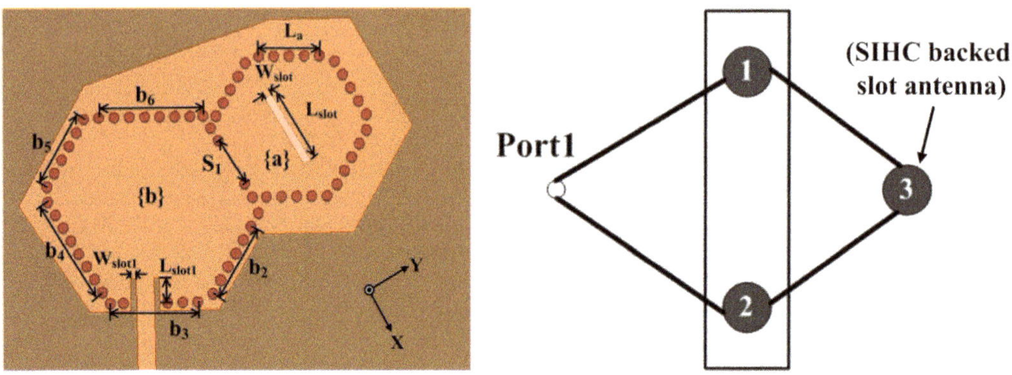

(a) (b)

Figure 8.7 (a) Schematic of the Design-I filtering antenna along with dimensions. L_a = 12.21 mm, L_{slot1} = 3.4 mm, W_{slot1} = 0.8 mm, L_{slot} = 11.45 mm, W_{slot} = 1.6 mm, b_6 = 14 mm, b_3 = 12.275 mm; b_2 = 12.6 mm, b_4 = 15.6 mm, b_5 = 12 mm, and S_1 = 7.325 mm. (b) Corresponding coupling topology.

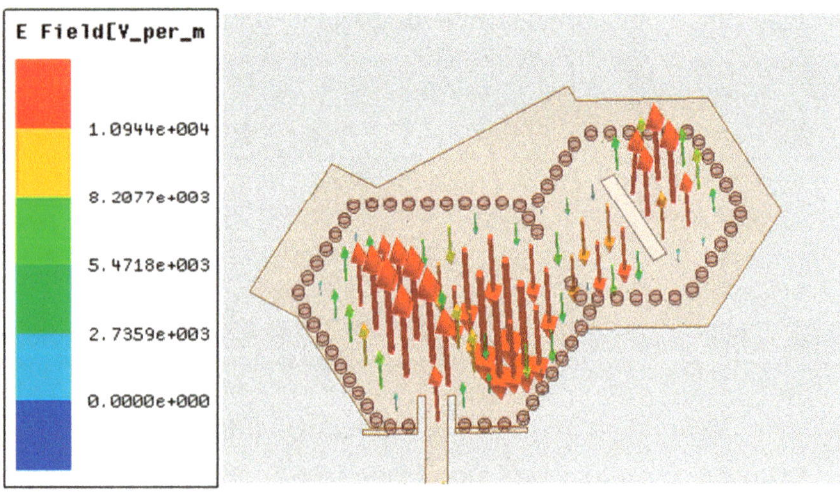

Figure 8.8 Vector E-field distribution in filtering antenna.

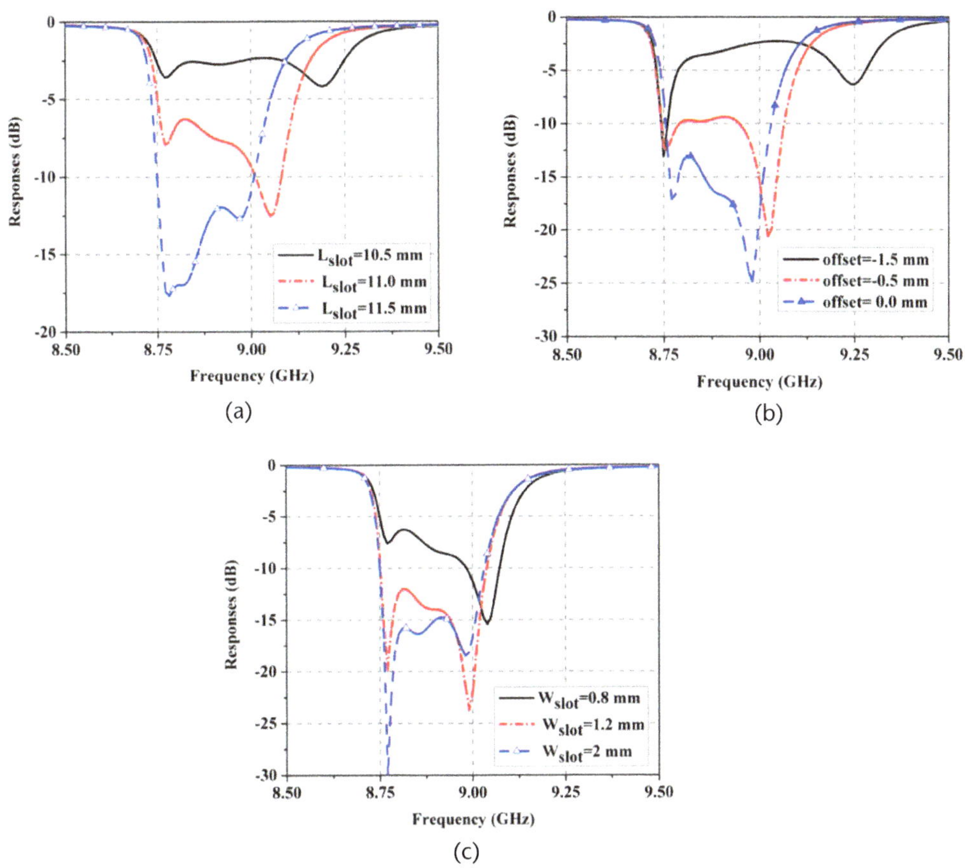

Figure 8.9 Effects of the antenna's design parameters on reflection coefficients. (a) Slot length (L_{slot}) variation. (b) Antenna offset position with respect to the center of the cavity (*offset*), as shown in Figure 5.3. (c) Slot width (W_{slot}) variation.

However, the variations in slot width (W_{slot}) hardly influence the resonant frequency of the radiating resonator; they only affect the input impedance matching, as shown in Figure 8.9(c). The calculated unloaded quality factor (Q_U) of the cavity {*a*} is around 600 for the first higher-order mode without loading the slot antenna. When the same cavity is loaded with a rectangular slot antenna, its Q decreases, and the loaded Q (Q_A) is around 44. The Q_A of a cavity-backed rectangular slot is extracted using EM simulation by applying very weak coupling to the SIHC-backed slot antenna. Q_A is extracted using the reactance slope [1].

$$Q_A = \frac{\omega_0}{2\mathrm{Re}\left(Z_{\mathrm{in}}(\omega_0)\right)} \left. \frac{d\left(\mathrm{Im}\left(Z_{\mathrm{in}}\right)\right)}{d\omega} \right|_{\omega=\omega_0} \tag{8.2}$$

A low Q cavity helps to achieve good impedance matching. Using extracted loaded Q, the filtering antenna is represented by lumped RLC elements, which help to carry out the filtering antenna synthesis through optimization [2]. Keysight ADS software is used for this purpose. The corresponding circuit schematic is shown in Figure 8.10. The equivalent circuit model and EM-simulated filtering antennas' S-parameters are compared with the measured filtering antennas' S-parameters in Figure 8.11. It shows good agreement between the measured and simulated responses of the filtering antenna. The measured center frequency of the filter is 8.89 GHz with a 290-MHz (–10-dB S_{11} impedance bandwidth) and the measured reflection coefficient is below 15 dB in the passband. Figure 8.12 exhibits measured simulated S-parameters and gain of the filtering antenna along with simulated S-parameter responses of the filter as shown in Figure 8.2. Antenna gain and radiation nulls are analogous to the filter insertion loss and transmission zeros, respectively.

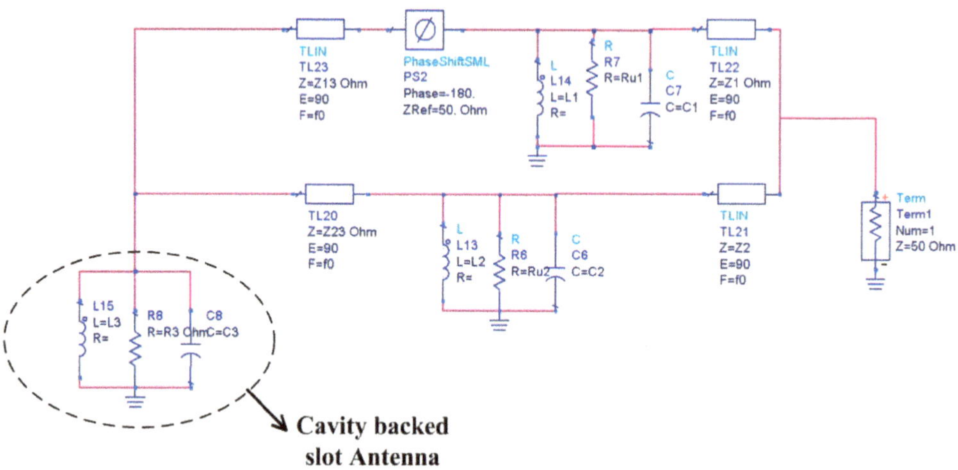

Figure 8.10 Equivalent lumped model of the Design-I filtering antenna.

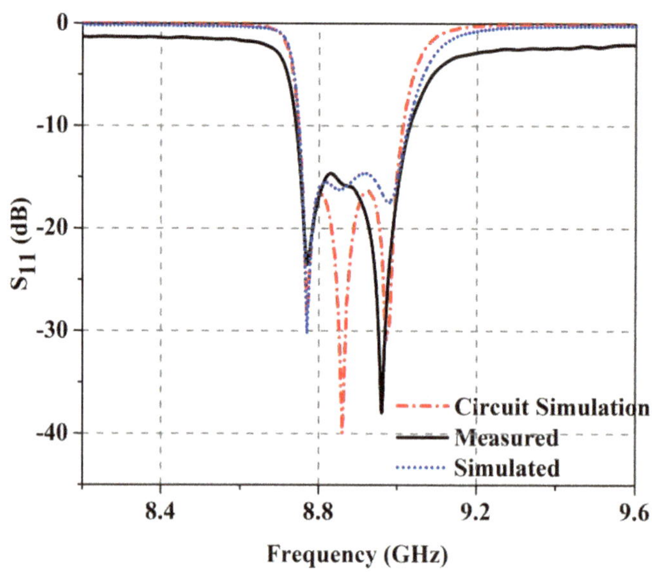

Figure 8.11 Measured, EM-simulated, and circuit-simulated responses of the filtering antenna.

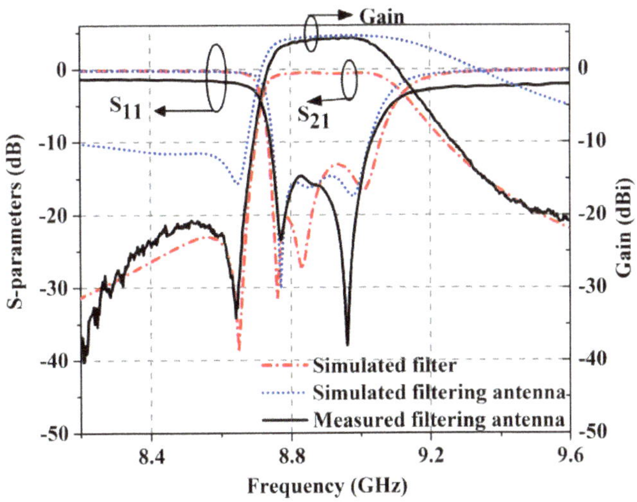

Figure 8.12 Measured and simulated gains and S-parameters of the filtering antenna and simulated S-parameters of the filter shown in Figure 8.1.

The filtering antenna is realized using the filter synthesis technique, which enables a flat gain in the operating band. The measured out-of-band filtering antenna gain is below 25 dB. The measured and simulated passband gains of the filtering antenna are 4.33 dBi and 4.57 dBi, respectively. It is also observed that the filtering antenna gain characteristic follows the transmission characteristic of the filter. The gain of the filtering antenna has one radiation null at 8.65 GHz, which is at the left side of the passband edge of the filter and thereby improves the filtering antenna's selectivity. This radiation null can easily be controlled by applying perturbation in the dual-mode SIHC. Thus, the design flexibility can be achieved. The normalized radiation patterns of the filtering antenna are shown in Figure 8.13. This is a unidirectional radiation pattern. Since a rectangular slot is placed on the back side of the cavity, the radiation appears at the $-Z$ direction. The measured cross-polarization level is below 18 dB and the measured front-to-back ratio (FTBR) is lower than 24 dB. The radiation pattern is similar throughout the operating passband. The top and bottom views of the fabricated prototype are provided in Figure 8.14(a, b), respectively.

8.1.2 Diplexing Antenna Design Using the SIHC

In the transceiver design, filtering in the upper and lower channels is implemented using a diplexer network. Traditionally, both the diplexer and antenna are designed separately. However, impedance mismatch or a detuned resonating frequency may degrade the system's performance. So diplexing antennas can provide a solution to the problem by designing both the antenna and the diplexer simultaneously, which improves selectivity and reduces losses. In this design, a diplexing antenna is realized using a dual-mode hexagonal cavity. In Chapter 6, diplexer synthesis and design procedures were discussed, where dual-mode hexagonal cavities are utilized to design diplexers. The same diplexer design technique is followed here to realize a diplexing antenna.

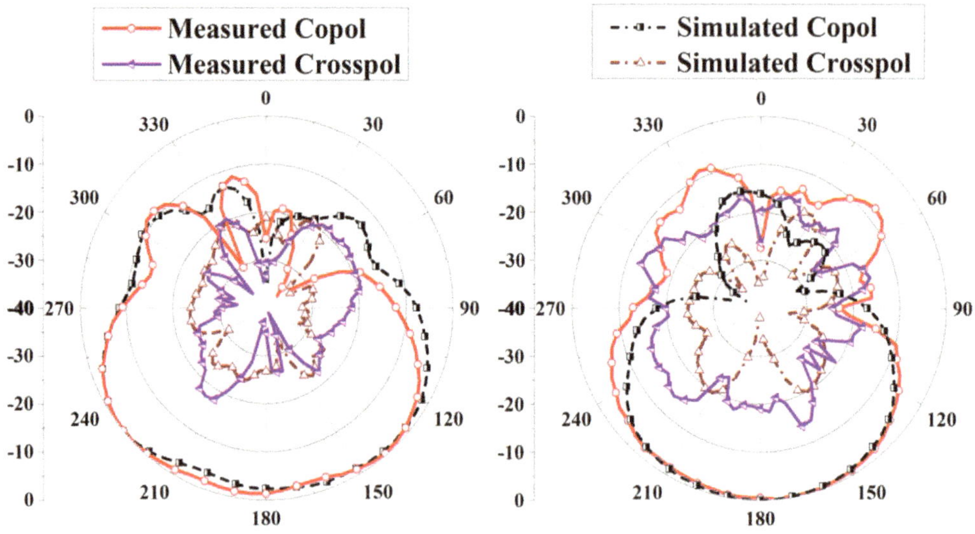

(a) (b)

Figure 8.13 Normalized radiation patterns of the filtering antenna. (a) E-plane and (b) H-plane radiation pattern at 8.89 GHz.

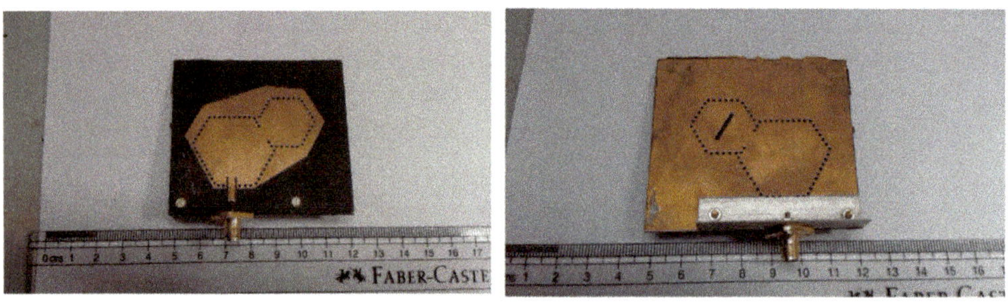

(a) (b)

Figure 8.14 Fabricated prototype of filtering antenna (Design-I): (a) top view and (b) bottom view.

8.1.2.1 Diplexing Antenna Design Using a Single Dual-Mode SIHC (Design II)

First, the diplexer is realized using design specifications and coupling topology. The design specifications of the diplexer are as follows:

- *Rx channel (lower band):* The passband edges are 9.13 to 9.33 GHz, and the return loss in the passband is 15 dB with one transmission zero at 9.5 GHz, which is above the Rx passband.
- *Tx channel (higher band):* The passband edges are 9.7 to 9.8 GHz, and the return loss in the passband is 15 dB with one transmission zero at 9.60 GHz, which is below the Tx passband.

Since both transmission zeros are placed between the two passbands, the diplexer selectivity improves. The coupling topology of the diplexer is shown in Figure 8.15,

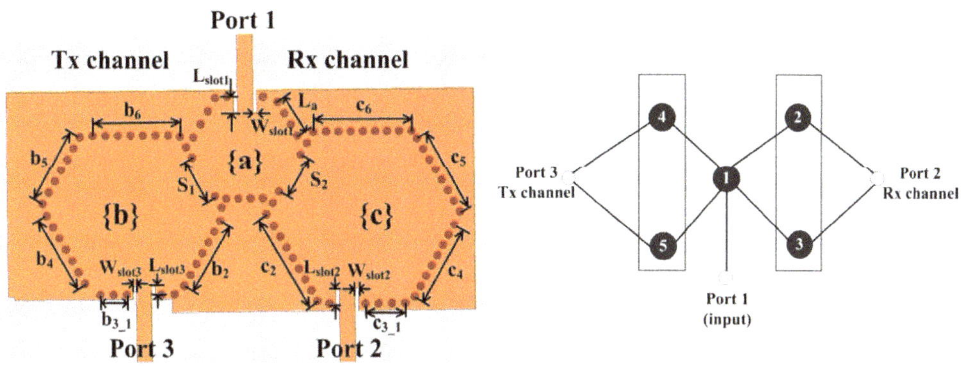

(a) (b)

Figure 8.15 Schematic of the diplexer along with dimensions. $S_2 = 6.775$ mm, $S_1 = 7.075$ mm, $L_{slot2} = 2.375$ mm, $W_{slot2} = 0.6$ mm, $L_{slot3} = 1.35$ mm, $W_{slot3} = 0.6$ mm, $L_{slot1} = 2.5$ mm, $W_{slot1} = 0.6$ mm, $c_6 = 15.3$ mm, $c_5 = 13.3$ mm, $c_4 = 12.6$ mm, $c_{3\_1} = 6.3$ mm, $c_2 = 15.5$ mm, $b_6 = 13.8$ mm, $b_5 = 10.0$ mm, $b_4 = 12$ mm, $b_{3\_1} = 4.2$ mm, $b_2 = 11.4$ mm, and $L_a = 9.7$ mm, and corresponding coupling topology.

in which Rx and Tx channels have one single dual-mode hexagonal cavity. Based on the design specifications and coupling topology, the coupling matrix of the diplexer is obtained using the generalized Chebyshev synthesis method given (8.3).

$$M = \begin{bmatrix} & P_1 & P_2 & P_3 & 1 & 2 & 3 & 4 & 5 \\ P_1 & 0 & 0 & 0 & 0.62 & 0 & 0 & 0 & 0 \\ P_2 & 0 & 0 & 0 & 0 & -0.336 & 0.405 & 0 & 0 \\ P_3 & 0 & 0 & 0 & 0 & 0 & 0 & -0.01 & 0.326 \\ 1 & 0.62 & 0 & 0 & 0.278 & 0.183 & 0.259 & 0.671 & 0.2048 \\ 2 & 0 & -0.336 & 0 & 0.183 & 0.406 & 0 & 0 & 0 \\ 3 & 0 & 0.405 & 0 & 0.259 & 0 & 0.935 & 0 & 0 \\ 4 & 0 & 0 & -0.01 & 0.671 & 0 & 0 & -0.461 & 0 \\ 5 & 0 & 0 & 0.326 & 0.248 & 0 & 0 & 0 & -0.891 \end{bmatrix} \quad (8.3)$$

The diplexer is designed using the Rogers RT-Duroid 5880 substrate with relative dielectric constants (ε_r) and a thickness (h) of 0.787 mm. Based on the coupling matrix, the diplexer is designed. The schematic and the dimensions of the diplexer are shown in Figure 8.15(a). Resonator $\{c\}$ is a dual-mode SIHC associated with the Rx channel, whereas resonator $\{b\}$ is associated with the Tx channel. In the Rx channel, transmission zero is placed above the passband, which is achieved by inward perturbation using side wall c_5. Similarly, in the Tx channel, side wall b_5 of $\{b\}$ is utilized for outward perturbation, which leads to a design-specified transmission zero below the Tx passband. Therefore, using the perturbation technique in dual-mode SIHC, the desired self-resonant frequencies of the orthogonal modes are achieved.

The self-resonant frequencies, external quality factors, and coupling coefficients obtained from (8.3) are as follows: $f_{01} = 9.37$ GHz, $f_{02} = 9.32$ GHz, $f_{03} = 9.15$ GHz, $f_{04} = 9.61$ GHz, $f_{05} = 9.75$ GHz, $K_{12} = 0.0128$, $K_{13} = 0.0181$, $K_{14} = 0.0468$, $K_{15} =$

0.0173, Q_E (P_1, 1) = 40, Q_E (P_2, 2) = 130, Q_E (P_2, 3) = 90, Q_E (P_3, 4) = 14,325, and Q_E (P_3, 5) = 130. Figure 8.16 shows the simulated and theoretical S-parameters responses of the diplexer. The simulated response shows good agreement with the theoretical response of the diplexer.

After successfully designing the diplexer, the next objective is to realize the diplexing antenna based on the designed diplexer mentioned above. Figure 8.17

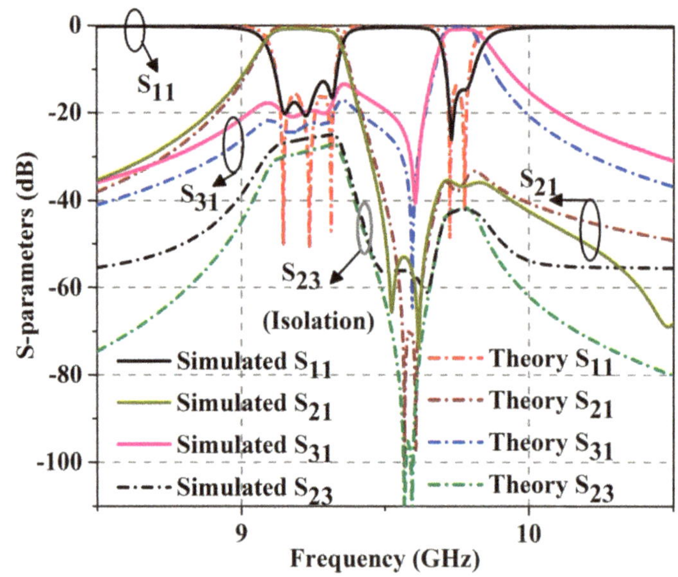

Figure 8.16 Simulated and theoretical (coupling matrix) diplexer responses.

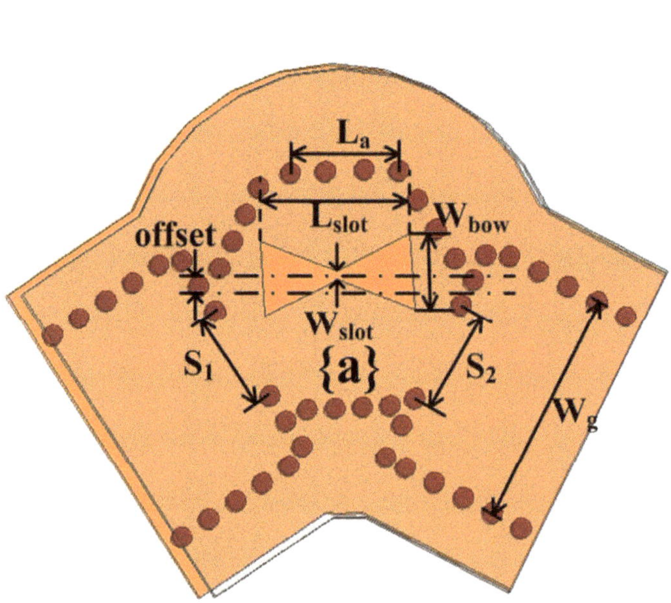

Figure 8.17 Schematic of the single-mode SIHC ({*a*})-backed bow-tie slot antenna coupled with an SIW feeding network. (W_{bow} = 5 mm, W_{slot} = 0.6 mm, L_{slot} = 10.3 mm, *offset* = −1.0 mm, S_1 = 7.075 mm, S_2 = 6.775 mm, W_g = 16 mm, and L_a = 9.7 mm.)

shows a schematic and its dimensions of a single-mode, SIHC-backed, bow-tie radiating slot that is magnetically coupled with SIW. Figure 8.18(a, c) exhibit the vector electric field distributions in the single-mode SIHC cavity without and with a radiating slot for the fundamental mode at 9.41 GHz and 8.98 GHz, respectively, when each side-wall length (L_a) of the SIHC is 9.7 mm. Figure 8.18(b, d) show the corresponding normalized input impedance.

In this design, the dimensions of the single-mode SIHC (L_a) and the coupling iris windows (S_1 and S_2) are kept the same as they are in the diplexer. However, due to the presence of a bow-tie slot on the back side of the cavity, the fundamental mode gets disturbed and shifted to 8.98 GHz. The dimensions and position of the bow-tie slot are chosen carefully to get proper impedance matching. The bow-tie slot length (L_{slot}) is chosen as a half-wavelength ($\lambda/2$) at the frequency of operation.

At 8.98 GHz, the electric field vector and normalized input impedance are shown in Figure 8.18(c, d), respectively. To obtain the desired resonant frequency of {a} loaded with a bow-tie slot at 9.41 GHz, each side-wall length (L_a) of the hexagonal cavity should be reduced. After L_a is made 9 mm, the resonant frequency of the radiating resonator is again tuned at 9.4 GHz, as shown in Figure 8.19. Thus, the single-mode SIHC can be replaced with SIHC loaded with a bow-tie radiating

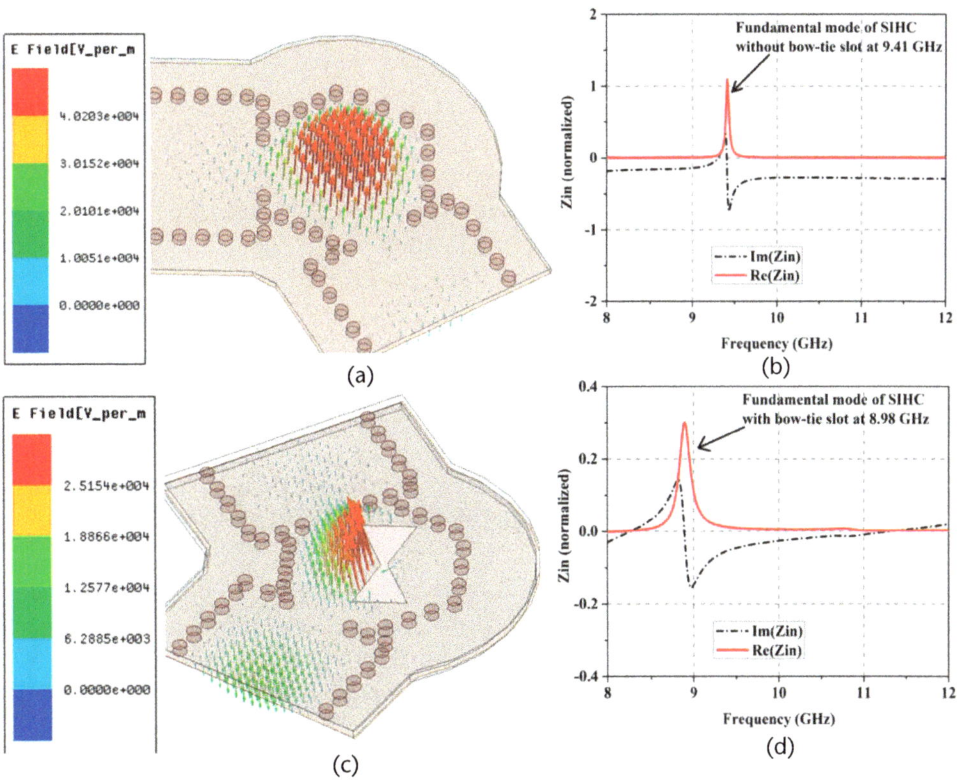

Figure 8.18 Plot of electric field vector and normalized input impedance of the single-mode SIHC with a dimension of each side wall (L_a) of 9.7 mm. (a) Electric field vector plot for fundamental mode at 9.41 GHz without introducing the slot. (b) Corresponding normalized input impedance (Z_{in}). (c) Electric field vector plot of cavity-backed bow-tie slot for fundamental mode at 8.98 GHz, keeping all the dimensions same and the antenna dimension is provided in Figure 5.20. (d) Corresponding normalized input impedance (Z_{in}).

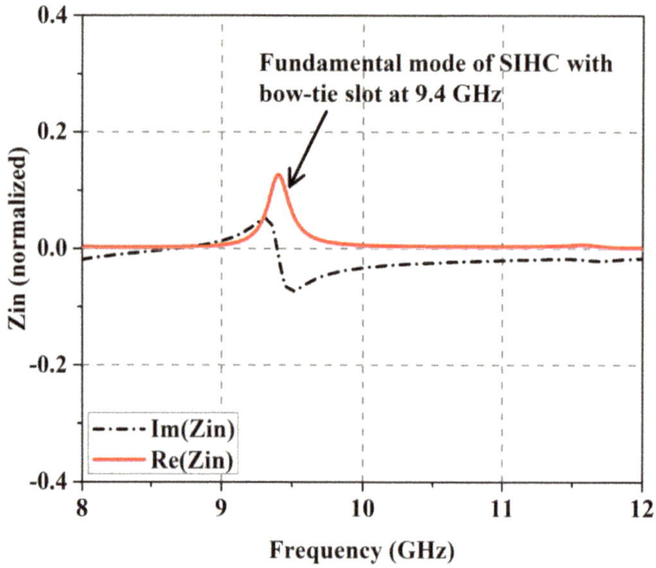

Figure 8.19 Normalized input impedance (Z_{in}) plot of the single-mode SIHC-backed bow-tie slot tuned at 9.4 GHz. Each side-wall dimension (L_a) for the SIHC is 9 mm.

resonator keeping the same self-resonant frequency. The bow-tie slot is adopted in realizing diplexing antennas instead of a rectangular slot because a bow-tie slot antenna has better impedance-matching performance as compared to a rectangular slot antenna [3].

The common resonator of the diplexer is replaced by a single-mode, SIHC-backed, bow-tie slot radiating resonator of which each side wall (L_a) is 9 mm. The coupling topology of the diplexing antenna is provided in Figure 8.20(a), where resonator 1 is a single-mode, SIHC-backed, bow-tie slot antenna. The geometry of the diplexing antenna is provided in Figure 8.20(b), along with the dimensions. The dual-mode SIHC {*c*} is attached to the Rx channel, which exhibits a lower-frequency band response, whereas {*b*} is attached to the Tx channel, which exhibits a higher-frequency band response.

In Figure 8.21, it is observed that the filtering antenna's response is highly sensitive to the variations of the bow-tie slot antenna's design parameters. Figure 8.21(a) shows an increase in the flaring width (W_{bow}) of the bow-tie slot, which improves impedance matching in both Rx and Tx channels. Other than this, it also affects the Rx and Tx bandwidth. Similarly, *offset* variations of the bow-tie slot shift the resonant frequency and change the input impedance of the radiating resonator *{a}*. As a consequence, matching gets disturbed, as shown in Figure 8.21(b). From the parametric variations, the optimum flare of the bow-tie and the offset length are chosen. The calculated unloaded quality factor (Q_U) of the resonator {*a*} is around 400 for the fundamental mode without loading with a bow-tie slot antenna. When the same cavity is loaded with bow-tie slot antenna, its Q decreases.

Using EM simulation, the loaded Q_A and equivalent lumped circuit element of the cavity-backed bow-tie slot are extracted by applying very weak coupling shown in Figure 8.22(a). The inductive iris window S_1 is kept as small as possible to avoid

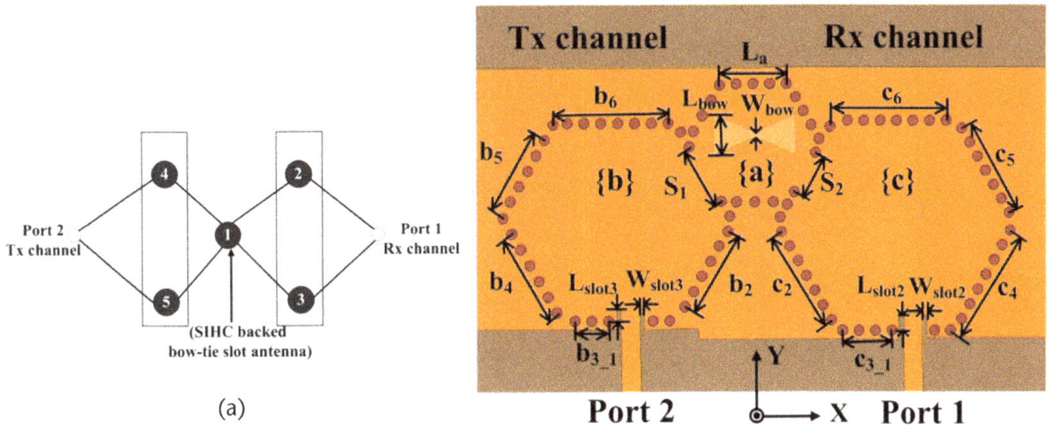

(a) (b)

Figure 8.20 (a) Coupling topology of the diplexing antenna. (b) Schematic of the diplexer antenna along with dimensions: $S_2 = 5.575$ mm, $S_1 = 7.725$ mm, $L_{slot} = 10.3$ mm, $L_{slot2} = 2.05$ mm, $W_{slot2} = 0.6$ mm, $L_{slot3} = 1.35$ mm, $W_{slot3} = 0.6$ mm, $W_{bow} = 5.0$ mm, $W_{slot} = 0.6$ mm, $c_6 = 14.5$ mm, $c_5 = 12$ mm, $c_4 = 14.4$ mm, $c_{3\,1} = 6.3$ mm, $c_2 = 12.6$, $b_6 = 14.4$ mm, $b_5 = 10.9$ mm, $b_4 = 11.25$ mm, $b_{3\,1} = 4.2$ mm, $b_2 = 10.75$ mm, and $L_a = 9.0$ mm (refer to Figure 8.17).

a loading effect. In the design, S_1 is 3.7 mm and the via's diameter and spacing are 1.3 mm and 2 mm, respectively. The iris window can be considered an admittance inverter J, which is attached to an equivalent lumped resonator of the cavity-backed bow-tie slot antenna, as shown in Figure 8.22(b). The obtained lumped parameter values are listed in Figure 8.22. The EM and lumped circuit simulated input impedances are shown in Figure 8.22(c). The extracted loaded Q (Q_A) is around 42.0. Q_A and the equivalent lumped model of the cavity-backed bow-tie slot are extracted using the reactance slope (8.4) as mentioned in [1].

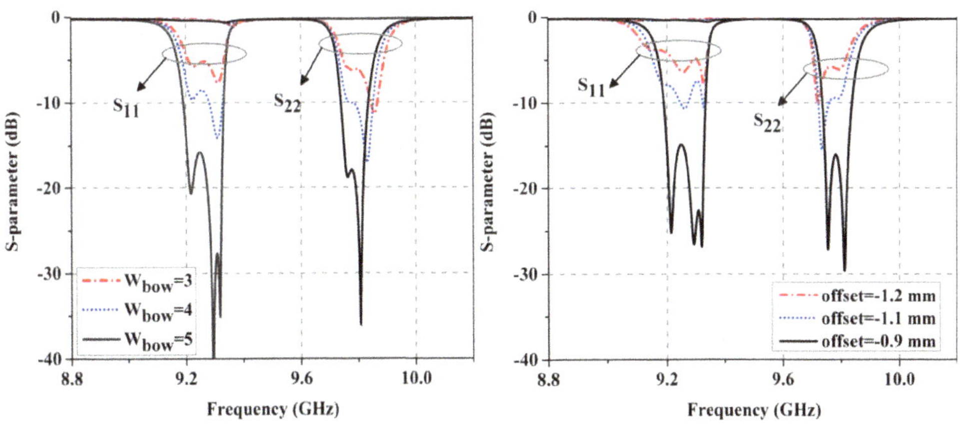

(a) (b)

Figure 8.21 Effects of the antenna's design parameters on the reflection coefficients of the diplexing antenna. (a) Flair width (W_{bow}) variation. (b) Variation of antenna offset position with respect to center line of the SIHC cavity shown in Figure 5.20.

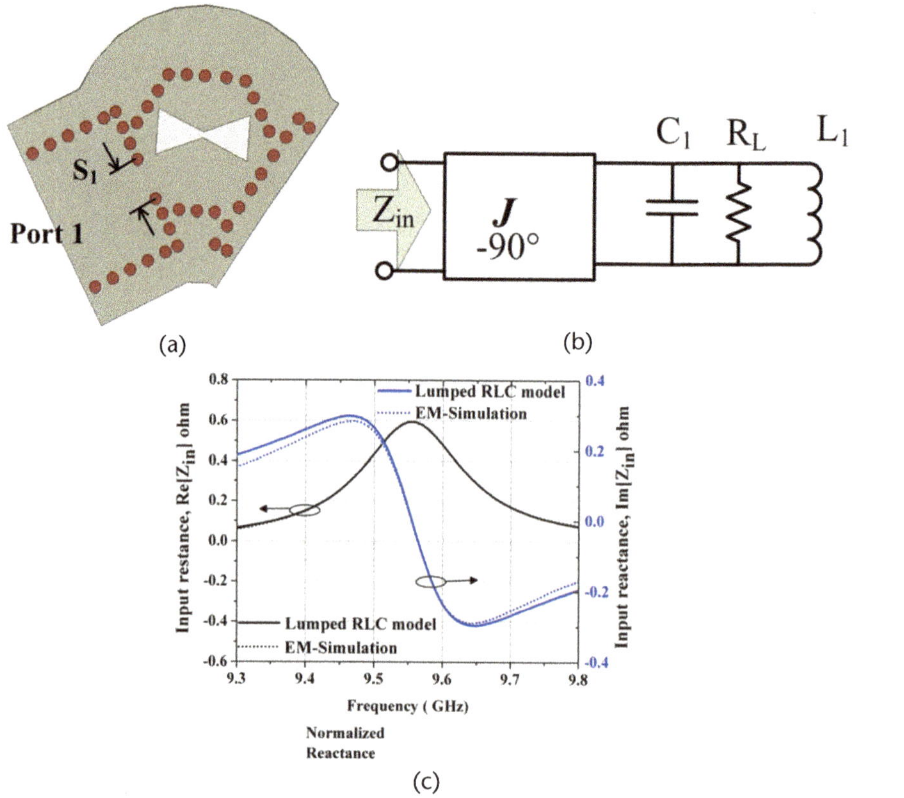

Figure 8.22 Analysis model for extracting Q of SIHC-backed bow-tie slot radiating resonator. (a) Single-mode, SIHC-backed bow-tie resonator weakly coupled with the SIW. Antenna dimensions are provided in Figure 8.17. Here $S_1 = 3.7$ mm, d (diameter of via) = 1.3 mm, and S (spacing between two vias) = 2 mm. (b) Equivalent lumped RLC circuit of the SIHC-backed, bow-tie slot radiating resonator. Extracted equivalent lumped circuit: $L_1 = 0.417$ nH, $C_1 = 0.665$ pF, $R_L = 0.596$, and extracted $Q_A = 42$. (c) Input impedance of the lumped circuit model and EM simulation.

$$
\begin{aligned}
Q_A &= \frac{\omega_0}{2\mathrm{Re}\left(Z_{\mathrm{in}}(\omega_0)\right)} \frac{d\left(\mathrm{Im}\left(Z_{\mathrm{in}}\right)\right)}{d\omega}\Bigg|_{\omega=\omega_0} \\
L_1 &= \frac{\omega_0}{2} \frac{d\left(\mathrm{Im}\left(Z_{\mathrm{in}}\right)\right)}{d\omega}\Bigg|_{\omega=\omega_0}
\end{aligned}
\tag{8.4}
$$

The lower Q cavity helps to achieve good impedance matching. The calculated loaded Q allows representing the filtering antenna using lumped RLC elements, which helps to carry out the diplexing antenna synthesis by optimization [2]. Keysight ADS software is used to carry out diplexing antenna synthesis by the optimization process. The corresponding circuit schematic is shown in Figure 8.23. Circuit and EM-simulated S-parameters of a diplexing antenna are compared with the measured ones, as shown in Figure 8.24(a). It shows good agreement between the measured and simulated responses of the diplexing antenna. The measured center frequencies

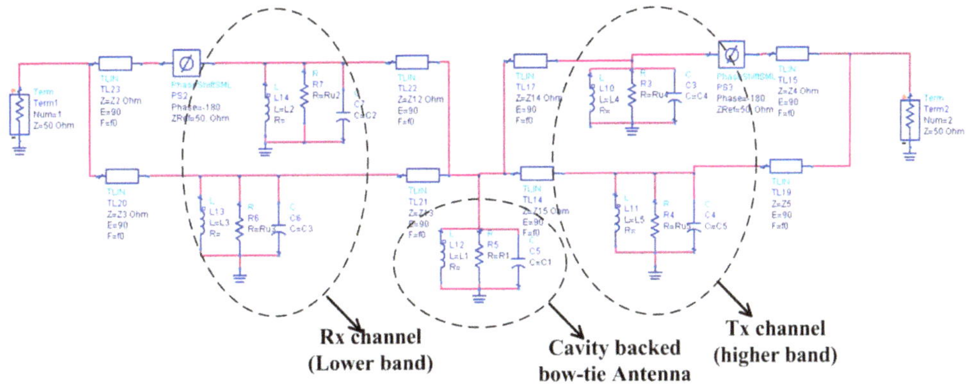

Figure 8.23 Circuit schematic model of the diplexing antenna.

of the Rx and Tx channels are 9.22 GHz and 9.8 GHz, respectively, with impedance bandwidths (–10-dB S_{11}) of 180 MHz (1.95%) and 120 MHz (1.24%), respectively. Besides that, the measured reflection coefficient is below 15 dB in the Rx and Tx channels. Figure 8.24(b) shows that the measured and simulated isolations are below 20 dB. Figure 8.25(a, b) compare the measured and simulated S-parameters and gain of diplexing antenna with diplexer's S-parameters to validate our design procedure. It is observed that the diplexing antenna follows the diplexer's responses, which is expected.

The diplexing antenna's gain and radiation null are analogous to the diplexer's insertion loss and transmission zero. Since the diplexing antenna design is based on the filter synthesis process, the diplexing antenna gains are flat in both the Rx and Tx passbands. The measured gains of the diplexing antenna are 3.57 dBi and 4.1 dBi in the Rx and Tx channels, respectively, whereas the simulated gains are 3.8 dBi and 4.32 dBi, respectively. In the Rx channel, the radiation null appears above the passband at 9.36 GHz, whereas, in the diplexer, transmission zero appears above the passband at 9.52 GHz. Similarly, in the Tx channel, the radiation null is at 9.61 GHz, whereas, in the diplexer, transmission zero appears below the passband at

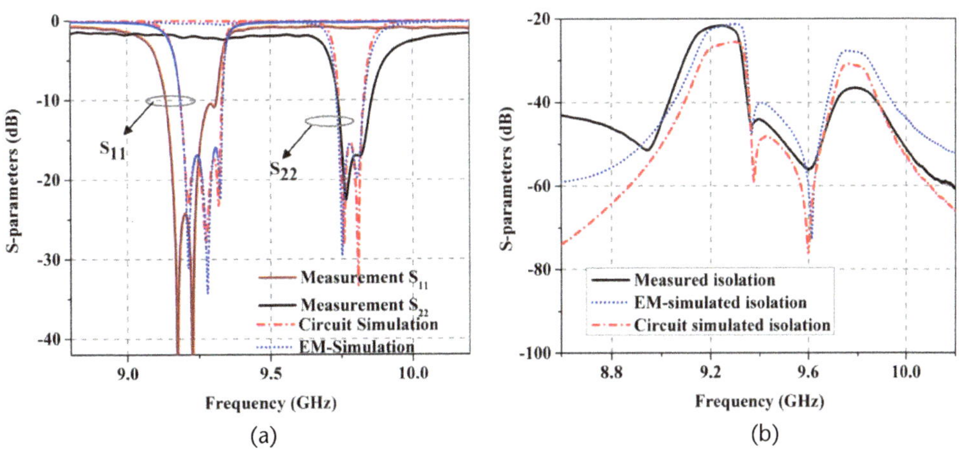

Figure 8.24 Measured, EM-simulated, and circuit-simulated responses: (a) reflection coefficients (dB), and (b) isolations of the diplexing antenna.

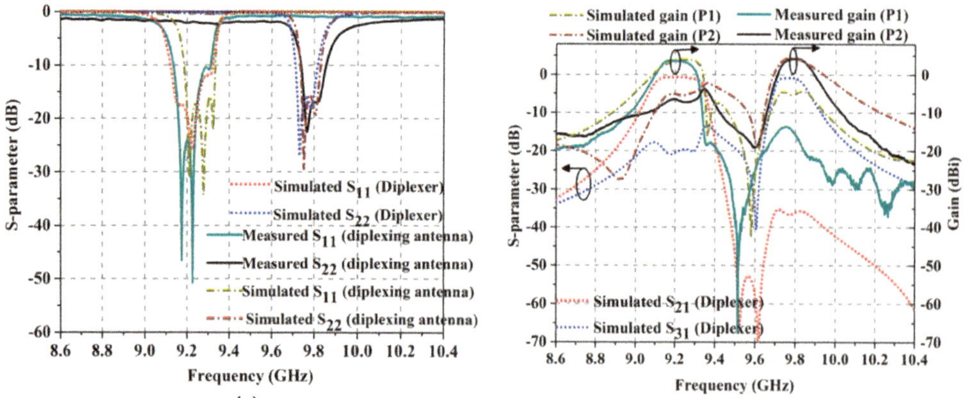

(a) (b)

Figure 8.25 (a) Measured and simulated S-parameters of the diplexing antenna and the simulated S-parameters of the diplexer. (b) Measured and simulated gains of the diplexing antenna and simulated transmission coefficients (dB) of the diplexer.

9.60 GHz. Figure 8.25(b) shows good agreement between diplexing antenna gain and simulated diplexer insertion loss. The Rx channel has more than one radiation null, which is because of the loading effect of the Tx channel over the Rx channel, as discussed in Chapter 6. Both Rx and Tx channels have a sharp band skirt and high selectivity.

Figure 8.26(a, b) exhibit the measured and simulated normalized radiation patterns of the E and H-planes of the Rx channel (Port 1) at 9.22 GHz. It is observed that the measured cross-polar level is below 20 dB and the measured FTBR is 15 dB. Similarly, Figure 8.26(c, d) present the measured and simulated normalized radiation patterns of the E and H-planes of the Tx channel (Port 2) at 9.8 GHz. It is observed that the measured cross-polar level is below 25 dB and the measured FTBR is 15.2 dB. Figure 8.27(a, b) exhibit the top view and bottom view, respectively, of the fabricated prototype.

8.1.2.2 Diplexing Antenna Design Using a Pair of Dual-Mode SIHCs (Design III)

The schematic and dimensions of the diplexer are shown in Figure 8.28. The coupling topology is provided in Figure 8.28(b), which shows that each channel has two dual-mode resonators and a common single-mode resonator 1. Similarly, from the schematic, it is observed that each channel has a pair of dual-mode SIHCs and the common resonator {*a*}. Resonators {*b*} and {*c*} are dual-mode SIHCs attached to the Rx channel, whereas reonators {*d*} and {*e*} are dual-mode SIHCs attached to the Tx channel. Rx and Tx filters are considered at lower and higher-frequency bands, respectively. The design specifications of the diplexer are as follows:

- *Rx channel (lower band):* The passband edges are 9.255 to 9.427 GHz, the return loss in the passband is 18 dB, with two transmission zeros at 9.18 GHz and 9.5 GHz, which are on both sides of the Rx passband.
- *Tx channel (higher band):* The passband edges are 9.635 to 9.815 GHz and the return loss in the passband is 15 dB with two transmission zeros at 9.57 GHz and 9.89 GHz, which are on both sides of the Tx passband.

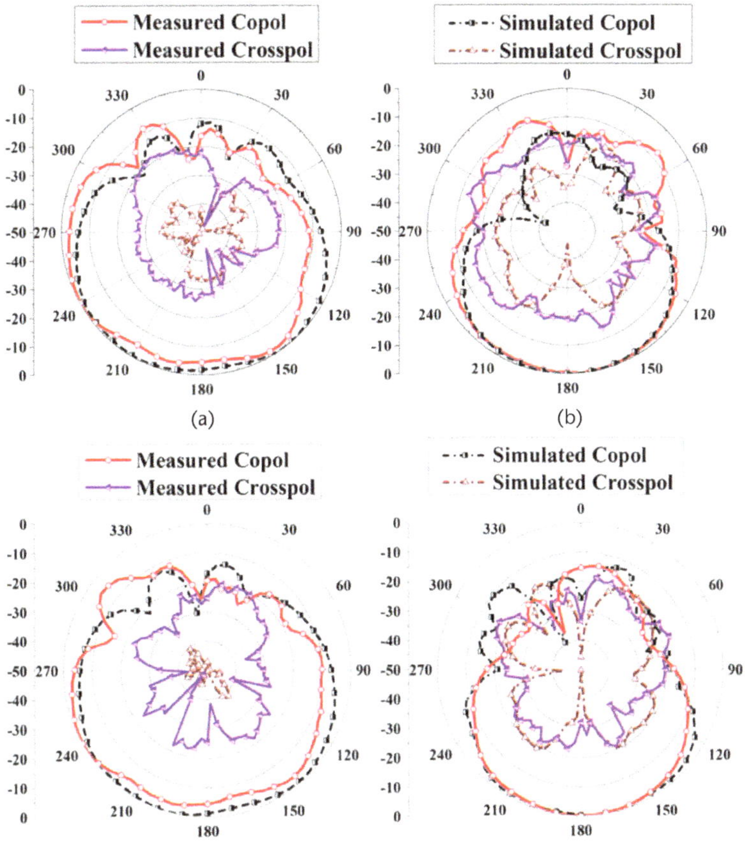

Figure 8.26 Normalized radiation pattern of the diplexing antenna. (a) E-plane and (b) H-plane radiation patterns at 9.22 GHz for the Rx channel. (c) E-plane and (d) H-plane radiation patterns at 9.8 GHz for the Tx channel.

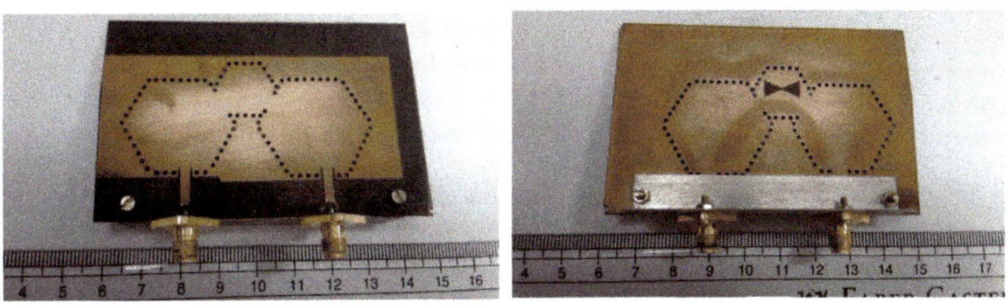

Figure 8.27 Fabricated prototype of the diplexing antenna using a single dual-mode SIHC (Design II): (a) top view and (b) bottom view.

Both Rx and Tx channels have two transmission zeros that are placed on both sides of their respective passbands, which therefore improves the diplexer's selectivity as well as isolation. Based on design specifications and coupling topology, the diplexer synthesis process is carried out using the generalized Chebyshev synthesis method. The details of diplexer design procedures are discussed in Chapter 6. The

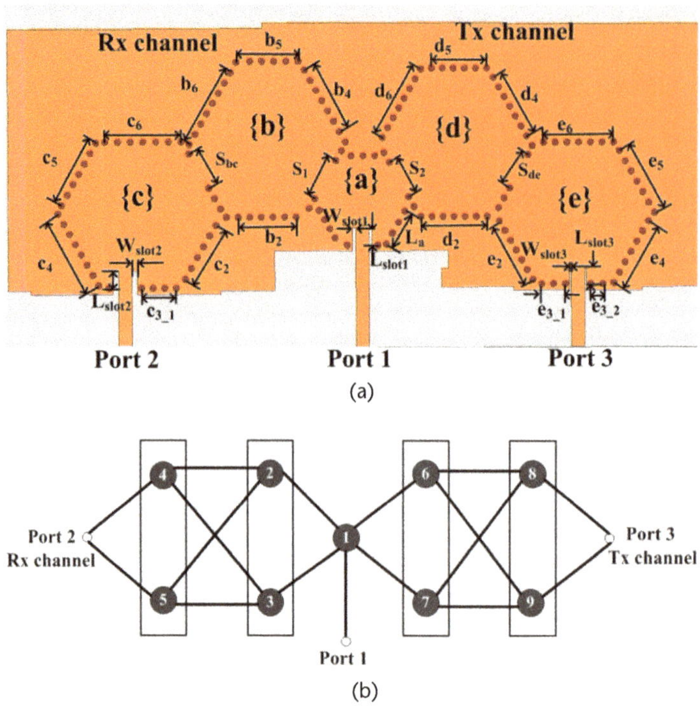

Figure 8.28 Schematic of the diplexer along with dimensions. W_{slot1} = 0.6 mm, L_{slot1} = 3 mm, L_{slot2} = 3 mm, W_{slot2} = 1.0 mm, W_{slot3} = 0.35 mm, L_{slot3} = 2.8 mm, b_5 = 11.3 mm; b_2 = 10.5 mm, c_6 = 13.8 mm, c_{3_1} = 6.3 mm, d_2 = 12.0 mm, d_5 = 10.0 mm, e_{3_1} = 4.2 mm, e_{3_2} = 2.2 mm, e_6 = 12.6 mm, S_1 = 7.675 mm, b_6 = 14.7 mm, c_2 = 11.0 mm, c_5 = 12.6 mm, d_6 = 14.0 mm, S_{de} = 6.725 mm, e_4 = 12.6 mm, S_{bc} = 7.0 mm, b_4 = mm, c_4 = mm, S_2 = 6.925 mm, d_4 = 12 mm, e_2 = 11.35 mm, e_5 = 12.6 mm, and L_a = 9.5 mm.

coupling matrix corresponding to the design specifications and coupling topology are given in (8.5).

$$
M = \begin{bmatrix}
 & P_1 & P_2 & P_3 & 1 & 2 & 3 & 4 & 5 & 6 & 7 & 8 & 9 \\
P_1 & 0 & 0 & 0 & 1 & 0 & 0 & 0 & 0 & 0 & 0 & 0 & 0 \\
P_2 & 0 & 0 & 0 & 0 & 0 & -0.181 & 0.562 & 0 & 0 & 0 & 0 & 0 \\
P_3 & 0 & 0 & 0 & 0 & 0 & 0 & 0 & 0 & 0 & 0 & -0.405 & 0.399 \\
1 & 1 & 0 & 0 & -0.062 & 0.818 & 0.056 & 0 & 0 & 0.084 & 0.486 & 0 & 0 \\
2 & 0 & 0 & 0 & 0.818 & 0.235 & 0 & 0.346 & -0.016 & 0 & 0 & 0 & 0 \\
3 & 0 & 0 & 0 & 0.056 & 0 & 0.677 & 0.194 & 0.334 & 0 & 0 & 0 & 0 \\
4 & 0 & -0.181 & 0 & 0 & 0.3460 & 0.194 & 0.57 & 0 & 0 & 0 & 0 & 0 \\
5 & 0 & 0.0562 & 0 & 0 & -0.016 & 0.334 & 0 & 0.704 & 0 & 0 & 0 & 0 \\
6 & 0 & 0 & 0 & 0.084 & 0 & 0 & 0 & 0 & -0.313 & 0 & 0.010 & -0.376 \\
7 & 0 & 0 & 0 & 0.4865 & 0 & 0 & 0 & 0 & 0 & -0.714 & 0.209 & 0.134 \\
8 & 0 & 0 & -0.405 & 0 & 0 & 0 & 0 & 0 & 0.01 & 0.209 & -0.536 & 0 \\
9 & 0 & 0 & 0.359 & 0 & 0 & 0 & 0 & 0 & -0.376 & 0.134 & 0 & -0.798
\end{bmatrix} \tag{8.5}
$$

Chapter 5 presented a detailed study on how perturbation brings two transmission zeros on both sides of the passband in each channel. A similar design approach is followed here for the realization of a diplexer response with transmission zeros on both sides of the passband. Theoretical and simulated responses of the diplexer are shown in Figure 8.29.

It shows good agreement between theoretical and simulated responses. From this figure, we see that the isolation is below 20 dB in both Rx and Tx passbands. After designing the diplexer, the next objective is to realize the diplexing antenna. The coupling topology of the diplexing antenna is shown in Figure 8.30. The self-resonant frequency of the single-mode SIHC {*a*} is 9.8 GHz, and the corresponding side-wall length of the hexagonal cavity (L_a) is 9.5 mm. When the bow-tie slot is etched on the bottom side of resonator {*a*}, the resonant frequency of the single-mode SIHC gets disturbed and shifts to a lower frequency of 9.03 GHz. Bow-tie dimensions are provided in Figure 8.31. To retain the desired self-resonant frequency (9.8 GHz), each side-wall length L_a of the hexagonal cavity is reduced to 9.41 mm. The length (L_{slot}) of the bow-tie slot antenna is considered $\lambda/2$ at the operating frequency. The only difference between the diplexer and diplexing antenna's coupling topology is that the single-mode resonator {*a*} (common resonator) with the input port in diplexer is now loaded with a bow-tie slot while the input port is removed. The geometry of the diplexing antenna is shown in Figure 8.31, where the dimensions of the diplexing antenna are also provided. Using full-wave EM simulation-based

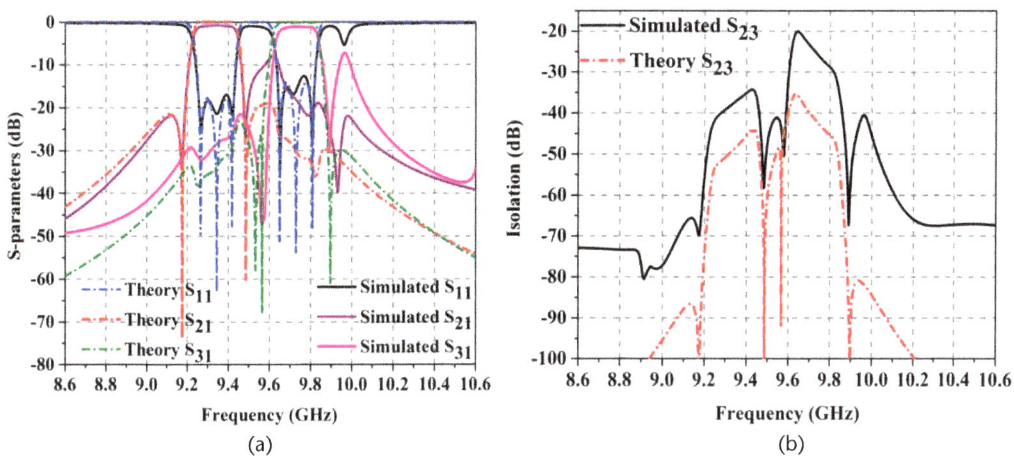

Figure 8.29 (a) Simulated and theoretical (coupling matrix) diplexer responses and (b) isolation.

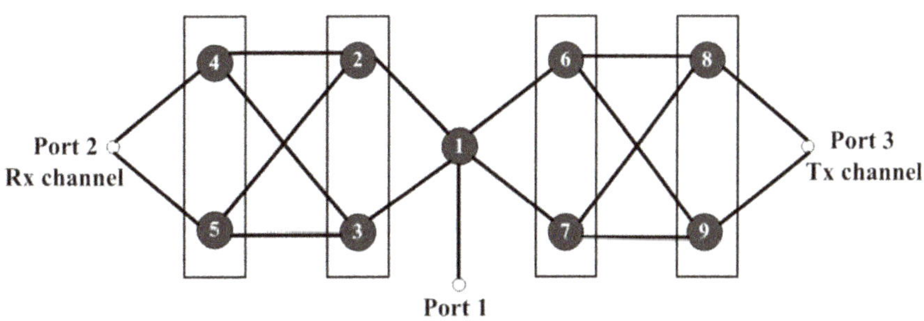

Figure 8.30 Coupling topology of diplexing antenna (Design III).

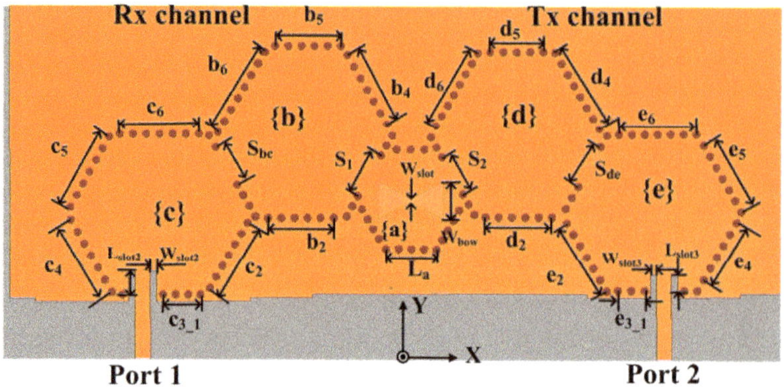

Figure 8.31 Schematic of the diplexing antenna along with dimensions: $L_{slot} = 10.3$ mm, $W_{slot} = 0.45$ mm, $W_{bow} = 5.8$ mm, $L_{slot2} = 3.8$ mm, $W_{slot2} = 1.0$ mm, $W_{slot3} = 0.95$ mm, $L_{slot3} = 2.5$ mm, $b_5 = 10.5$ mm, $b_2 = 10.5$ mm, $c_6 = 12.6$ mm, $c_{3\_1} = 6.0$ mm, $d_2 = 10.5$ mm, $d_5 = 8.4$ mm, $e_{3\_1} = 4.2$ mm, $e_6 = 12.3$ mm, $S_1 = 7.325$ mm, $b_6 = 14.7$ mm, $c_2 = 12.0$ mm, $c_5 = 12.6$ mm, $d_6 = 13.3$ mm, $S_{de} = 7.175$ mm, $e_4 = 10.5$ mm, $S_{bc} = 6.775$ mm, $b_4 = 12.6$ mm, $c_4 = 13.32$ mm, $S_2 = 6.675$ mm, $d_4 = 14$ mm, $e_2 = 12.6$ mm, $e_5 = 12.0$ mm, and $L_a = 9.41$ mm.

parametric studies, the other bow-tie slot parameters are chosen. Figure 8.32(a, b) demonstrates the influences of W_{bow} and W_{slot} on the reflection coefficients. The variations of W_{bow} and W_{slot} only affect the impedance matching of both ports, but they do not affect the impedance bandwidth of Rx and Tx channels.

Figure 8.33(a, b) compare the measured and simulated diplexing antenna's reflection coefficients and gain with the diplexer's reflection coefficients and transmission coefficients, respectively. The measured center frequencies of Rx and Tx channels are 9.35 GHz and 9.70 GHz, respectively, with impedance bandwidths (−10-dB S_{11}) of 205 MHz (2.21%) and 184 MHz (1.97%), respectively. It is also observed that the diplexing antenna follows the diplexer's responses, which is expected. Since diplexing antenna design is based on the diplexer synthesis process, the diplexing antenna gains are flat in both the Rx and Tx passbands. The measured gains are 4.41 dBi and 4.0 dBi, whereas simulated gains are 4.7 dBi and 4.3 dBi in the Rx and Tx

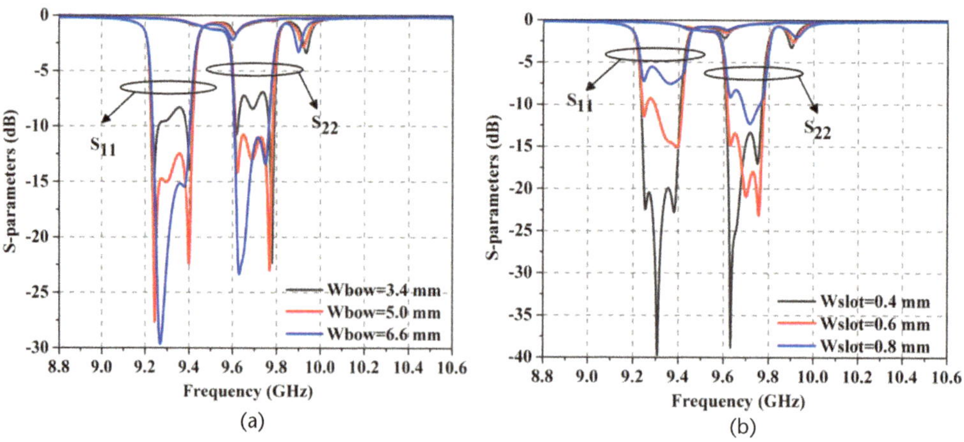

Figure 8.32 Effects of antenna design parameters on reflection coefficients. (a) Variation of reflection coefficient with a change in the amount of flaring W_{bow}. (b) Changes in reflection coefficients with a bow-tie length (W_{slot}) variation.

channels, respectively. In the Rx channel, the radiation null appears on both sides of the passband at 9.19 GHz and 9.5 GHz, whereas, in the diplexer, transmission zeros appear on both sides of the passband at 9.18 GHz and 9.49 GHz. Similarly, in the Tx channel, radiation nulls are at 9.58 GHz and 9.89 GHz, which are on both sides of the passband, whereas in the diplexer, transmission zeros appear at 9.57 GHz and 9.89 GHz. Figure 8.33(b) shows good agreement between diplexing antenna gain and simulated diplexer insertion loss.

Each Rx and Tx channel in the (−*Z*) direction has two radiation nulls (or transmission zeros) on both sides of the band edges, thereby improving selectivity. The radiation nulls of the diplexing antenna can be easily controlled by applying perturbation in dual-mode SIHC. Thus, design flexibility can be achieved. Figure 8.34 shows that the measured and simulated isolations of diplexing antennas are below 20 dB. Thus, design flexibility can be achieved. Figure 8.34 shows that

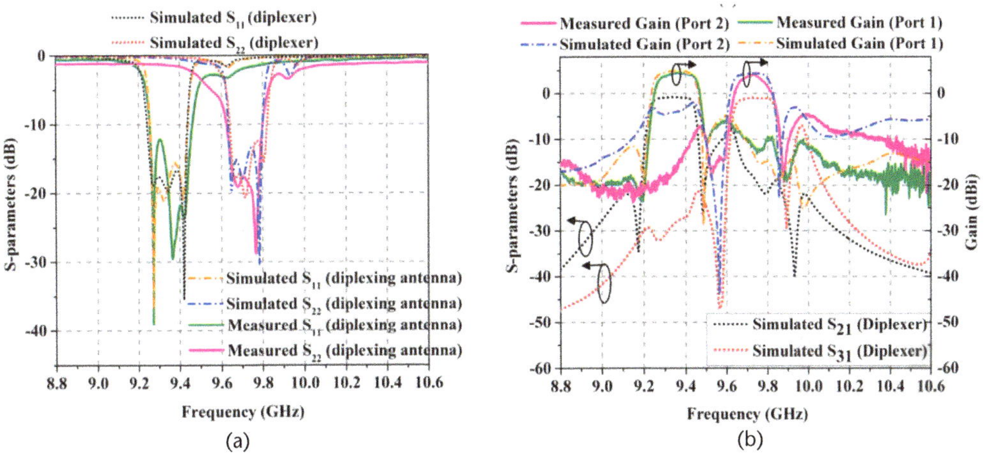

Figure 8.33 (a) Measured and simulated S-parameters of the diplexing antenna and simulated S-parameters of the diplexer. (b) Measured and simulated gains of the diplexing antenna and simulated transmission coefficients (dB) of the diplexer.

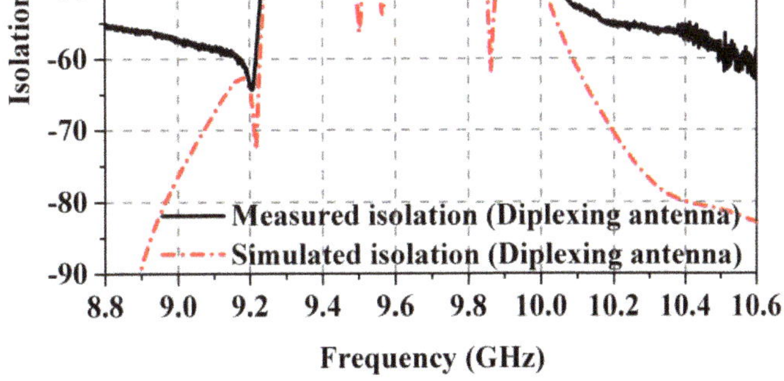

Figure 8.34 Measured and simulated isolations of the diplexing antenna.

the measured and simulated isolations of diplexing antennas are below 20 dB. It is observed that the measured cross-polar level is below 20 dB. Figure 8.35(a, b) exhibit the measured and simulated normalized radiation patterns of the E-plane and H-plane of the Rx channel (Port 1) at 9.35 GHz, where the measured FTBR is 15 dB. Similarly, Figure 5.3(c, d) present the measured and simulated normalized radiation patterns of the E-plane and H-plane of the Tx channel (Port 2) at 9.7 GHz. It is observed that the measured cross-polar level is below 20 dB and the measured FTBR is 15 dB. Figure 8.36(a, b) exhibit the top view and bottom view, respectively, of the fabricated prototype.

8.1.3 Triplexing Antenna Design Using the SIHC (Design IV)

In Chapter 6, a triplexer is realized using a combination of dual-mode and single-mode SIHCs. The triplexer is a four-port network where Port 2, Port 3, and Port 4

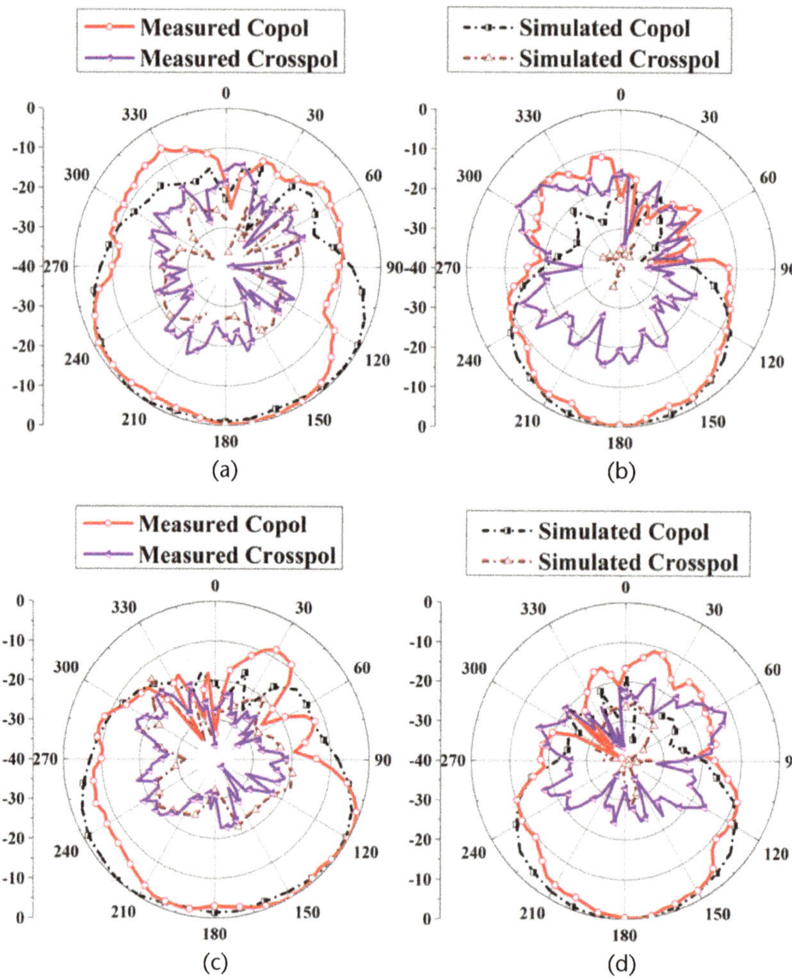

Figure 8.35 Normalized radiation pattern of the diplexing antenna: (a) E-plane and (b) H-plane radiation patterns at 9.35 GHz for Rx channel, and (c) E-plane and (d) H-plane radiation pattern at 9.7 GHz for Tx channel.

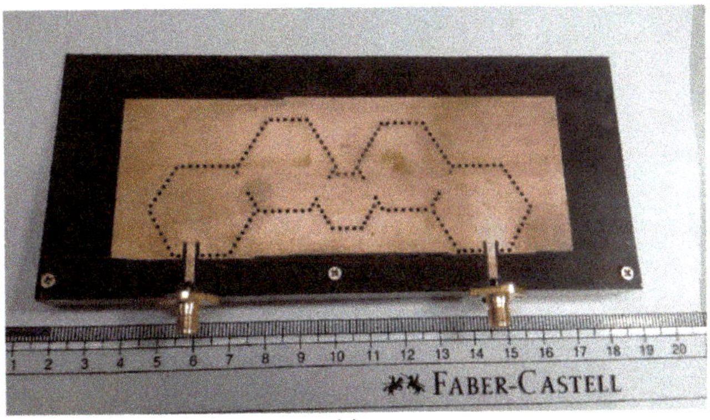

(a)

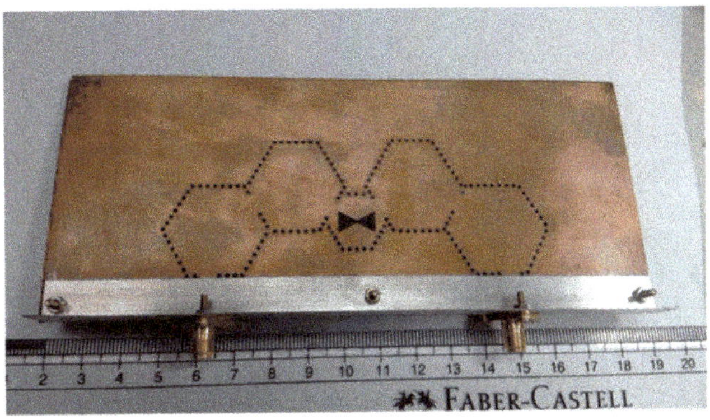

(b)

Figure 8.36 Fabricated prototype of diplexing antenna design using a pair of dual-mode SIHCs (Design III): (a) top view and (b) bottom view.

are considered Rx (lower), Mx (middle), and Tx channels (higher-frequency band), respectively. The schematic of the triplexer along with its dimensions are provided in Figure 6.33, and the corresponding coupling topology is also presented in Figure 6.34. Based on the designed triplexer, a triplexing antenna is realized as shown in Figure 8.37, and its coupling topology is provided in Figure 8.38, where it is observed that the common resonator 1 of the triplexer is acting as a radiating resonator. A triplexing antenna has three channels: Rx, Mx, and Tx. Two resonators are attached to each channel. Out of two resonators, one is a single-mode SIHC, and the other is a dual-mode SIHC. The schematic of the triplexing antenna is provided in Figure 8.37, along with its dimensions. A bow-tie slot is placed on the back side of the single-mode resonator {*a*}. The length (L_{slot}) of the bow-tie slot is half-wavelength ($\lambda/2$) at the operational frequency. The other dimensions of the bow-tie are chosen carefully in order to act as an efficient radiating resonator. The loaded quality factor of the cavity-backed bow-tie slot resonator is extracted using the weak coupling method, as was done in the previous design.

The triplexer's coupling coefficients, given in (6.31), are considered the initial coupling values of the triplexing antenna in circuit simulation. After a few optimizations in circuit simulation considering the external quality factor, the desired

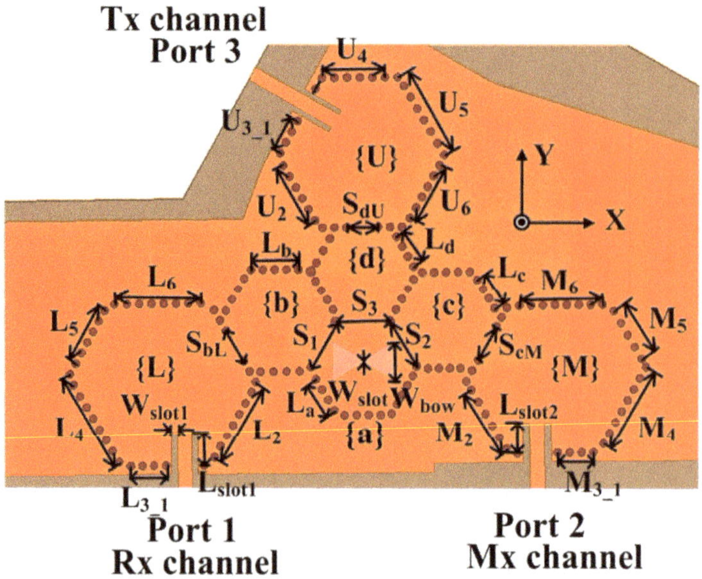

Figure 8.37 Schematic of the triplexing antenna along with dimensions: $L_a = 9.4$ mm, $L_b = 10.43$ mm, $L_c = 9.72$ mm, $L_d = 9.59$ mm, $W_{bow} = 6.6$ mm, $W_{slot} = 0.4$ mm, $S_3 = 8.9$ mm, $S_{dU} = 5.225$ mm, $L_6 = 14.7$ mm; $M_6 = 14.0$ mm, $U_4 = 12.6$ mm, $L_{3\_1} = 6$ mm, $M_{3\_1} = 6$ mm, $L_{slot} = 10.3$ mm, $L_{slot1} = 5.5$ mm, $W_{slot1} = 1.1$ mm, $L_{slot1} = 5.1$ mm, $S_1 = 8.6$ mm, $S_{cM} = 6.325$ mm, $L_2 = 12.6$ mm, $L_5 = 10.0$ mm, $M_4 = 14.7$ mm, $U_6 = 10.0$ mm, $S_2 = 8.63$ mm, $S_{bL} = 5.775$ mm, $L_4 = 17$ mm, $M_2 = 13.2$ mm, $M_5 = 9.2$ mm, $U_2 = 10.5$ mm, and $U_5 = 15$ mm.

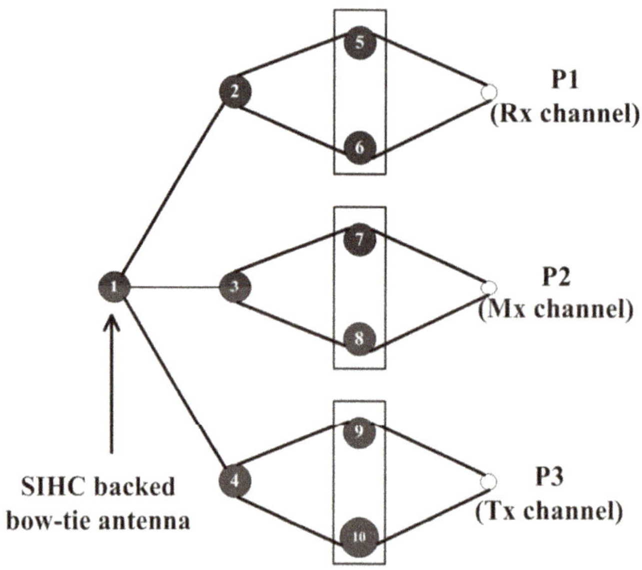

Figure 8.38 Coupling topology of triplexing antenna.

coupling values of the triplexing antenna are achieved in order to obtain the desired frequency response. The dimensions of the triplexing antenna are provided in Figure 8.37. The measured and simulated reflection coefficients of the triplexing antenna are shown in Figure 8.39.

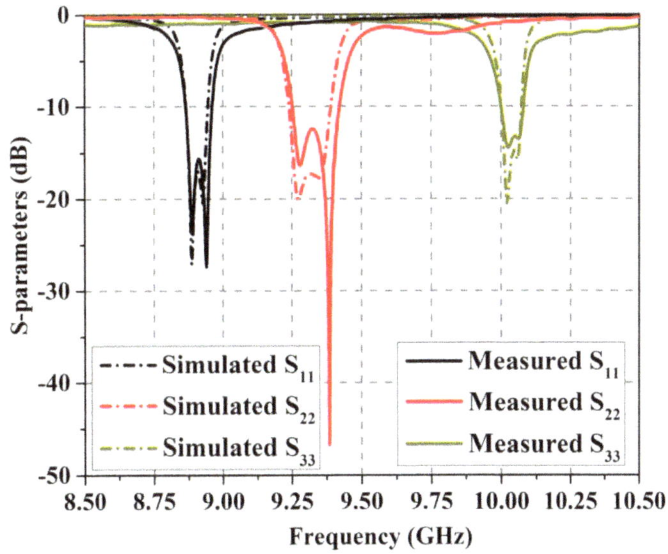

Figure 8.39 Plot of the measured and simulated reflection coefficients of the triplexing antenna.

- *Measured response of Rx channel:* Center frequency at 8.9 GHz with a −10-dB impedance bandwidth of 100 MHz (1.12%) and reflection coefficient is below 15 dB.
- *Measured response of Mx channel:* Center frequency at 9.35 GHz with a −10-dB impedance bandwidth of 171 MHz (1.83%) and reflection coefficient is below 15 dB.
- *Measured response of Tx channel:* Center frequency at 10.0 GHz with a −10-dB impedance bandwidth of 80 MHz (0.79%) and reflection coefficient is below 15 dB.

Figure 8.40 shows the measured and simulated triplexing antennas' gain and reflection coefficients. Since the triplexer synthesis technique is adopted to realize a triplexing antenna, the triplexing antenna exhibits better band-edge selectivity and flat passband response. The measured Rx, Mx, and Tx channels' gains are 3.24 dBi, 4.06 dBi, and 2.48 dBi, respectively, whereas the simulated gains are 3.56 dBi, 4.27 dBi, and 2.65 dBi, respectively. Since dual-mode cavities are employed to design triplexing antenna, the triplexing antenna provides improved skirt selectivity with stopband suppression. The measured isolations between all three channels shown in Figure 8.41 are below 20 dB.

The measured normalized radiation patterns of the triplexing antenna are shown in Figures 8.42 and 8.43. This is a unidirectional radiation pattern. Since the bow-tie slot is placed on the back side of the cavity, the radiation appears in the *z*-direction.

- Figure 8.42(a, b) present the measured and simulated normalized radiation patterns in the E and H-planes of the Rx channel (Port 1) at 8.91 GHz. It is observed that measured cross-polar level is below 20 dB and the FTBR is 12.5 dB.

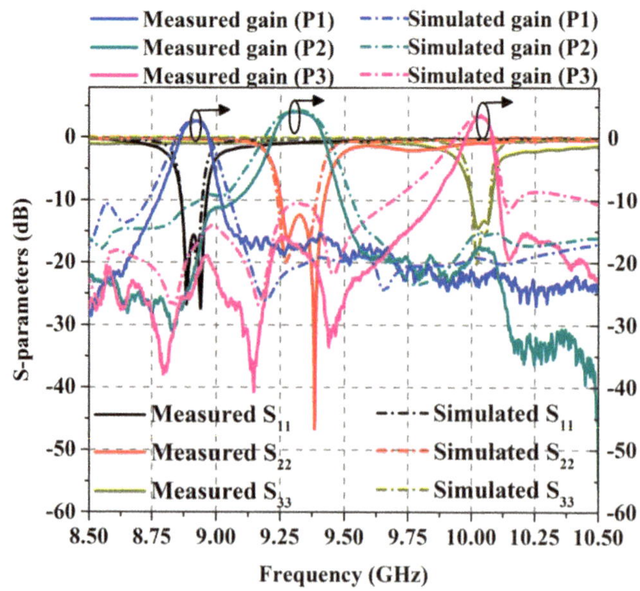

Figure 8.40 Measured and simulated gains and S-parameters of the triplexing antenna.

- Figure 8.42(c, d) show the measured and simulated normalized radiation pattern in the E and H-planes of the Mx channel (Port 2) at 9.33 GHz. It is observed that measured cross-polar level is below 30 dB and the FTBR is 12 dB.
- Similarly, Figure 8.43(a, b) show the measured and simulated normalized radiation patterns of the E and H-planes of the Tx channel (Port 3) at 10.0

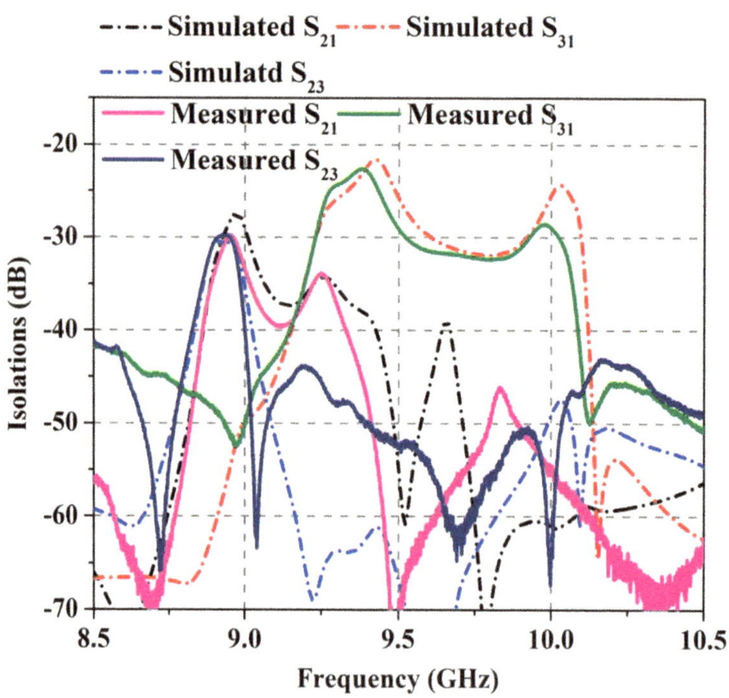

Figure 8.41 Measured and simulated isolations of the triplexing antenna.

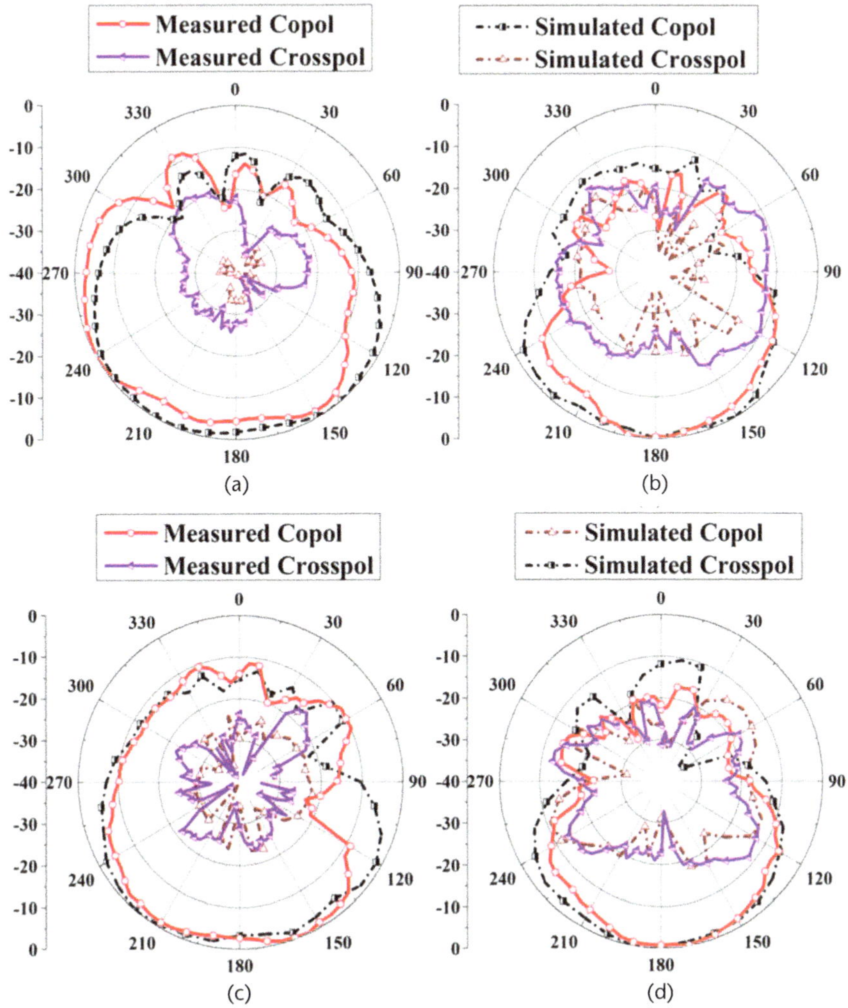

Figure 8.42 Normalized radiation patterns of the triplexing antenna at the Rx and Mx channels. (a) E-plane and (b) H-plane radiation patterns at 8.90 GHz for the Rx channel. (c) E-plane and (d) H-plane radiation patterns at 9.35 GHz for the Mx channel.

GHz. It is observed that measured cross-polar level is below 15 dB and the FTBR is 18.0 dB.

The radiation pattern is similar throughout the operating passband of Rx, Mx, and Tx channel. Figure 8.44 shows the fabricated prototype of the triplexing antenna (Design IV), where Figure 8.44(a, b) exhibit the top and bottom views.

8.1.4 Balun Filtering Antenna (Design V)

The schematic of the balun filtering antenna is shown in Figure 8.45, along with its dimensions. The design of the balun filter and its performance were discussed in Section 6.4.1. The same balun filter is employed to realize the balun filtering antenna. This is a fifth-order balun filter with two transmission zeros on both sides

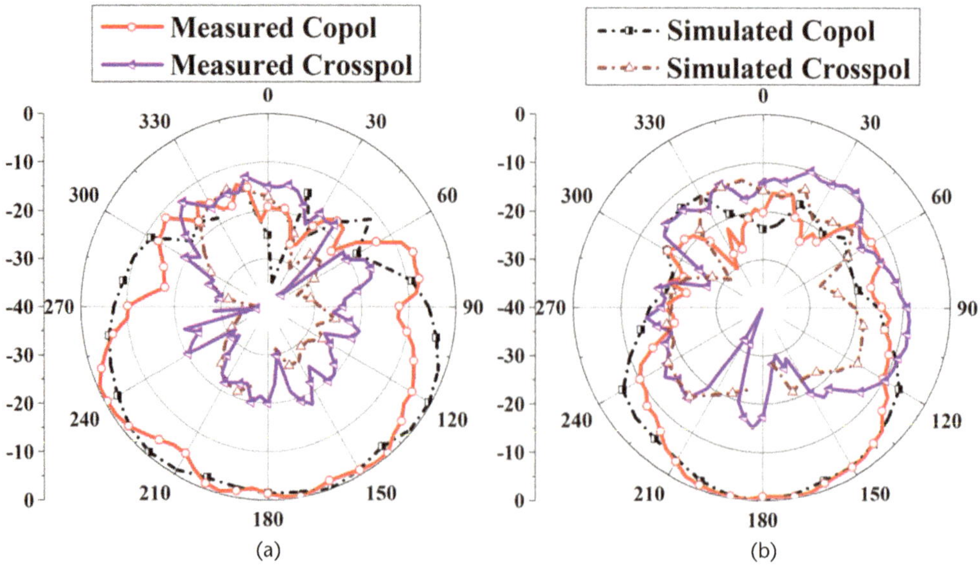

Figure 8.43 Normalized radiation patterns of the triplexing antenna at the Tx channel: (a) E-plane and (b) H-plane radiation pattern at 10.0 GHz.

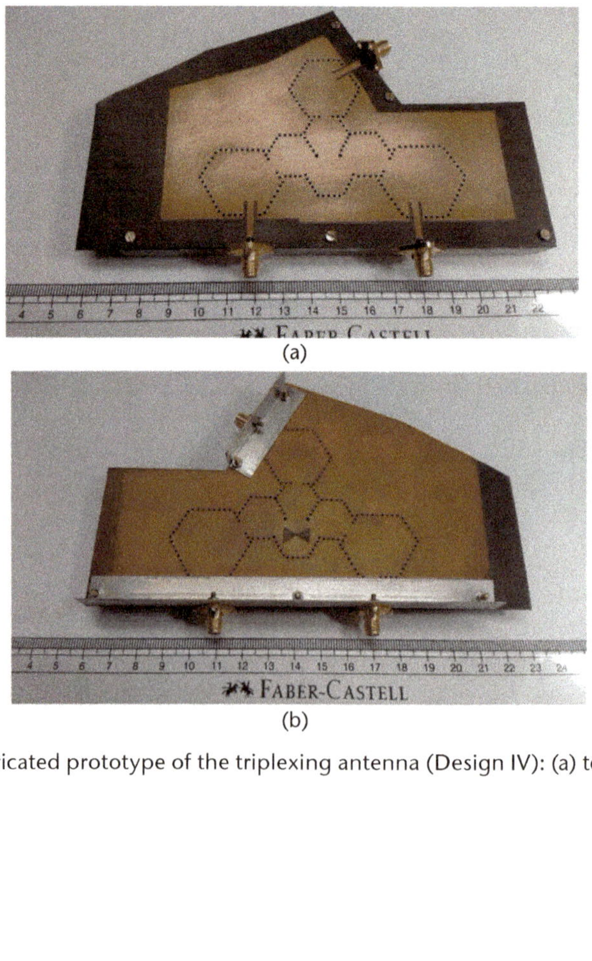

Figure 8.44 Fabricated prototype of the triplexing antenna (Design IV): (a) top view and (b) bottom view.

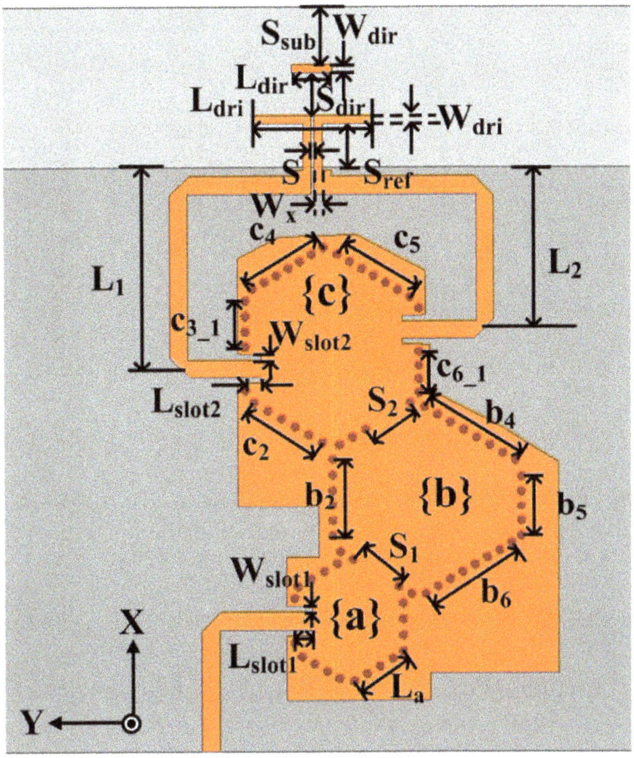

Figure 8.45 Balun filter dimensions were provided in Chapter 6. Other than balun filter dimensions, all quasi-Yagi antenna-related dimensions of the SIHC-based balun filtering antenna are given: $L_1 = 27.65$ mm, $L_2 = 22.35$ mm, $S = 0.53$ mm, $W_{dri1} = 1.1$ mm, $S_{ref} = 5.9$ mm, $L_{dri} = 16.6$ mm, $S_{dir} = 6.0$ mm, $W_{dir} = 1.1$ mm, $L_{dir} = 5.6$ mm, and $S_{sub} = 8$ mm.

of the passband. Figure 8.46 presents the coupling topology of the balun filtering antenna. Figure 8.47(a, b) demonstrate the effect of a quasi-Yagi antenna's design parameter variations on the performance of the balun filtering antenna. The variations of L_{dri} and L_{dir} only affect the input impedance matching without affecting the −10-dB impedance bandwidth. This shows the load insensitivity of the balun filter.

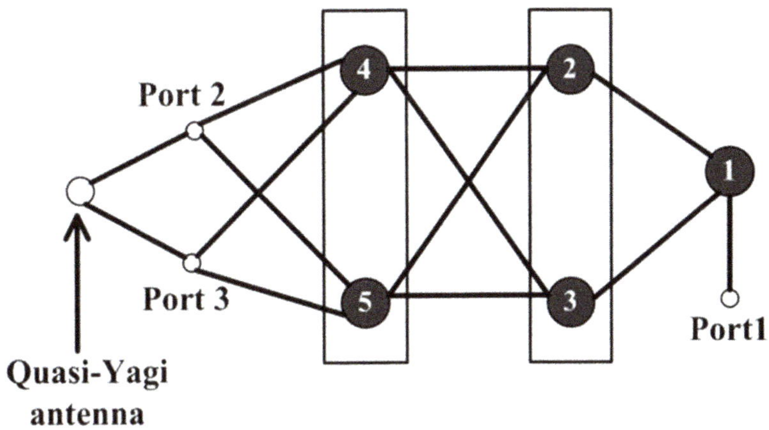

Figure 8.46 Coupling topology of the balun filtering antenna.

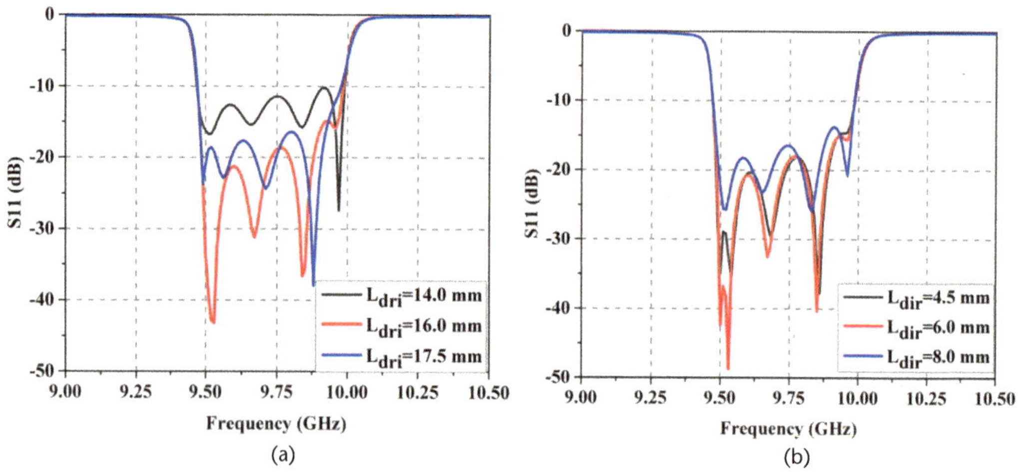

Figure 8.47 Effect of the quasi-Yagi antenna design parameter variations on the reflection coefficient of the balun filtering antenna.

Figure 8.48 exhibits the measured and simulated responses of the balun filtering antenna. It shows good agreement between the measured and simulated responses. The antenna gain and radiation null are analogous to the filter's insertion loss and transmission zero, respectively. In the balun filter's passband, the balun filtering antenna has a flat gain along the +*X* direction; it has two radiation nulls appearing at 9.4 GHz and 10.1 GHz on both sides of the balun filter's passband at the same frequency of the transmission zeros of the balun filter. The designed balun filtering antenna possesses a sharp band skirt and good selectivity. The measured and simulated center frequencies are 9.71 GHz and 9.72 GHz with 512-MHz and 509-MHz ($S_{11} = -10$ dB) impedance bandwidths, respectively. The measured and simulated gains are 4.7 and 5.0 dBi, respectively, as shown in Figure 8.48. Figure 8.49 exhibits measured and simulated gains and S-parameters of the balun filtering

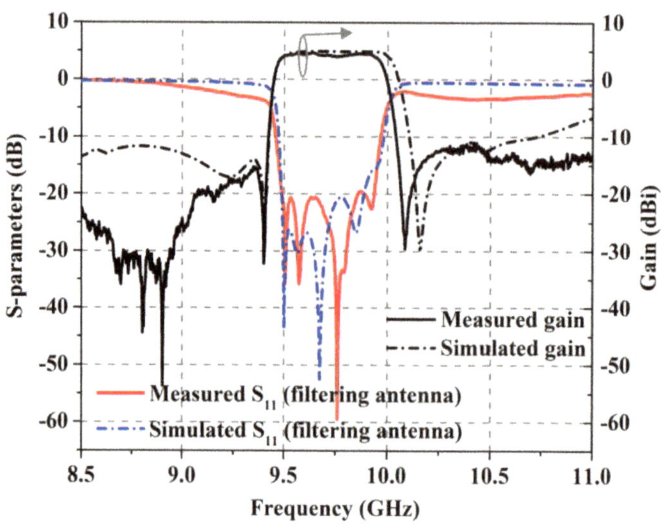

Figure 8.48 Measured and simulated gains and S-parameters of the balun filtering antenna.

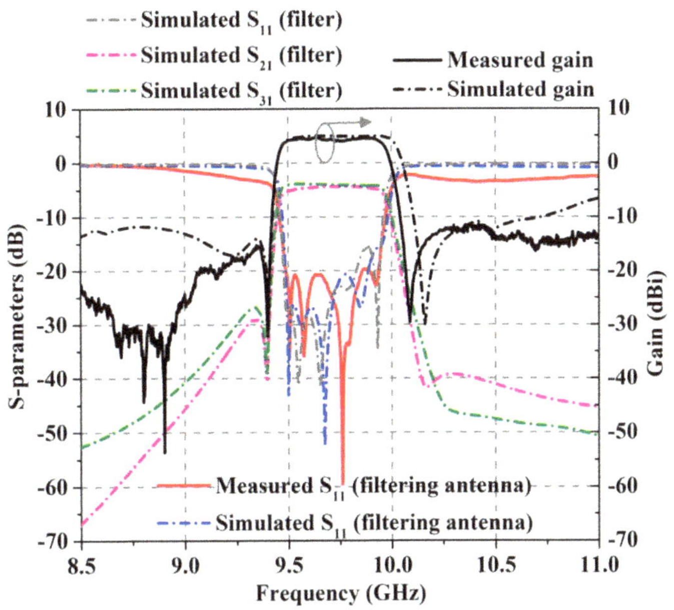

Figure 8.49 Measured and simulated gains and S-parameters of the balun filtering antenna and the S-parameters of balun filter.

antenna, along with the simulated S-parameter responses of the balun filter. As expected, it is observed that the balun filtering antenna follows the same characteristics as the balun filter.

Figure 8.50(a, b) are normalized radiation patterns at 9.55 GHz in the E and H-planes, respectively. Figure 8.50(c, d) are normalized radiation patterns at 9.7 GHz in the E and H-planes, respectively. Similarly, Figure 8.51(a, b) illustrates normalized radiation patterns at 9.9 GHz in the E and H-planes, respectively. It is observed that the measured cross-polar level is below 16 dB and the FTBR is 16.4 dB. Figure 8.52(a, b) exhibit the top and bottom views of the balun filtering antenna.

8.1.5 Dual-Band Filtering Antenna Using the SIHC (Design VI)

Here, a balun diplexer is adopted to feed two wideband quasi-Yagi antennas. The design comprises a balun diplexer, a pair of wideband quasi-Yagi antennas, and a connecting microstrip line. In coupling topology, as shown in Figure 8.53, it has been observed that both the Rx and Tx sections of the balun diplexer are attached to a quasi-Yagi antenna. The schematic of the dual-band filtering antenna is shown in Figure 8.54, along with its dimensions. The design of the second-order balun diplexer and its performance were discussed in Section 6.4.2. The same balun diplexer is employed to realize the dual-band filtering antenna. The dual-mode SIHCs {*b*} and {*c*} are attached to the Rx and Tx channels, respectively. In the Rx channel, the design-specified transmission zero is above the passband, whereas, in the Tx channel, it is below the passband.

Hence, it improves the diplexer's selectivity. The balanced outputs of the resonators {*b*} and {*c*} are acting as feed lines to the quasi-Yagi antenna.

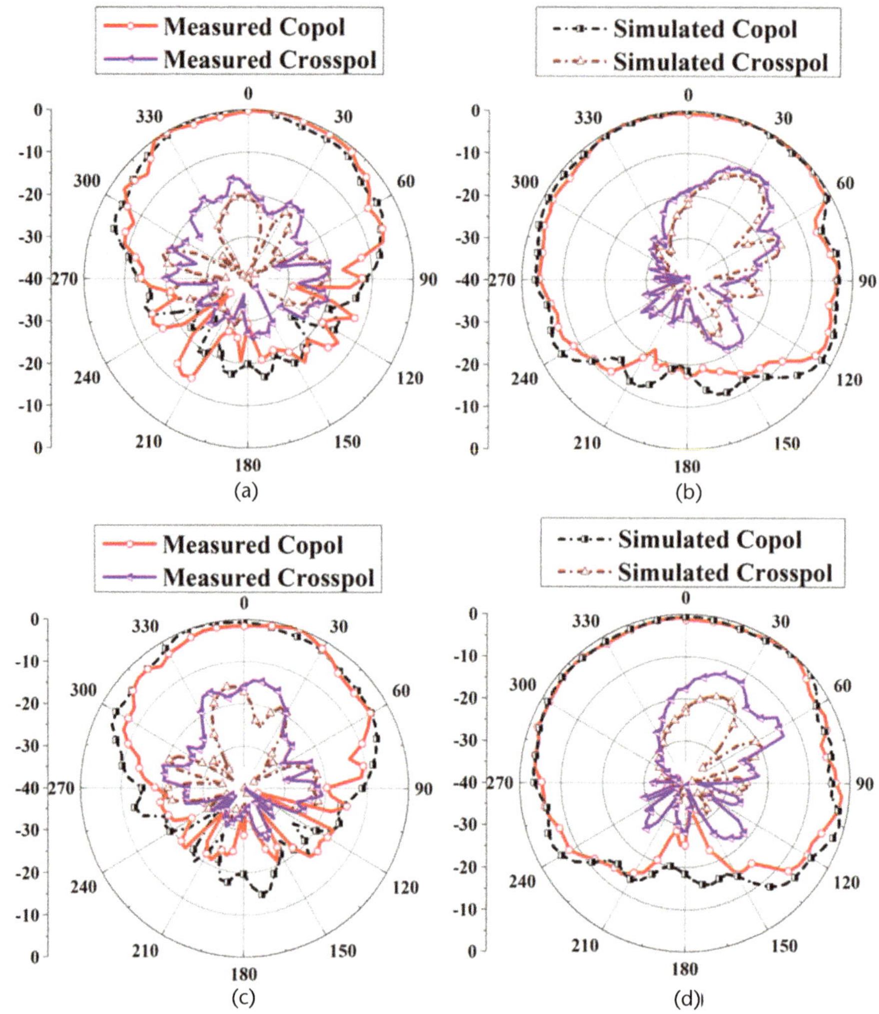

Figure 8.50 Normalized radiation patterns of the balun filtering antenna. (a) E-plane and (b) H-plane radiation patterns at 9.55 GHz. (c) E-plane and (d) H-plane radiation patterns at 9.7 GHz.

The measured gain and measured and simulated S-parameters are exhibited in Figure 8.55. In the lower band, the radiation null leads to a steep skirt at the right-side band edge, whereas, in the higher band, it leads to a steep skirt at the left-side band edge. The measured center frequencies of the lower and higher bands are at 9.46 GHz and 9.95 GHz, with 177-MHz (1.87%) and 220-MHz (2.21%) impedance bandwidths, respectively. The lower-band (Rx) measured and simulated gains are 5.2 dBi and 5.6 dBi, respectively, and the higher-band (Tx) measured and simulated gains are 4.98 dBi and 5.3 dBi, respectively. Figure 8.56 gives the performance comparison between dual band filtering antenna and balun diplexer in terms of measured gain and S-parameters. In both bands, the design shows flat gain response since the diplexer synthesis technique is adopted to realize a dual-band filtering antenna. Measured radiation nulls appear at 9.66 GHz for both the bands. The balun diplexer's transmission zeros are fairly matched with the dual-band filtering antenna's radiation nulls. Figure 8.57(a, b) shows the lower-band radiation pattern

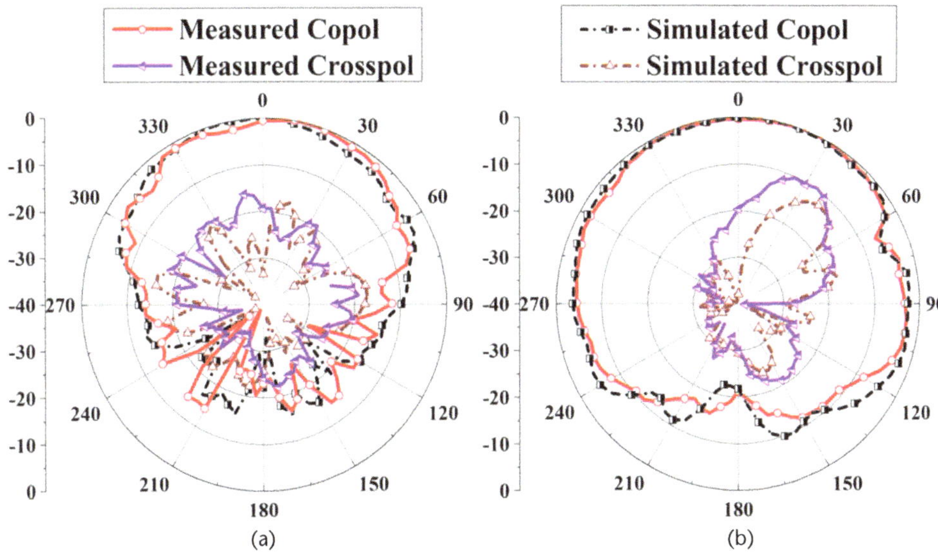

Figure 8.51 Normalized radiation patterns of the balun filtering antenna. (a) E-plane and (b) H-plane radiation patterns at 9.9 GHz.

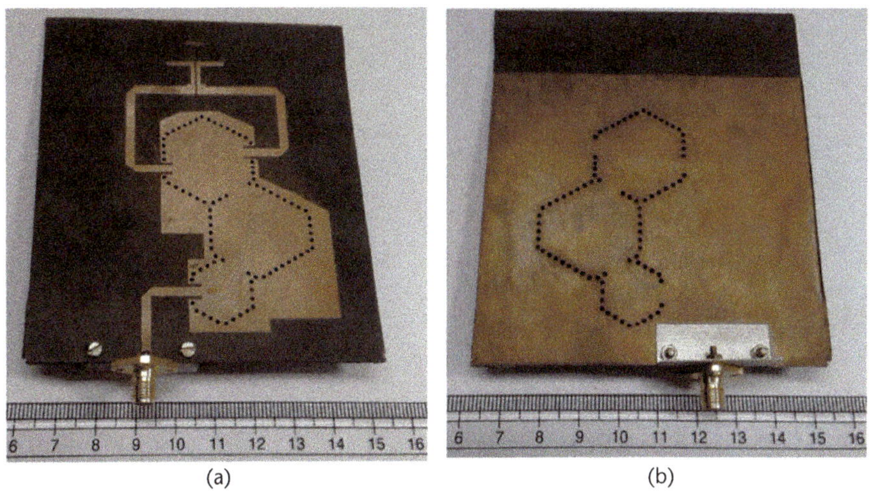

Figure 8.52 Fabricated prototype of balun filtering antenna (Design V). (a) Top view. (b) Bottom view.

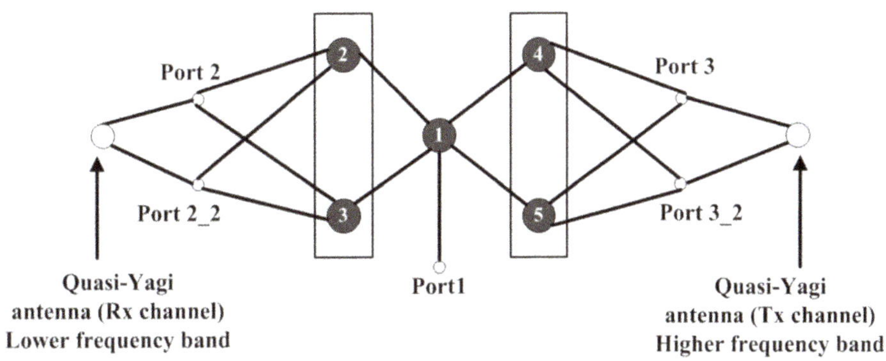

Figure 8.53 Coupling topology of the dual-band filtering antenna.

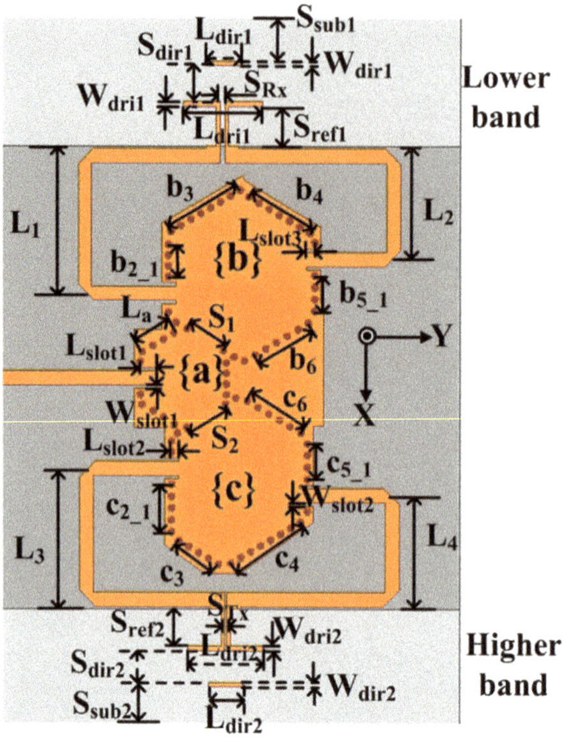

Figure 8.54 Balun diplexer's dimensions are provided in Chapter 4, Design II. Other than the balun diplexer's dimensions, all the dimensions of the dual-band filter antenna using a balun diplexer are: L_1 = 24.68 mm, L_2 = 19 mm, L_3 = 23.52 mm, L_4 = 19 mm, S_{Rx} = 0.86 mm, S_{Tx} = 0.42 mm, W_{dri1} = 0.75 mm, S_{ref1} = 6.94 mm, L_{dri1} = 14.0 mm, S_{dir1} = 6.2 mm, W_{dir1} = 0.75 mm, L_{dir1} = 6.1 mm, S_{sub1} = 7.5 mm, W_{dri2} = 0.75 mm, S_{ref2} = 6.5 mm, L_{dri2} = 13.4 mm, S_{dir2} = 5.25 mm, W_{dir2} = 0.75 mm, L_{dir2} = 5.9 mm, and S_{sub2} = 6.65 mm.

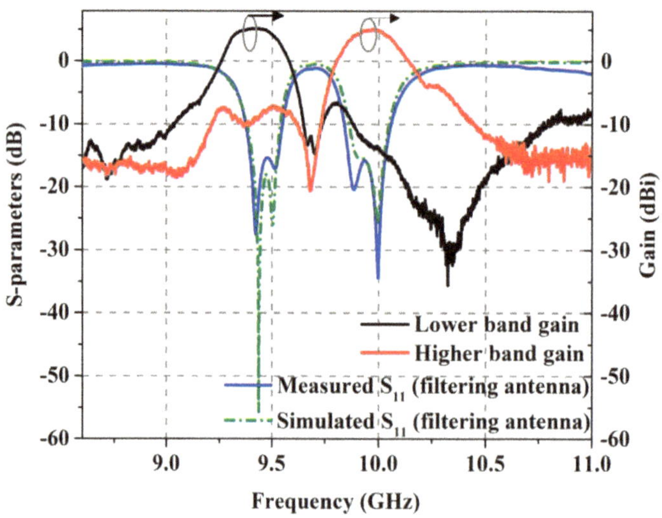

Figure 8.55 Measured and simulated gains and S-parameters of the dual-band filtering antenna.

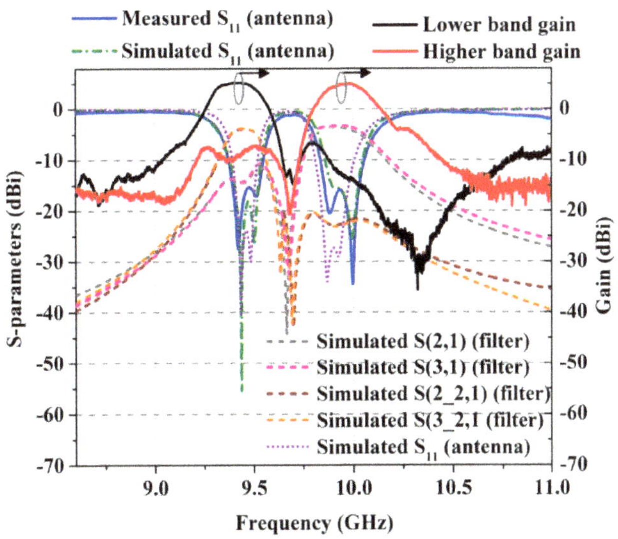

Figure 8.56 Measured and simulated gains and S-parameters of the dual-band filtering antenna and S-parameter of the balun diplexer.

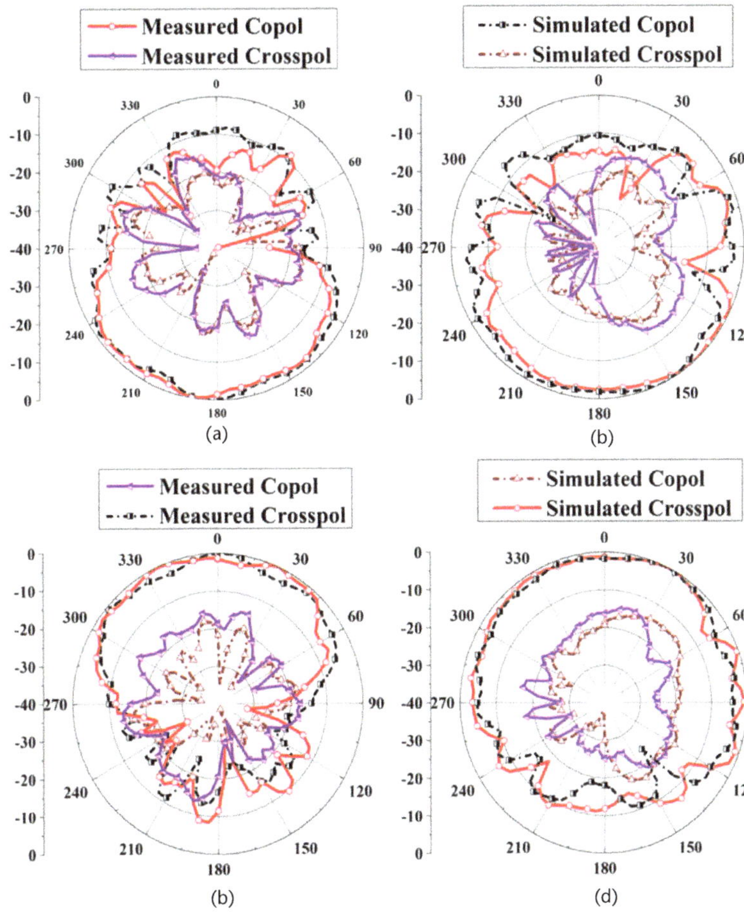

Figure 8.57 Normalized radiation patterns of the dual-band filtering antenna. (a) E-plane and (b) H-plane radiation patterns at 9.46 GHz (lower band). (c) E-plane and (d) H-plane radiation patterns at 9.95 GHz (higher band).

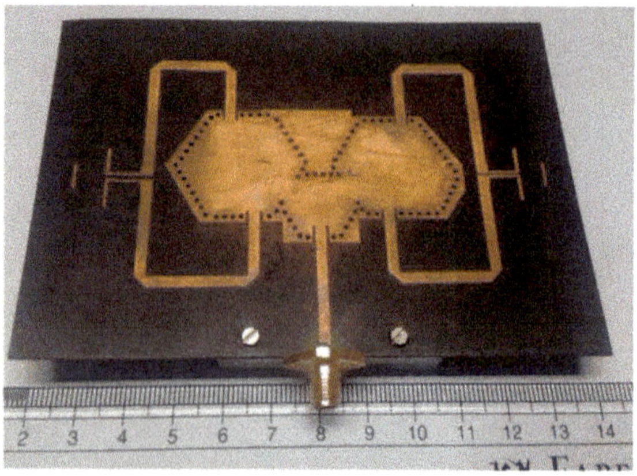

(a)

(b)

Figure 8.58 A fabricated prototype of a dual-band filtering antenna design using a dual-mode SIHC-based balun diplexer (Design VI). (a) Top view. (b) Bottom view.

in the E and H-planes at 9.46 GHz, respectively. It is noticed that the measured cross-polar level is below 15 dB and the FTBR is 12.7 dB. Figure 8.57(c, d) shows the higher-band radiation pattern in the E and H-planes at 9.95 GHz, respectively. It is noticed that the measured cross-polar level is below 16 dB and the FTBR is 10.1 dB. Figure 8.58(a, b) exhibits the top and bottom views of the fabricated prototype.

8.2 SIW AIA

The AIA consists of an active device and a radiating element integrated on the same substrate. With rapidly growing modern communication systems, AIAs have received much attention because of their compactness and low cost. There are many practical applications of AIAs, such as power combining, beam steering/switching, retro-directive arrays, and high-efficiency power amplifier designs. At the millimeter-wave frequency, higher transmission line loss, limited source power,

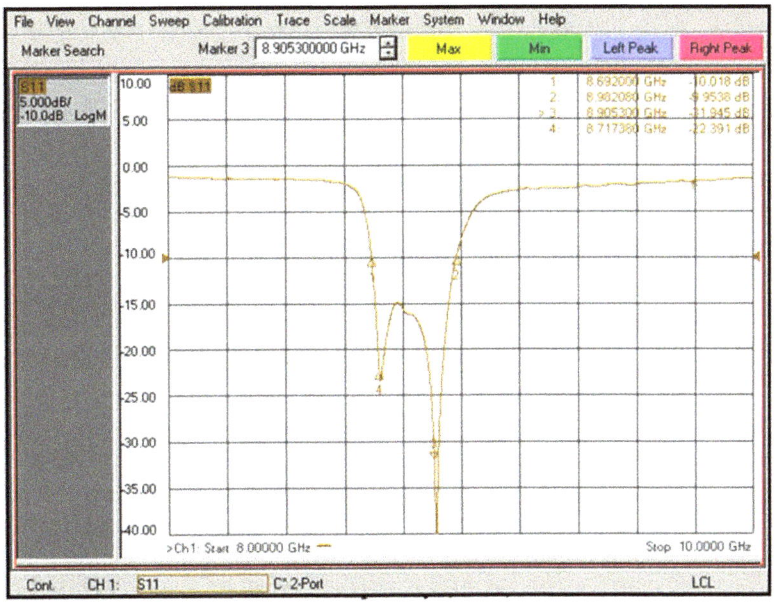

(a)

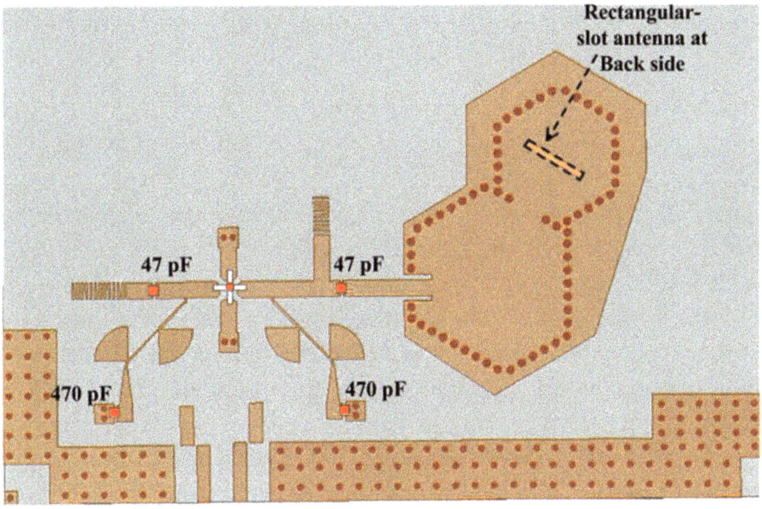

(b)

Figure 8.59 (a) Measured S-parameters of the filtering antenna. (b) Schematic of the negative resistance oscillator (NRO) integrated with filtering antenna. The dimensions related to filtering antenna are provided in Section 8.1.1.

and reduced antenna efficiency make AIA an attractive candidate [4]. AIAs can be divided into four different categories: low noise amplifier (LNA) active antennas [5–7], power-amplifying active antennas [8, 9], self-oscillating active antennas [10], and self-mixing active antennas [11]. A novel design of a Class C push-pull power amplifier integrated with a uniplanar quasi-Yagi antenna is presented in [8], where the corrugation along with the truncated ground plane allows for the combination of out-of-phase power as well as second-harmonic tuning.

8.2.1 SIW-Based Active Integrated Filtering Antenna Design

The measured S-parameter responses of the filtering antenna and the schematic of the active integrated filtering antenna (AIFA) are shown in Figure 8.59(a, b), respectively. The dimensions of the third-order filtering antenna are provided in Figure 8.7. The measured gain and radiation pattern of the filtering antenna are provided in Figures 8.12 and 8.13, respectively. The AIFA is realized on the Rogers RT/Duroid 5880 substrate with a thickness of 0.787 mm and relative dielectric constant of 2.2. The filtering antenna is integrated with an oscillator. The oscillator is designed using 1-port negative-resistance topology. The pHEMT NE3210S01 device is used for the design. Both the source ports of the pHEMT device are connected with a microstrip short-circuited stub that ensures maximum negative resistance at the desired frequency. The desired oscillation frequency is chosen at the center frequency of the filtering antenna, which is at 8.89 GHz. The bias conditions of the oscillator are as follows: drain voltage (V_{ds}) 2.5V and drain current (I_{ds}) 20 mA. The radiated power from the AIFA is received by a reference double-ridged horn antenna (gain = 12.3 dBi) at a distance of 30.48 cm. Figure 8.60(a) shows the received output power spectrum at the frequency of 8.85 GHz with −25.34-dBm output power. The measured cable insertion loss is 2.6 dB at 8.85 GHz as shown in Figure 8.61. The effective isotropic radiated power (EIRP) can be calculated using (8.6).

$$EIRP = P_t G_t = \frac{P_r}{G_r L_r}\left(\frac{4\pi R}{\lambda_0}\right)^2 \tag{8.6}$$

where P_r is the measured received power, P_t is the oscillation output power, L_r is the insertion loss of the connecting cable, G_t is the gain of the filtering antenna, which is 4.33 dBi, and R is the distance between transmitting and receiving antenna (30.48 cm). Using (8.6), the calculated EIRP of the active integrated filtering antenna is found to be 6.0 dBm. From EIRP, the calculated oscillation output power (P_t) is 1.7 dBm. The DC power consumption of the AIFA is 50 mW and the DC to RF conversion efficiency is about 5.40%.

Figure 8.60(b) shows the measured phase-noise from the received output power spectrum of the AIFA. The measured SSB phase-noise is −97.5 dBc/Hz at 1-MHz offset for the oscillation frequency 8.85 GHz. The calculated FOM is −159.5 dBc/Hz at 1-MHz offset. The fabricated prototype of the AIFA is shown in Figure 8.62. Figure 8.62(a, b) are the top view and bottom view of the AIFA.

8.2.2 SIW-Based Voltage-Controlled Active Integrated Balun Filtering Antenna

The schematic of the voltage controlled active integrated balun filtering antenna (VCAIBFA) is provided in Figure 8.63. The dimensions related to the balun filtering antenna are provided in Figure 8.45. The measured S-parameters and gain of the passive balun filtering antenna are provided in Figure 8.48. The measured radiation patterns of the antenna are provided in Figures 8.50 and 8.51. The VCAIBFA is realized on Rogers RT/Duroid 5880 substrate with thickness of 0.787 mm and relative dielectric constant of 2.2.

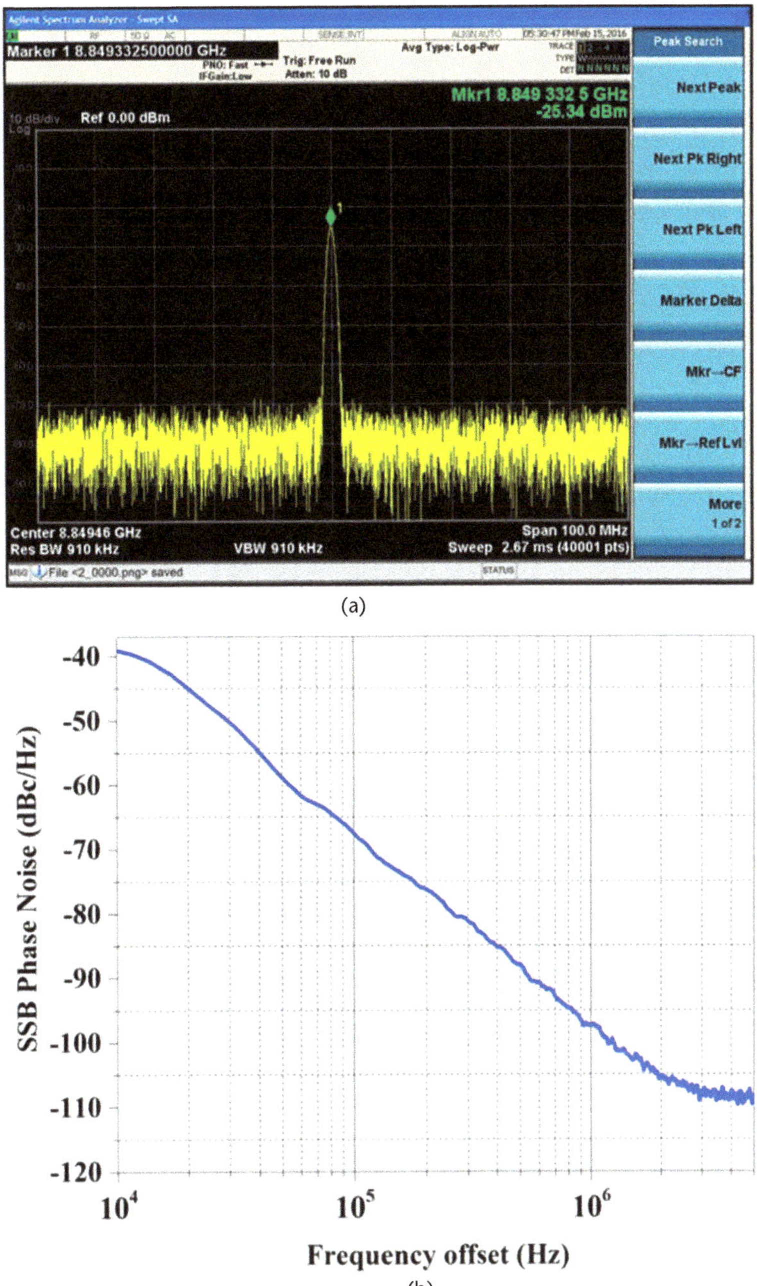

Figure 8.60 (a) Spectrum of radiated power from active integrated filtering antenna (AIFA) with the receiving antenna located 30.48 cm away from AIFA. (b) Measured phase noise of the AIFA.

The filtering antenna is integrated with an oscillator. The oscillator is designed using one port negative resistance topology. The pHEMT NE3210S01 device is used for the design. Both the source ports of the pHEMT device are connected with microstrip short circuited stub that ensures maximum negative resistance at the desired frequency. A varactor diode (MA46H070-1056) is placed at the gate

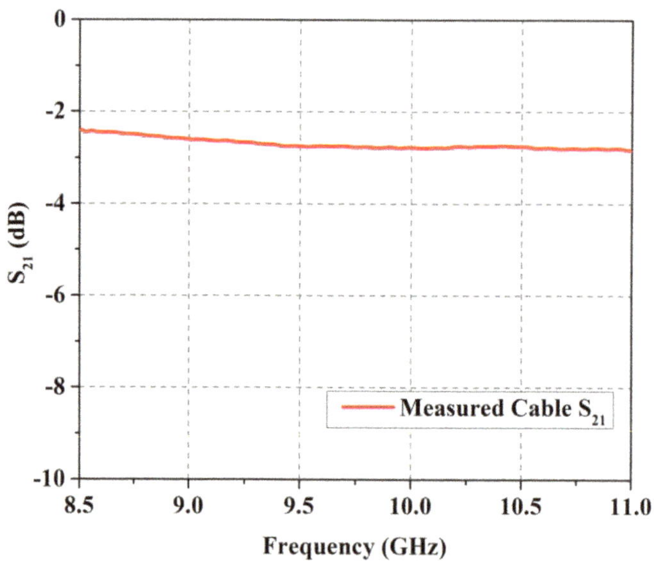

Figure 8.61 Measured cable transmission coefficient in dB.

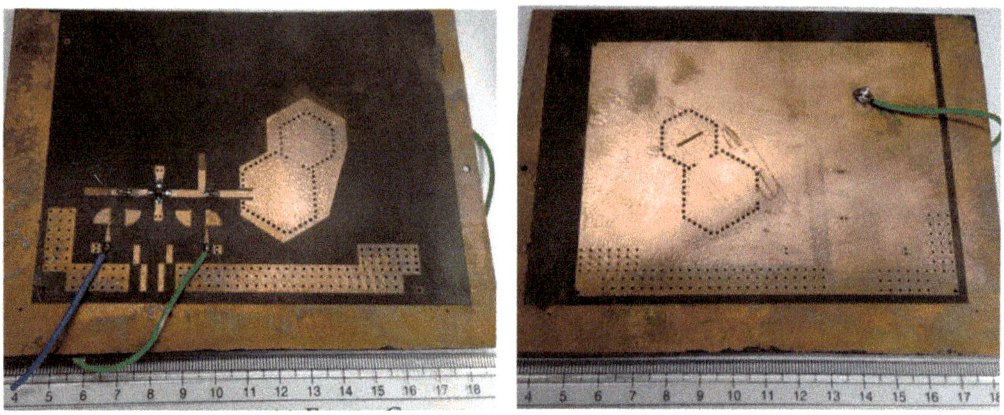

(a) (b)

Figure 8.62 (a) Top view of the AIFA. (b) Bottom view of the AIFA.

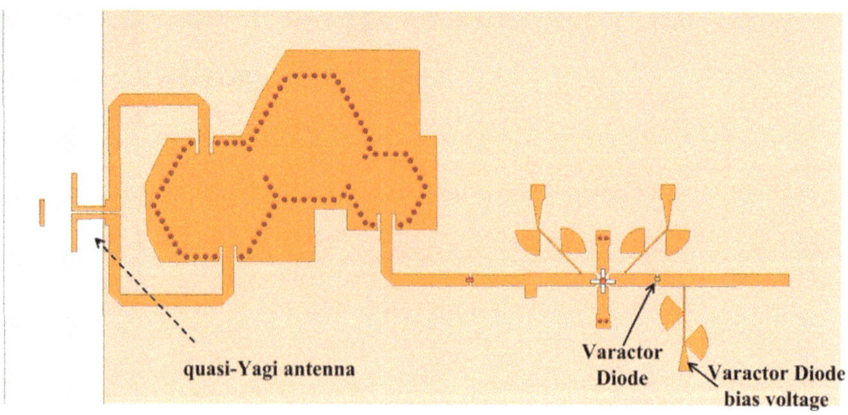

Figure 8.63 Schematic of the VCAIBFA. The dimensions related to quasi-Yagi-based filtering antenna are provided in Figure 8.45.

end of the active device. The varactor diode biasing range is from 0V to 20V and the junction capacitance varies from 1.1 pF to 0.15 pF. The bias conditions of the oscillator are: drain voltage (V_{ds}) 2.5V and drain current (I_{ds}) 20 mA. The radiated power from the VCAIBFA is received by a reference double-ridged horn antenna (gain = 12.3 dBi) at a distance of 30.48 cm. The received power levels are −27.21

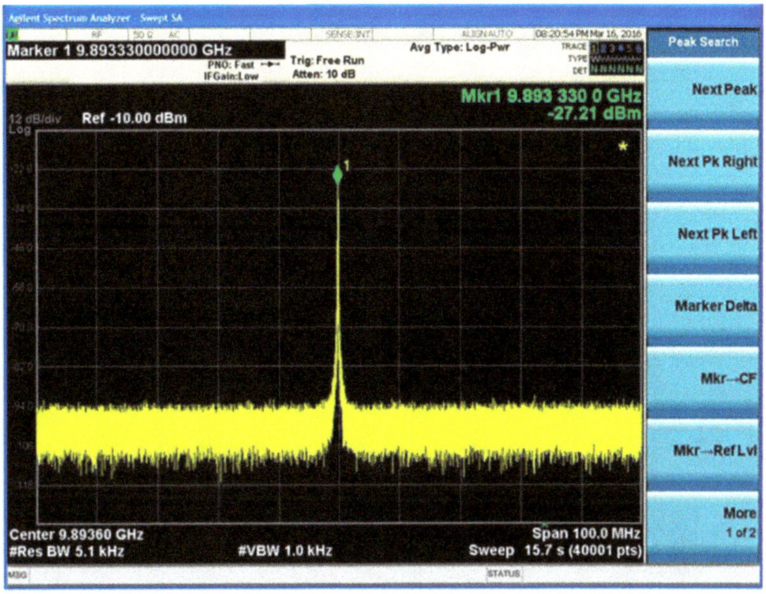

(a)

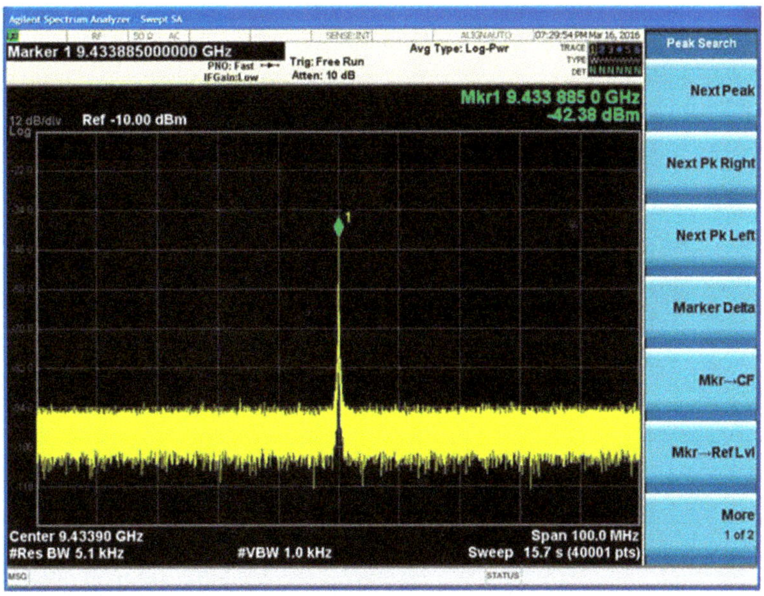

(b)

Figure 8.64 Radiated power from the voltage-controlled active integrated filtering antenna, with the receiving antenna located 30.48 cm away from the VCAIBFA. (a) Varactor diode voltage (V_C) = 20V. (b) V_C = 0.8V.

dBm and –42.38 dBm with varactor diode bias voltage (V_C) at 20V and 0.8V, respectively, as shown in Figure 8.64.

Figure 8.65(a) exhibits the variations of the received output power spectrum with the changes of varactor diode control voltage (V_C). Figure 8.65(b) shows the measured SSB phase noise from the received power of the VCAIBFA when V_C is kept at 9.87V. The measured SSB phase noises are –77 dBc/Hz and –106 dBc/Hz at 100-kHz and 1-MHz offset for the oscillation frequency of 9.65 GHz. The calculated FOM is –168.7 dBc/Hz at 1-MHz offset. Figure 8.66(a) shows the measured tuning frequency range and received output power variations with the change in V_C. The VCAIBFA is tuned from 9.42 GHz to 9.89 GHz. Since the radiation null of balun filtering antenna is present at 9.4 GHz, as mentioned in Section 8.1.4, a sudden decrease in the received output power at a nearby frequency is observed. The

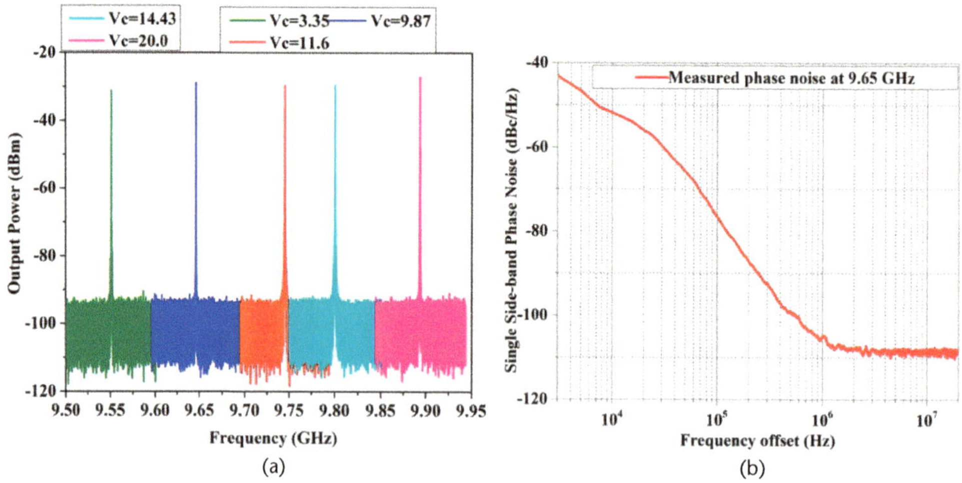

Figure 8.65 (a) Measured radiation power spectrum with the variations of V_C of the varactor diode. (b) Measured phase noise of the VCAIBFA at 9.65 GHz with V_C = 9.87V.

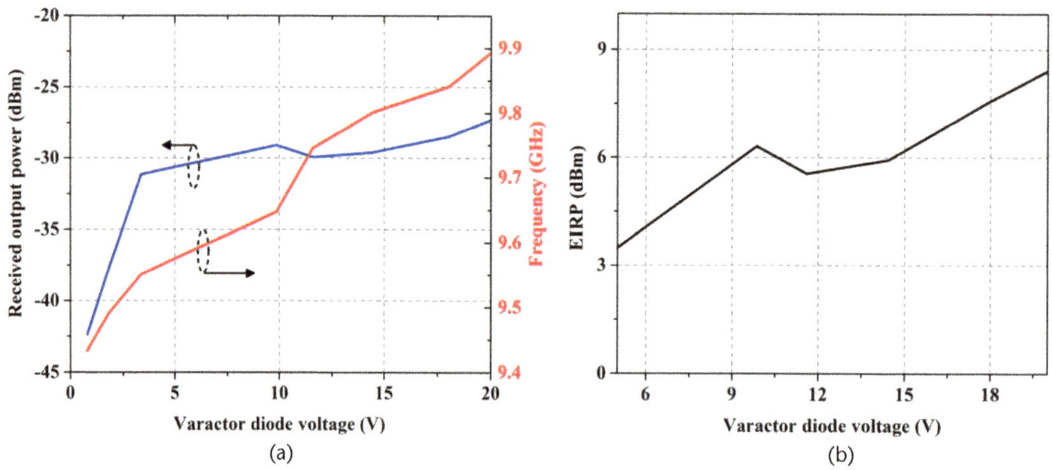

Figure 8.66 (a) Received output power and frequency tuning of VCAIBFA with varactor diode voltage variations. (b) Measured EIRP of the VCAIBFA with varactor diode voltage variations.

received output power variation is 3.5 dB from 9.55 GHz to 9.89 GHz. The EIRP can be calculated using (8.6). The variation of EIRP with the change of varactor diode control voltage (V_C) is shown in Figure 8.66(b). The measured cable insertion loss is 2.8 dB at 9.65 GHz as shown in Figure 8.61. The distance (R) between the transmitting and receiving antennas is 30.48 cm.

The calculated EIRP of the design is 5.61 dBm at 9.89 GHz. From EIRP, the oscillation output power (P_t) can be determined using a passive balun filtering antenna gain (G_t), which is 4.47 dBi at 9.89 GHz. The calculated P_t is 1.14 dBm. The DC power consumption of the VCAIBFA is 50 mW and the DC-to-RF conversion efficiency is about 2.60% at 9.89 GHz. The fabricated prototype of the VCAIBFA is shown in Figure 8.67(a, b).

The design has highly selective frequency response because of two radiation nulls at 9.4 GHz and 10.09 GHz. The design has potential applications in highly selective, low-power, and low-cost transmitters such as X-band RFID systems.

8.2.3 SIHC-Based Transceiver Design Using a Diplexing Antenna

In this section, two different SIHC-based transceiver designs are discussed. The receiver section of Design I consists of the Rx channel of the diplexing antenna with the subharmonic SOM. The transmitting section of Design I is formed by integrating the Tx channel of the diplexing antenna with an oscillator, which is designed using a 1-port negative-resistance method. However, the receiver section of Design II consists of the Rx channel of the diplexing antenna with a fundamental SOM. The Tx channel of transceiver Design II is same as the Tx channel of the transceiver Design I.

8.2.3.1 Transceiver Design I

The schematic of the SIW-based transceiver design is shown in Figure 8.68. The transceiver consists of Rx and Tx modules. The Rx section comprises a diplexing antenna integrated with a SOM, whereas the Tx section comprises an NRO integrated with a diplexing antenna. The diplexing antenna using a pair of a dual-mode SIHC is discussed in Section 8.1.2.2. The detailed dimensions of the diplexing antenna are provided in Figure 8.31.

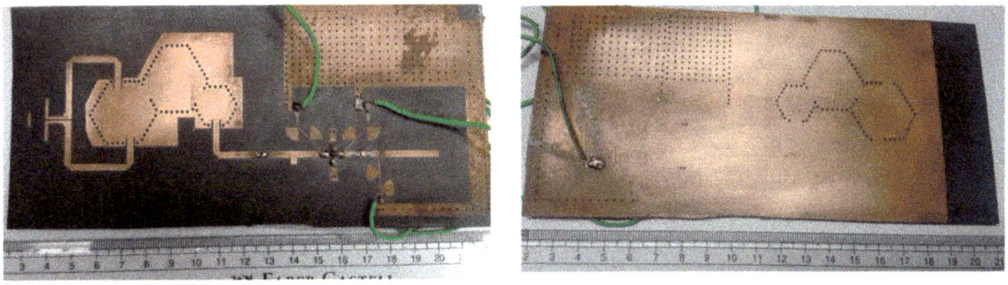

(a) (b)

Figure 8.67 (a) Top view of the VCAIBFA. (b) Bottom view of the VCAIBFA.

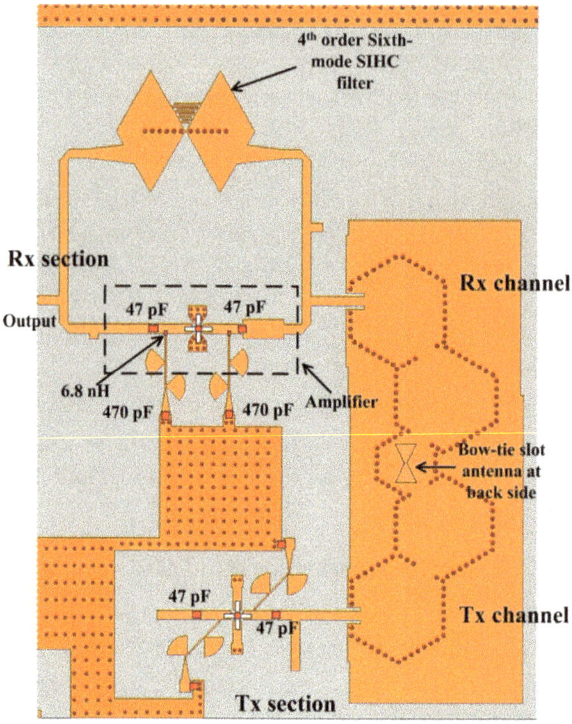

Figure 8.68 Schematic of the SIW-based transmitting (Tx) and receiving (Rx) sections. The Rx section consisting of subharmonic SOM integrated with diplexing antenna and the Tx section has NRO integrated with a diplexing antenna.

Rx Section

The Rx channel of the diplexing antenna is from 9.25 to 9.45 GHz with impedance bandwidth (–10-dB S_{11}) of 200 MHz. The Rx channel of diplexing antenna is integrated with the SOM through gate terminal of the pHEMT (NE3210S01) device. SOM is designed using feedback oscillator. Fourth-order, sixth-mode SIW-based BPF is employed in the feedback loop as a frequency stabilization element. The center frequency of BPF filter is kept at 4.0 GHz with fractional bandwidth (FBW) of 12.5%. Sixth-mode SIW-based bandpass filter is discussed in Chapter 5. The schematic of the filter is shown in Figure 8.69(a) and the measured and simulated S-parameters of the filter are provided in Figure 8.69(b). The dimensions related to the BPF are provided in Figure 5.35. The sixth-mode SIW-based BPF exhibits miniaturization in the filter size as compared to that of the SIHC-based fourth-order bandpass filter.

To design an amplifier at 4.2 GHz using a common source configuration, the simultaneous conjugate matching network is designed for 4.2 GHz. The sixth-mode SIW-based highly selective BPF is placed in the feedback path of the amplifier using microstrip line of electrical length of 360° or multiple of 360°. The output signal is collected from the drain of the pHEMT device. The fundamental oscillation frequency is kept at 4.2 GHz (f_{LO}). The second harmonic of oscillating signal (f_{2LO}) is mixed with the RF signal received by the Rx channel of the diplexing antenna that generates the desired IF signal. The output signal from designed SOM is collected

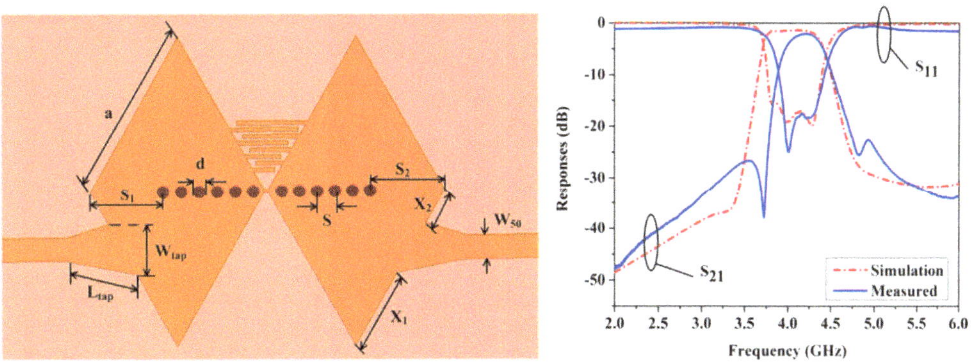

(a) (b)

Figure 8.69 (a) Schematic of a half-mode BPF employed to stabilize the fundamental oscillation frequency of the subharmonic SOM. The dimensions of the filter are provided in Figure 5.35. (b) Measured and simulated S-parameters of the filter.

through external lowpass filter (LPF). The schematic of the designed lowpass filter and its measured S-parameter responses are shown in Figure 8.70(a, b). The output spectrum of SOM through external LPF is shown in Figure 8.71. Figure 8.71(a) shows the output spectrum in absence of RF signal.

When an RF signal of 9.33 GHz (f_{RF1}) is received through a diplexing antenna, the output power spectrum is shown in Figure 8.71(b). The fundamental oscillation signal is at 4.2 GHz (f_{LO}). The second harmonic of the fundamental oscillation frequency appears at 8.4 GHz (f_{2LO}) that mixes with the RF signal of 9.33 GHz (f_{RF1}) and generates a desired IF signal appearing at 900 MHz (f_{IF1}). The IF signal output power is obtained at −27.46 dBm. Similarly, Figure 8.72(a) shows output spectrum in the absence of the RF signal. When an RF signal of 9.41 GHz (f_{RF2}) is received through a diplexing antenna, the output power spectrum is shown in Figure 8.72(b). The desired IF signal appears at 983.5 MHz (f_{IF2}) with −27.8-dBm output power. Figure 8.73(a) exhibits the output power spectrum of desired IF signal (f_{IF1}) at 887 MHz when RF signal is kept at 9.3 GHz (f_{RF1}). Figure 6.53(a) shows the desired IF signal (f_{IF2}) at 987 MHz when RF signal is kept at 9.4 GHz (f_{RF2}). Figure 8.74(a) shows the output power spectrum of second harmonic of the oscillating frequency,

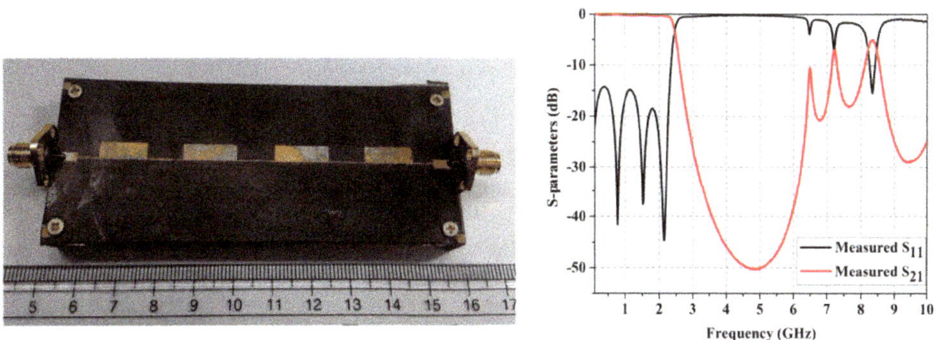

(a) (b)

Figure 8.70 (a) Fabricated photograph of the lowpass filter (LPF) externally attached to the output of the SOM. (b) Measured S-parameters of the LPF.

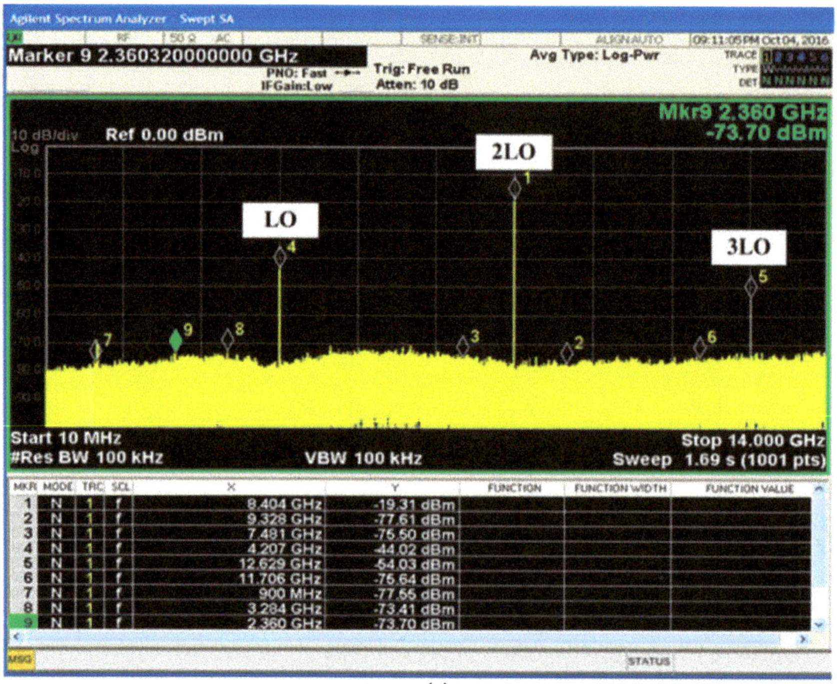

(a)

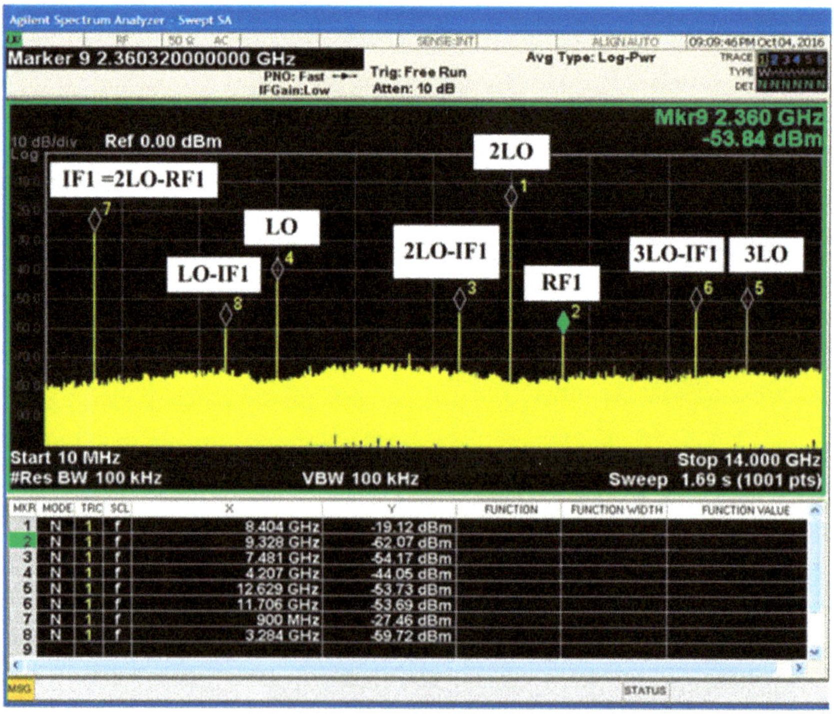

(b)

Figure 8.71 Output spectrum of SOM when the RF is kept at 9.33 GHz (f_{RF1}).

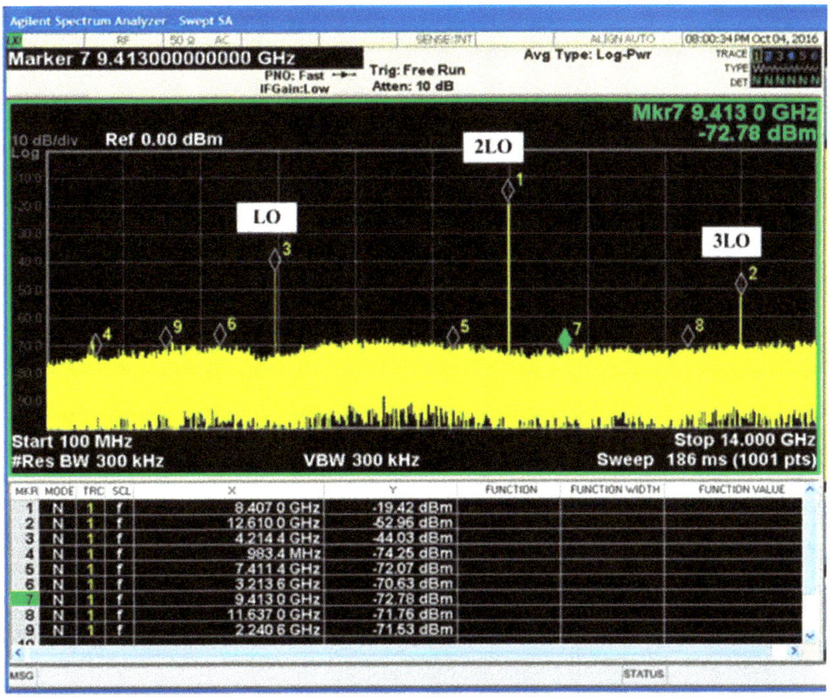

(a)

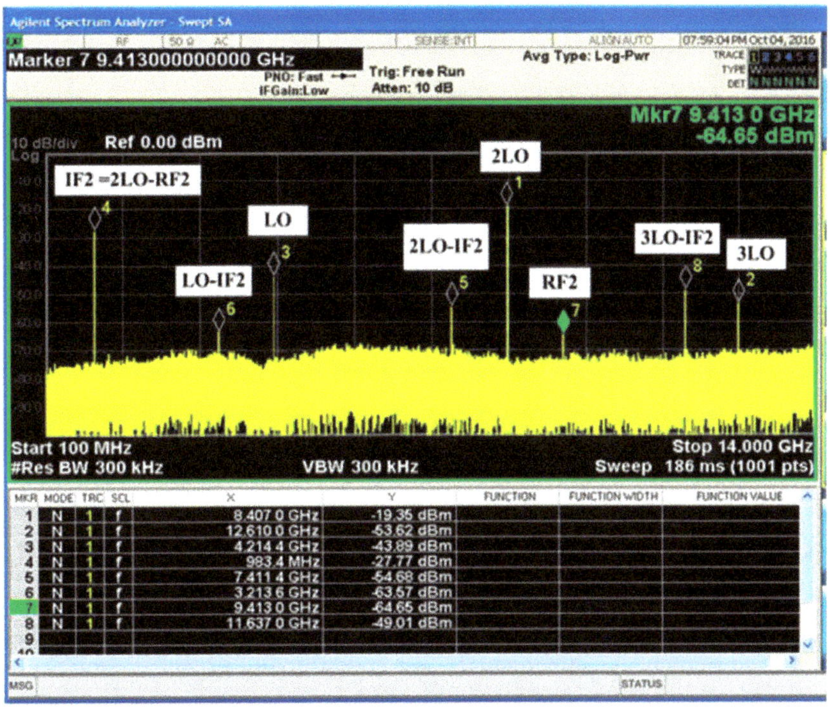

(b)

Figure 8.72 Output spectrum of SOM when the RF is kept at 9.41 GHz (f_{RF2}).

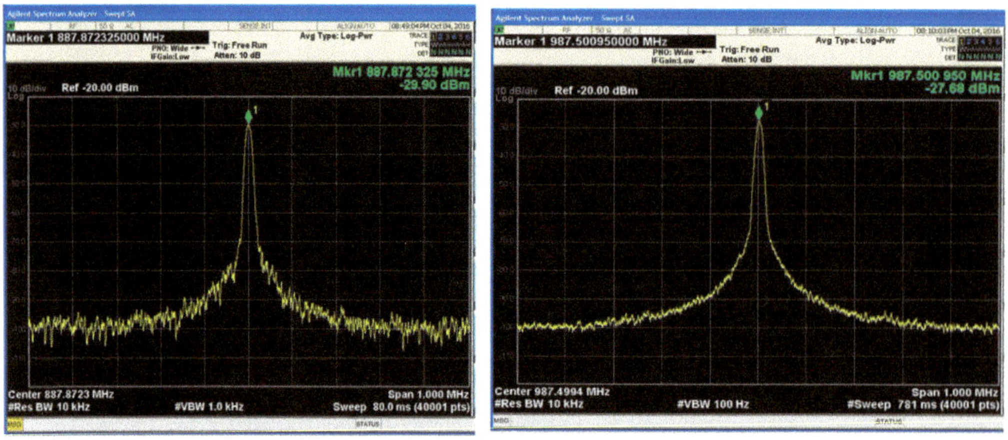

(a) (b)

Figure 8.73 Output spectrum of the IF signal. (a) Spectrum of IF signal (f_{IF1}) when RF signal 9.3 GHz (f_{RF1}). (b) Spectrum of IF signal (f_{IF2}) when RF signal 9.4 GHz (f_{RF2}).

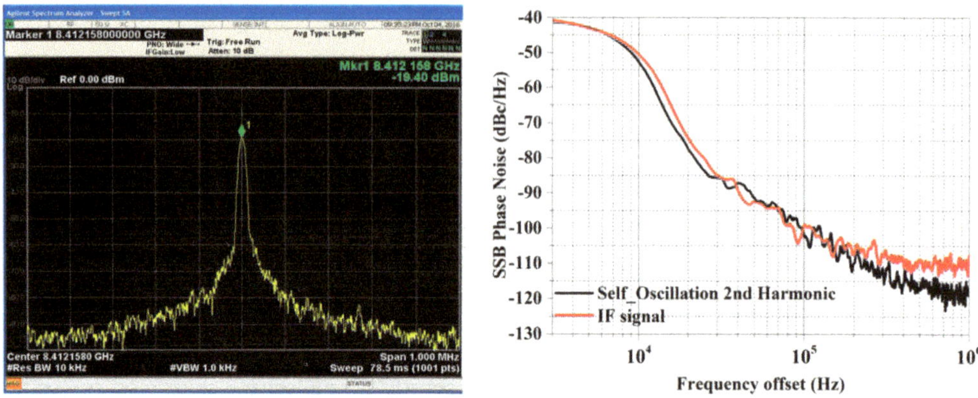

(a) (b)

Figure 8.74 (a) Spectrum of second harmonic of oscillating signal of SOM. (b) Measured SSB phase noise of IF signal and second harmonic of the oscillating signal of SOM.

which is at 8.412 GHz (f_{2LO}). Figure 8.74(b) exhibits SSB phase noise of IF signal and second harmonic of oscillating signal of SOM. The measured IF and second harmonic of oscillating signal phase noises are −99.5 dBc/Hz and −102 dBc/Hz at 100 kHz and −110.3 dBc/Hz and −117.3 dBc/Hz at 1-MHz offset, respectively. The DC bias conditions are V_{ds} = 2V and I_{ds} = 20 mA. The calculated FOM based on measured phase noise of both IF and second harmonic of oscillating signal are 142.5 dBc/Hz and −165.4 dBc/Hz at 100-kHz offset, respectively. The measured IF output power variation with RF is exhibited in Figure 8.75(a). The IF output power is measured keeping the RF transmitting power from a double-ridged horn antenna at 0 dBm. The distance between transmitting double-ridged horn antenna and receiving SIHC-based diplexing antenna is 30.48 cm. The transmitting double-ridged horn antenna and receiving diplexing antenna gains are 12.3 dBi and 4.37 dBi, respectively. Figure 8.75(b) shows the measured conversion gain of desired IF signal of the subharmonic SOM. The IF conversion gain varies from 1.4 dB to 0 dB with an RF variation from 9.25 to 9.42 GHz.

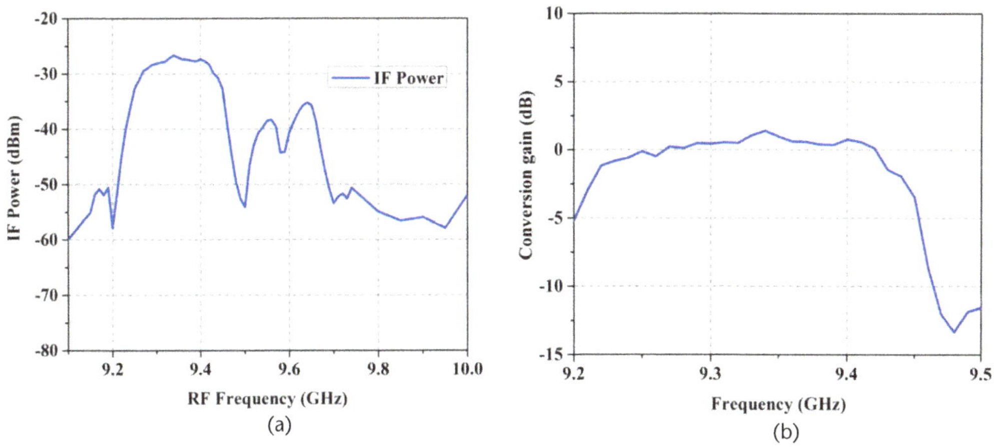

Figure 8.75 (a) IF power variation with RF-for-RF input power 0 dBm. (b) Measured conversion gain of the subharmonic SOM.

Tx Section

The Tx channel of the diplexing antenna is integrated with the oscillator, which is designed using 1-port negative-resistance topology. pHEMT NE3210S01 device is used for the design. Both the source ports of the pHEMT device are connected with microstrip short-circuited stub that ensures maximum negative resistance at the desired frequency. The Tx channel of the diplexing antenna is from 9.62 to 9.81 GHz. The desired oscillation frequency is chosen at 9.75 GHz. The bias conditions of the oscillator are: drain voltage (V_{ds}) 2.5V and drain current (I_{ds}) 20 mA. The radiated power from the Tx channel is received by a reference double-ridged horn antenna (gain = 12.3 dBi) at a distance of 30.48 cm. Figure 8.76 shows the received

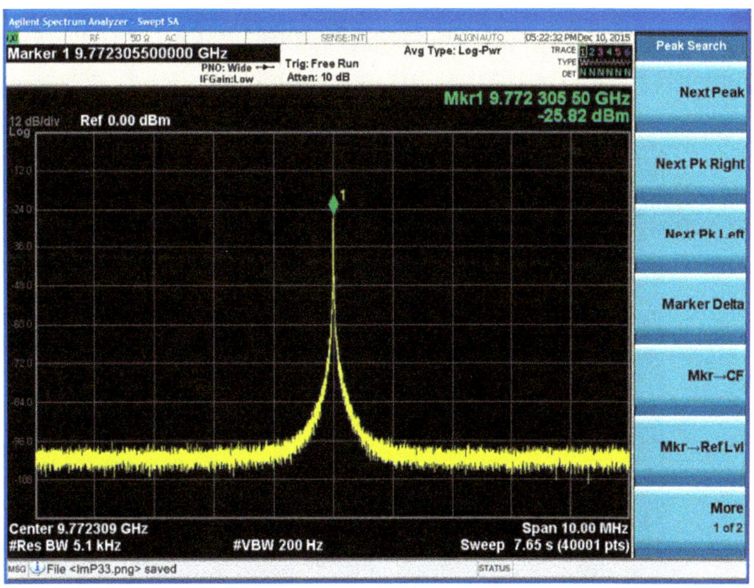

Figure 8.76 Output spectrum of received power from Tx section of the SIW-based transceiver module.

output power spectrum at the frequency of 9.772 GHz with −25.82-dBm output power. The measured cable insertion loss is 2.8 dB at 9.77 GHz as shown in Figure 8.61. The EIRP can be calculated using (8.6). The gain of the diplexing antenna is 4.1 dBi. The calculated EIRP of the active integrated filtering antenna is 6.5 dBm. From EIRP, the oscillation output power (P_t) can be estimated as 2.2 dBm. The DC power consumption of the Tx channel is 50 mW and the DC-to-RF conversion efficiency is about 5.0%. Figure 8.77 shows the measured phase noise from the received output power spectrum of the Tx section of the SIW-based transceiver. The measured SSB phase noises are −84 dBc/Hz and −110.5 dBc/Hz at 100-kHz and 1-MHz offset, respectively, for the oscillation frequency of 9.772 GHz. The calculated FOM is −169.5 dBc/Hz at 1-MHz offset frequency. The fabricated prototype of the SIW-based transceiver is shown in Figure 8.78.

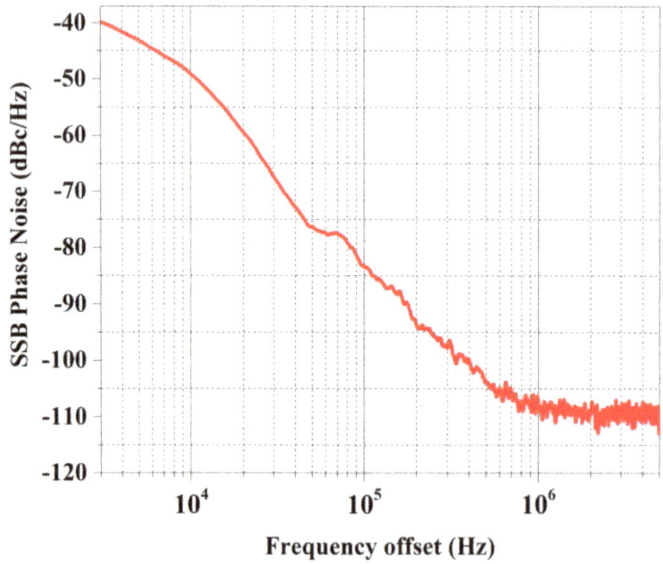

Figure 8.77 Measured SSB phase noise of the NRO integrated with a diplexing antenna.

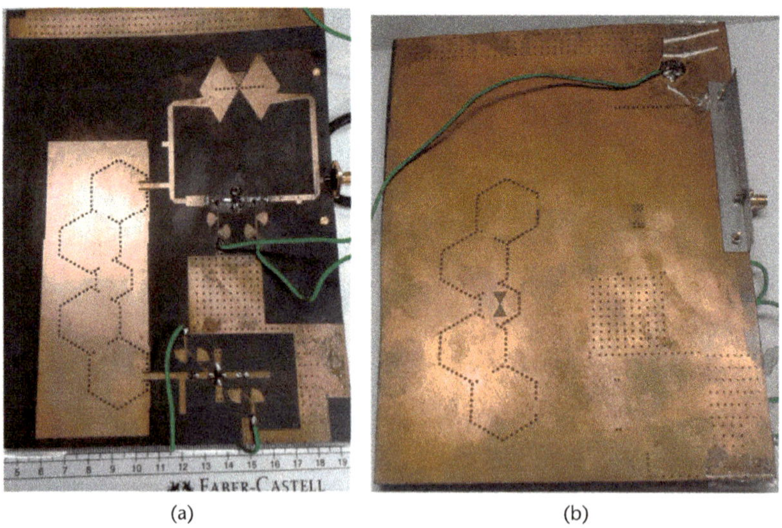

(a) (b)

Figure 8.78 Fabricated prototype of the subharmonic SOM. (a) Top view. (b) Bottom view.

8.2.3.2 Transceiver Design II

The schematic of the SIW-based transceiver design is shown in Figure 8.79. The SIW-based transceiver design consists of Rx and Tx sections. The Rx section comprises a diplexing antenna integrated with a SOM, whereas the Tx section comprises an NRO integrated with a diplexing antenna. The design of a diplexing antenna using a pair of dual-mode SIHC is discussed in Section 8.1.2.2. Dimensions related to the diplexing antenna are provided in Figure 8.31.

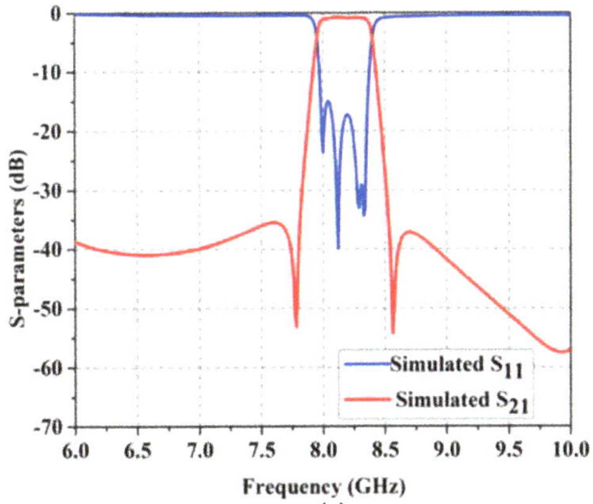

(a)

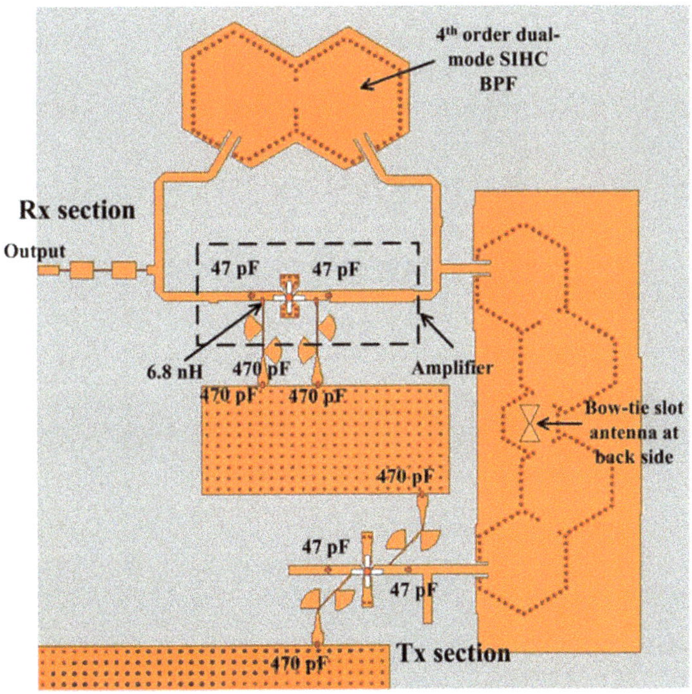

(b)

Figure 8.79 (a) Simulated S-parameters of the SIW-based fourth-order bandpass filter employed in a feedback loop to stabilize fundamental self-oscillation frequency. (b) Schematic of the SIW-based transmitting (Tx) and receiving (Rx) sections. The Rx section consists of SOM integrated with a diplexing antenna and a Tx section consists of NRO integrated with diplexing antenna.

Rx Section

The Rx channel of the diplexing antenna is from 9.25 GHz to 9.45 GHz, with an impedance bandwidth (–10-dB S_{11}) of 200 MHz. The Rx channel of the diplexing antenna is integrated with the SOM through the gate terminal of the pHEMT (NE3210S01) device. SOM is designed using a feedback oscillator. Fourth-order dual-mode SIHC-based BPF is employed in the feedback loop as a frequency stabilization element. The design specifications of the filter are as follows: center frequency = 8.2 GHz; FBW = 5%; and transmission zeros = 7.78 GHz and 8.56 GHz. The dimensions of its design parameters are provided in Figure 7.31. The simulated S-parameters of the filter are shown in Figure 8.79(a). The schematic of the SIW-based transceiver design is shown in Figure 8.79(b). The Rx channel of the diplexing antenna is attached to the SOM, which acts as a receiving section and is termed the Rx section. The oscillator is designed using a bandpass filter, which is employed in the feedback path of the amplifier designed at 8 GHz using a microstrip line of electrical length of 360° or a multiple of 360°. The output signal is collected from the drain of the pHEMT device. The fundamental oscillation frequency is kept at 8 GHz, which mixes with the RF signal received by the Rx channel of the diplexing antenna that generates the desired IF signal.

Figure 8.80 exhibits the output spectrum of SOM when the RF signal is at 9.45 GHz and received through a diplexing antenna. The fundamental oscillating frequency of the SOM mixes with an RF signal of 9.45 GHz and generates the desired IF signal at 1.45 GHz. The IF output power is –32.66 dBm. The DC bias is kept at V_{ds} = 2V and I_{ds} = 10 mA to obtain a self-oscillating mixing signal. Figure 8.81(a) shows the IF output spectrum at 1.32 GHz with an output power of –31.45 dBm when the RF is at 9.3 GHz.

Figure 8.81(b) shows the IF output spectrum at 1.47 GHz with an output power of –31.87 dBm when the RF is at 9.45 GHz. In both cases, the RF signal is

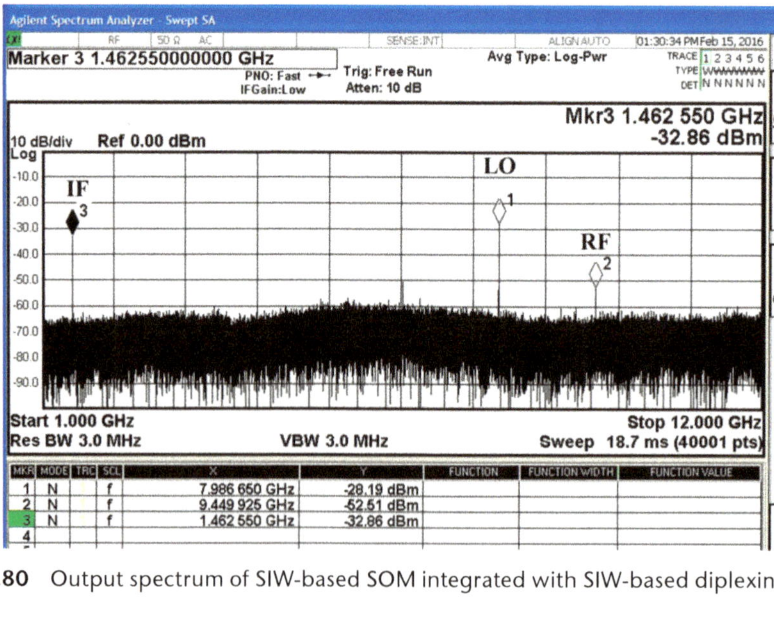

Figure 8.80 Output spectrum of SIW-based SOM integrated with SIW-based diplexing antenna.

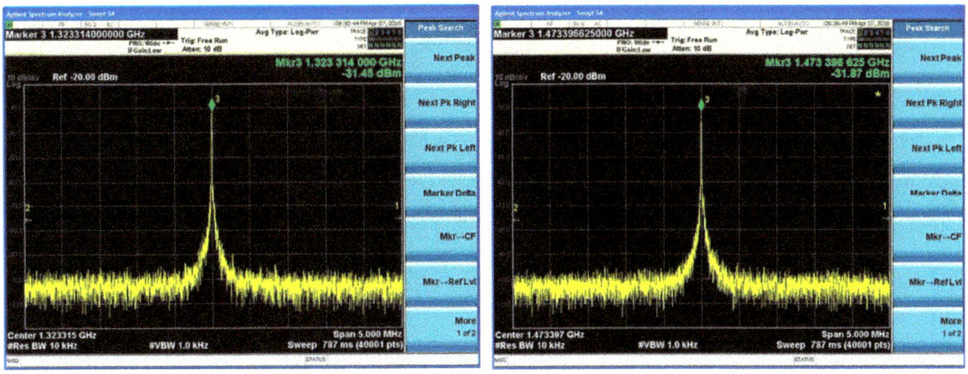

(a) (b)

Figure 8.81 Spectrum of IF signal: (a) spectrum of IF signal when RF signal at 9.3 GHz, and (b) spectrum of IF signal when RF signal at 9.45 GHz.

transmitted through a double-ridged horn antenna, which is kept 30.48 cm away from the diplexing antenna. The double-ridged horn antenna has a gain of 12.3 dBi and transmitted power of 0 dBm. Figure 8.82(a) exhibits the IF output power spectrum with RF variations from 9.3 GHz to 9.45 GHz. From the graph, it is observed that almost the same IF output power is obtained with a variation in RF. Figure 8.82(b) shows SSB IF phase noise. The measured IF phase noises are −96.5 dBc/Hz and −110.7 dBc/Hz at 100-kHz and 1-MHz offset frequencies, respectively. The DC bias conditions are V_{ds} = 2V and I_{ds} = 10 mA. The calculated FOM-based on the measured phase noise of the IF signal is −162.9 dBc/Hz at the 100-kHz offset frequency. Measured IF output power variations with changes in RF are exhibited in Figure 8.83(a). The gain of the receiving antenna is 4.37 dBi. Figure 8.83(b) shows the measured conversion gain of the desired IF signal of the SOM. The IF band conversion gain varies from −2 dB to −3.5 dB when the RF varies from 9.25 to 9.42 GHz. The detailed analysis of the Tx section is provided in Section 8.2.3.1. The fabricated prototype of the SIW-based transceiver is shown in Figure 8.84.

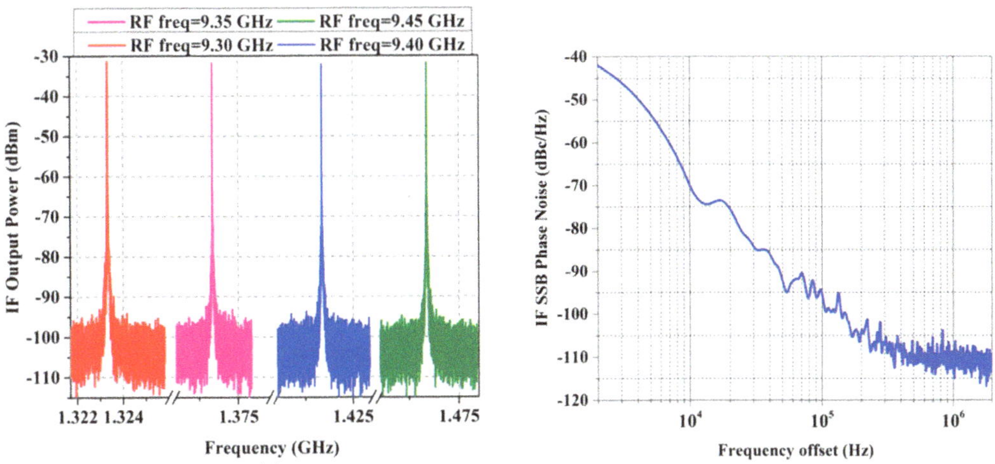

(a) (b)

Figure 8.82 (a) Spectrum of IF signal with the variations of RF. (b) Measured SSB phase noise of IF signal.

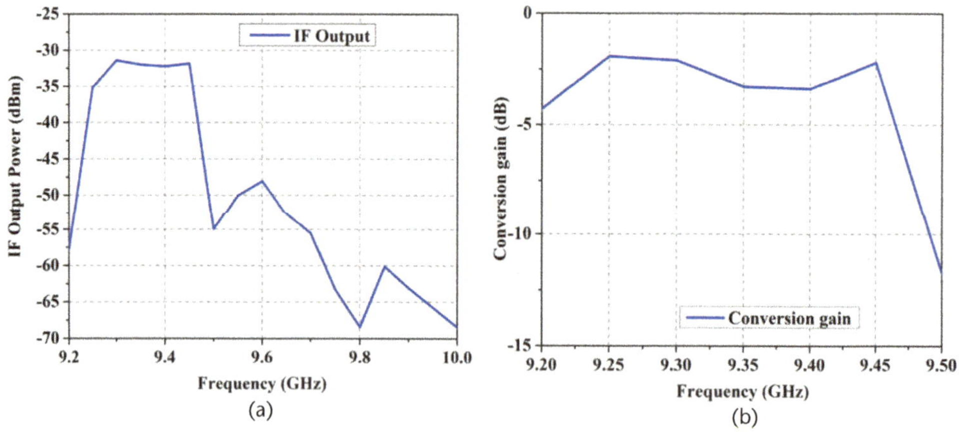

Figure 8.83 (a) IF output power variations with the RF. (b) Measured conversion gain of the SOM.

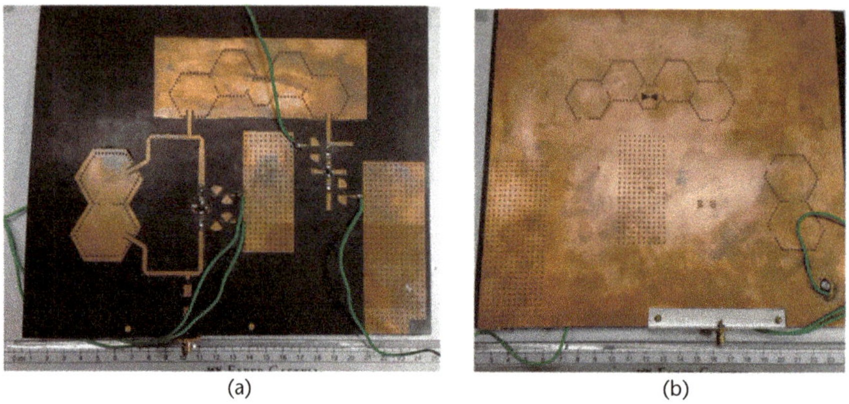

Figure 8.84 (a) Top view of the fabricated SOM. (b) Bottom view of the fabricated SOM.

8.3 SIW-Based RFIC Systems

In the previous sections, the integration of active components with SIW-based circuits has been shown. Such circuit implementation surely establishes the potential of SIW-based transceiver system for application in the commercial, space, and defense sectors. The major advantage of using SIW-based transceiver becomes prominent while using submillimeter-wave/millimeter-wave band where conventional technologies performance limitation can be overcome by the help of the SIW technology. However, the size of the system becomes comparatively large. One of the alternative solutions involve implementation of the RFIC-based system where all the major components can be integrated in the RFIC itself. For effective application scenario, the antenna gain requires to be high enough, which cannot be achieved by an on-chip antenna. An intermediate solution can be implementing the crucial components in PCB-based SIW technology while other components can be kept in the RFIC environment. Such a system will take benefit from both technology and will be compact in nature. However, an efficient integration between SIW and RFIC is required. Since

SIW is a single conductor line, it cannot be directly connected to RFIC. Therefore, a small section of microstrip-based balanced transmission line section is introduced in between the SIW and the RFIC. The SIW section offers an efficient means to implement passive feeding section (including a power divider and hybrid coupler) to the antenna. The antenna section can be implemented in the SIW environment itself. To achieve high gain, a waveguide-based antenna can be directly connected to the SIW section. In this case, the SIW is utilized as an efficient circuit component for smooth transition from planar PCB environment to nonplanar waveguide environment. The effectiveness of such system becomes prominent at mm-wave frequencies where the performance of other planar technologies is not satisfactory.

A design example of complete system implemented in single substrate can be in Figure 8.65. The SIW is realized using advanced ceramic-based photo-imageable thick-film technology where a trench-based metallic side wall is used instead of a via wall [7]. The slotted array antenna is realized in the SIW section, which was directly connected to the RFIC components realized on the same substrate. The transceiver system works at 60 GHz and the circuit dimension is 22.5 × 5.4 × 0.3 mm^3. Another design example is discussed in [8] where a 61-GHz RFID front end is realized using SIW-based circuits. Here the RFIC is directly connected to the SIW using balanced microstrip line transition followed by SIW-based hybrid coupler and SIW turnstile junction to open waveguide aperture antenna. The system generates high EIRP of 23.4 dBm and can communicate with the tag up to 4 mm. The readers are encouraged to explore similar works in [9, 10].

8.4 Application of the SIW Systems and Market Trends

Throughout this book, the potential of SIW technology has been explored and discussed in details. The SIW technology can be a very effective candidate for implementing low-cost circuits at submillimeter-wave and millimeter-wave systems with better performance than other planar technologies. Over the past few years, several reports demonstrated various uses of SIW-based systems at 24 GHz and above frequencies. In this section, a few examples are highlighted for the reader.

Over the years, many researchers have demonstrated the effective design strategy to implement mm-wave antenna systems and associated RF front end using SIW technology. One such example is shown in Figure 8.67. The system demonstrates SIW-based single-layer multibeam (24 beam) slot array antenna for 360° azimuth coverage for supporting MIMO communication [11]. The antenna system consists of modified Butler matrix feeding the slot array antenna. The MIMO operation is achieved by selecting 6 beams (and hence the dedicated input port) with higher signal to noise ratio (SNR) to construct 6 × M MIMO where M is the number of omnidirectional antennas at mobile terminal. With the recent development of 5G and beyond communication technologies, such system can be very much useful for low-cost implementation of indoor wireless access point, millimeter-wave fixed wireless access (FWA) applications.

Apart from millimeter-wave antenna applications, another important direction is millimeter-wave radar sensor for sensing applications. In Figure 8.68, a D-band (140-GHz) frequency-modulated continuous-wave (FMCW) radar using SIW-based

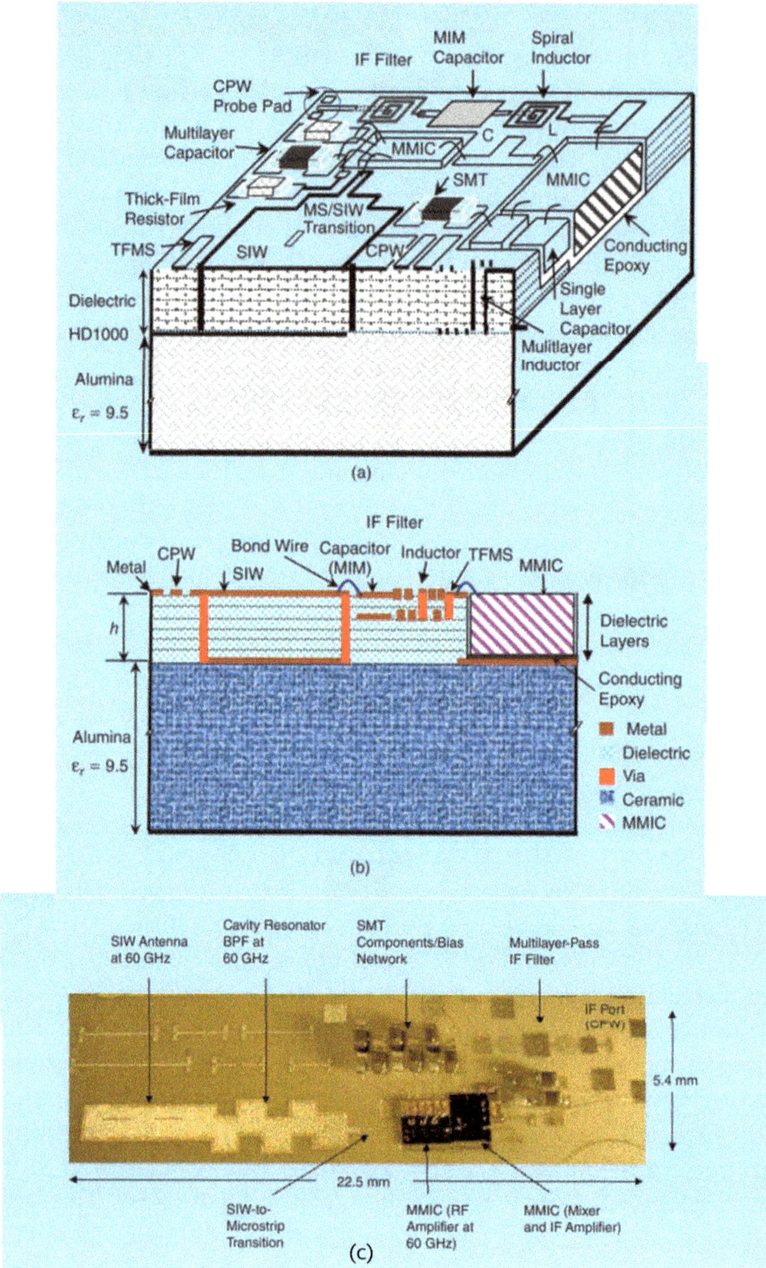

Figure 8.85 A 60-GHz transceiver including MMIC and SIW-based front end on single substrate. (© 2016 IEEE [12].)

circuits along with SiGe TRX chip is demonstrated [12]. The RFIC chip is connected to SIW via a balanced transition. The SIW section is used as an efficient interconnect between the chip and the high gain waveguide-based horn antenna. As application of such system, tube extrusion of plastic and imaging of fiber composite material is demonstrated. Many other design examples have been demonstrated in [13–15]. The readers are encouraged to explore the works to understand the application scenario of the SIW-based circuits.

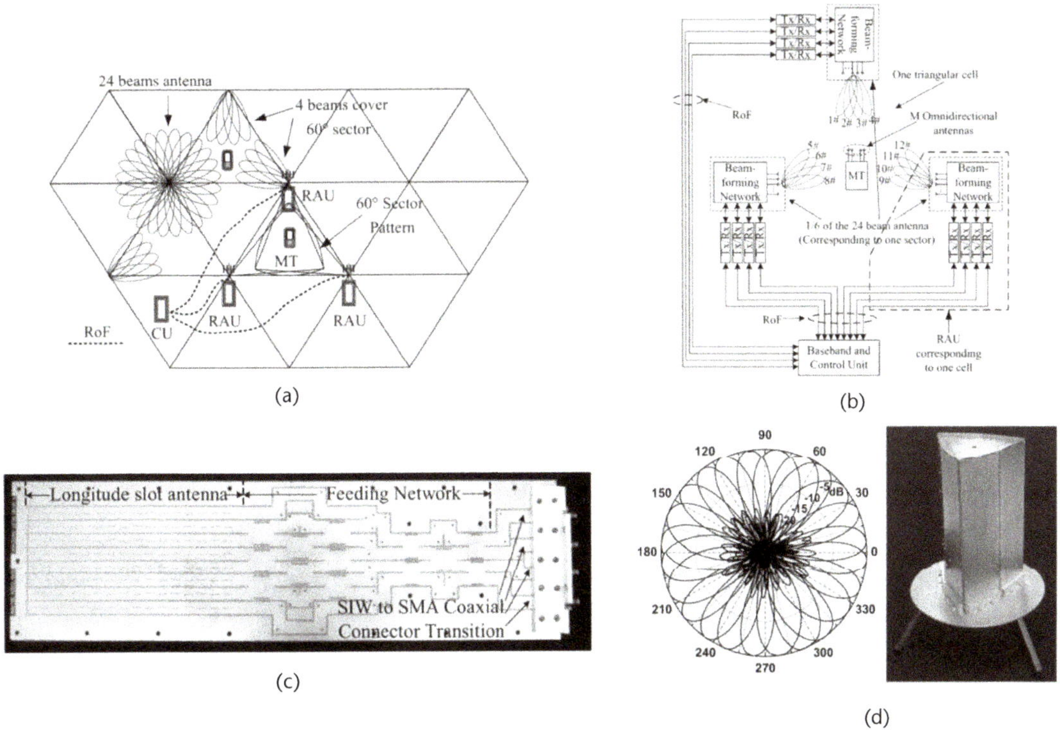

Figure 8.86 MIMO wireless network and block diagram of different component in a triangular cell, SIW multibeam antenna, and final 24-beam antenna design. (© 2009 IEEE [16].)

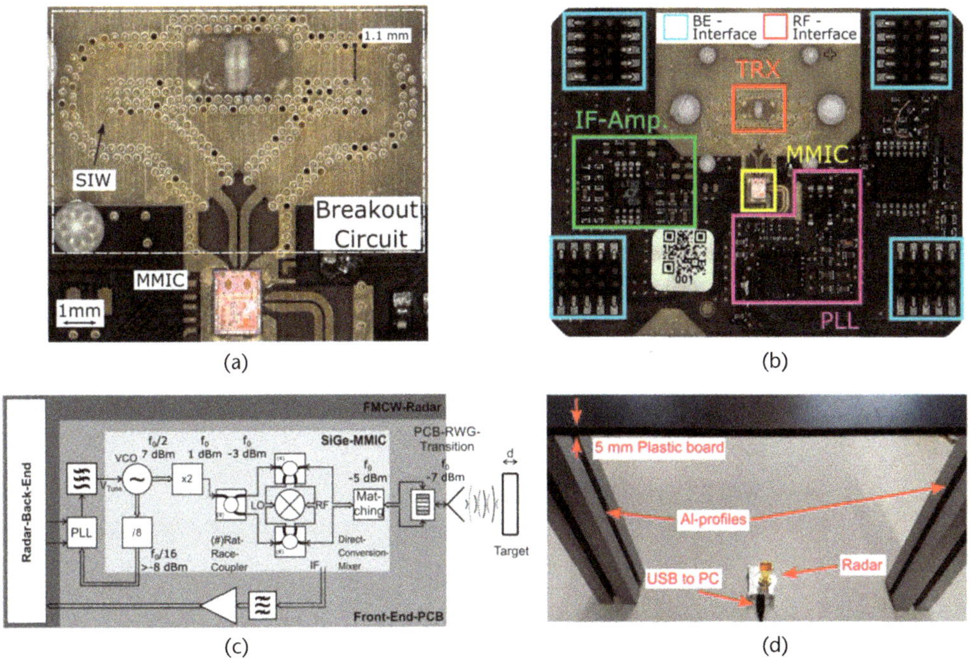

Figure 8.87 D-band MMIC to PCB interface, SIW section and transition to rectangular waveguide, complete radar front-end board, block diagram of the system, and measurement setup to determine the thickness of the plastic board. (© 2022 IEEE [17].)

8.5 Conclusions

Several designs of compact integrated filtering antenna networks were presented in this chapter. The designs utilize SIW technology-based high-*Q* filters to introduce radiation nulls into the antenna response. The designs also introduce diplexing and triplexing antennas for the realization of a multiport filter antenna network that can be used for simultaneous operation of multiband transceiver systems. Moreover, integration of balun in the filter antenna network for efficient feeding to a quasi-Yagi antenna-based transceiver system was also presented in this chapter. The efficient use of a higher-order mode of the SIW cavity to realize balun helps to realize a compact balun filtering antenna, making it suitable for low-profile wireless communication applications.

Several designs of low-profile active integrated antennas are also demonstrated, where cavity loaded with antenna is excited by CW oscillators. Further extension of the active integrated antenna using VCO is also demonstrated, which exhibits continuous tunability of the CW signal. The SIHC-based high-*Q* resonators are utilized to implement low phase-noise oscillators where highly selective bandpass filters act as frequency stabilization elements that help to achieve SSB phase noise below −130 dBc/Hz at 1-MHz offset frequency.

Also, various SIW-based self-oscillating mixer designs are explored. The designs use a SIW based high-*Q* filter network that is integrated with a quasi-Yagi antenna to implement a low-profile portable receiver network. All the designs are fabricated and measured using a standard horn antenna. Finally, the transmitter and receiver networks are integrated to implement a compact, low-profile transceiver network on a single substrate. The design uses a SIW-based diplexing antenna network along with a miniaturized SIW bandpass filter to realize the desired response of the transceiver. The designs are useful for low-profile X-band applications.

References

[1] Yusuf, Y., and X. Gong, "Compact Low-Loss Integration of High-Q 3-D Filters with Highly Efficient Antennas," *IEEE Transactions on Microwave Theory and Techniques*, Vol. 59, No. 4, April 2011, pp. 857–865.

[2] Hong, J., and M. Lancaster, *Microstrip Filters for RF/Microwave Applications*, Second Edition, New York: Wiley, 2011.

[3] Yu, C., and W. Hong, "37–38 GH Substrate Integrated Filtenna for Wireless Communication Application," *Microwave and Optical Technology Letters*, Vol. 54, No. 2, February 2012, pp. 346–351.

[4] Lin, Y. Y., C. H. Wu, and T. G. Ma, "Miniaturized Self-Oscillating Annular Ring Active Integrated Antennas," *IEEE Transactions on Antennas and Propagation*, Vol. 59, No. 10, October 2011, pp. 3597–3606.

[5] Segovia-Vargas, D., et al., "Broadband Active Receiving Patch with Resistive Equalization," *IEEE Transactions on Microwave Theory and Techniques*, Vol. 56, No. 1, January 2008, pp. 56–64.

[6] Andrenko, A. S., et al., "Active Broad X-Band Circular Patch Antenna," *IEEE Antennas and Wireless Propagation Letters*, Vol. 5, No. 1, December 2006, pp. 529–533.

[7] Dhar, S. K., et al., "An Active Integrated Ultra-Wideband MIMO Antenna," *IEEE Transactions on Antennas and Propagation*, Vol. 64, No. 4, April 2016, pp. 1573–1578.

[8] Hang, C. Y., et al., "High-Efficiency Push-Pull Power Amplifier Integrated with Quasi-Yagi Antenna," *IEEE Transactions on Microwave Theory and Techniques*, Vol. 49, No. 6, June 2001, pp. 1155–1161.

[9] Kim, H., and Y. J. Yoon, "Wideband Design of the Fully Integrated Transmitter Front-End with High Power-Added Efficiency," *IEEE Transactions on Microwave Theory and Techniques*, Vol. 55, No. 5, May 2007, pp. 916–924.

[10] Flynt, R. A., et al., "Low Cost and Compact Active Integrated Antenna Transceiver for System Applications," *IEEE Transactions on Microwave Theory and Techniques*, Vol. 44, No. 10, October 1996, pp. 1642–1649.

[11] Montiel, C. M., L. Fan, and K. Chang, "A Novel Active Antenna with Self-Mixing and Wideband Varactor-Tuning Capabilities for Communication and Vehicle Identification Applications," *IEEE Transactions on Microwave Theory and Techniques*, Vol. 44, No. 12, December 1996, pp. 2421–2430.

[12] Samanta, K. K., and I. D. Robertson, "Surfing the Millimeter-Wave: Multilayer Photoimageable Technology for High Performance SoP Components in Systems at Millimeter-Wave and Beyond," *IEEE Microwave Magazine*, Vol. 17, No. 1, January 2016, pp. 22–39.

[13] Funke, D. A., et al., "A 61-GHz RFID Frontend with SiGe Transceiver MMIC and SIW Coupling Network," *2020 50th European Microwave Conference (EuMC)*, Utrecht, Netherlands, 2020, pp. 165–168.

[14] Sene, B., et al., "An Automotive D-Band FMCW Radar Sensor Based on a SiGe-Transceiver MMIC," *IEEE Microwave and Wireless Components Letters*, Vol. 32, No. 3, March 2022, pp. 194–197.

[15] Corsi, J., et al., "A Millimeter-Wave Substrate Integrated Waveguide Filter in Si-BCB Technology," *2021 51st European Microwave Conference (EuMC)*, London, United Kingdom, 2021, pp. 849–852.

[16] Chen, P., et al., "A Multibeam Antenna Based on Substrate Integrated Waveguide Technology for MIMO Wireless Communications," *IEEE Transactions on Antennas and Propagation*, Vol. 57, No. 6, June 2009, pp. 1813–1821.

[17] Hansen, S., et al., "A SiGe-Chip-Based D-Band FMCW-Radar Sensor with 53-GHz Tuning Range for High Resolution Measurements in Industrial Applications," *IEEE Transactions on Microwave Theory and Techniques*, Vol. 70, No. 1, January 2022, pp. 719–731.

[18] Kueppers, S., et al., "A Compact 24 × 24 Channel MIMO FMCW Radar System Using a Substrate Integrated Waveguide-Based Reference Distribution Backplane," *IEEE Transactions on Microwave Theory and Techniques*, Vol. 68, No. 6, June 2020, pp. 2124–2133.

[19] Cheng, Y. M., et al., "Substrate-Integrated-Waveguide Beamforming Networks and Multibeam Antenna Arrays for Low-Cost Satellite and Mobile Systems," *IEEE Antennas and Propagation Magazine*, Vol. 53, No. 6, December 2011, pp. 18–30.

[20] Yang, B., et al., "Digital Beamforming-Based Massive MIMO Transceiver for 5G Millimeter-Wave Communications," *IEEE Transactions on Microwave Theory and Techniques*, Vol. 66, No. 7, July 2018, pp. 3403–3418.

Acronyms and Abbreviations

ADS	Advanced design system
AIA	Active integrated antenna
AIFA	Active integrated filtering antenna
BPF	Bandpass filter
CBCPW	Conductor-backed coplanar waveguide
CBSA	Cavity-backed slot antenna
CP	Circularly polarized
CPW	Coplanar waveguide
CRLH	Composite right/left handed
CSRR	Complementary split-ring resonator
DBS	Direct broadcasting from satellite
EIRP	Effective isotropic radiated power
EM	Electromagnetic
FBW	Fractional bandwidth
FCC	Federal Communication Commission
FEM	Finite element method
fIFH	Higher intermediate frequency
fIFL	Lower intermediate frequency
FIR	Frequency invariant reactive element
FOM	Figure of merits
FR	Frequency ratio
FTBR	Front-to-back ratio
GCPW	Ground coplanar waveguides
GPS	Global Positioning System
GR	Receiving antenna gain
GSM	Global System for Mobile Communication
GT	Transmitting antenna gain
HMBLC	Half-mode branch line coupler
HMSIW	Half-mode substrate integrated waveguide
HPBW	Half-power beamwidth

HiperLAN 2	High Performance Radio Local Area Network 2
IDC	Interdigital capacitor
IF	Intermediate frequency
IoT	Internet of Things
LHCP	Left-handed circularly polarized
LP	Linearly polarized
LTCC	Low temperature cofired ceramic
LTI	Linear time invariant
MIMO	Multiple-input multiple-output
Mx	Middle band
NFAP	National Frequency Allocation Plan
NRO	Negative resistance oscillator
PCB	Printed circuit board
PCS	Personal communication service
PEC	Perfect electric conductor
PMC	Perfect magnetic conductor
PTH	Plated through hole
Q	Quality factor
QE	External quality factor
Ql	Loaded quality factor
QMSIW	Quarter-mode substrate integrated waveguide
QU	Unloaded quality factor
quasi-TEM	Quasi-transverse electromagnetic
RF	Radio frequency
RHCP	Right-handed circularly polarized
RX	Receiver
SIHC	Substrate integrated hexagonal cavity
SIW	Substrate integrated waveguide
SMA	Subminiature version A
SOM	Self-oscillating mixer
SSB	Single sideband
TE	Transverse electric
TEM	Transverse electromagnetic
TM	Transverse magnetic
TX	Transmitter
UMTS	Universal Mobile Telecommunication System
UWB	Ultrawideband
VCAIBFA	Voltage-controlled active integrated balun filtering antenna

VNA	Vector network analyzer
WBAN	Wireless body area network
WiBro	Wireless broadband
Wi-Fi	Wireless fidelity
WiMax	Worldwide Interoperability for Microwave Access
WLAN	Wireless local area network
WPC	Wireless Planning Commission

About the Authors

Soumava Mukherjee received a BTech degree in electronics and communication engineering from the West Bengal University of Technology in 2009, an ME degree in electronics and telecommunication engineering from Bengal Engineering and Science University (BESUS) with specialization in microwave and communication engineering, Shibpur in 2011, and a PhD degree from the Department of Electrical Engineering, IIT Kanpur (IITK), in 2016. Since 2023, he has served as an associate professor in the Electrical Engineering Department, IIT Jodhpur, Rajasth, India, where he served as an assistant professor from 2016 to 2023. His current research interests include substrate integrated coaxial line antennas and circuits, substrate integrated waveguide (SIW) antennas, and multifrequency and broadband antennas. Dr. Mukherjee was a recipient of the European Microwave Student Grant in EuMC 2013 and EuMC 2014. He also received the International Travel Grant from the Department of Science and Technology (DST), Government of India. He is also a recipient of the Young Scientist Award AP-RASC-2019 (URSI Asia-Pacific Radio Science Conference) instituted by the International Union of Radio Science.

Prasun Chongder received an MTech degree in radio frequency design technology (RFDT) from CARE Center of IIT Delhi in 2009, and a PhD degree in RF & microwave specialization from the electrical engineering department of IIT Kanpur in 2017. While pursuing his PhD, he was involved in designing an advanced microwave transceiver system using substrate integrated waveguide (SIW) technology for radar communication system. Since 2018, he has been with the Department of Electronics & Communication Engineering, NIT Rourkela, where he is currently an assistant professor. His current research interests include substrate integrated circuits (SICs) for microwave and millimeter-waves applications, multiband filter design, and reconfigurability and advanced multiband reconfigurable filtering antenna design for radar communication systems.

Animesh Biswas received a BE degree from Bengal Engineering College, Shibpur, in electronics and telecommunication in 1980, and an MTech degree in microwave and radar engineering from the IIT Kharagpur, India, in 1982. He received a PhD degree in electrical engineering from the IIT Delhi, New Delhi, India, in 1989. From 1989 to 1990, he was a postdoctoral fellow at Oregon State University, where he was involved in characterizing multiconductor lines in layered medium. Since 1992, he has been on the faculty of the Department of Electrical Engineering, IIT Kanpur, Kanpur, India, where he is currently a professor. He was a technical consultant for M/S COMDEV Europe and was involved in the development of multimode DR filters and diplexers. He also was an RFIC designer for M/s. Ashvattha Semiconductors for a year and contributed significantly to the area of on-chip LNA and power amplifier development. His current research includes modeling of microwaves circuits, RF integrated circuits (RFICs), and numerical methods for solving electromagnetic problems. He has authored or coauthored over 220 papers in various peer-reviewed international journals and conference proceedings. Prof. Biswas is a Fellow, West Bengal Academy of Science and Technology, the Institution of Electronics and Telecommunication Engineers, India, and a Senior Member of IEEE. Recently, he received the prestigious Ram Lal Wardha award from IETE, India, for outstanding original contributions for the last 10 years in RF and microwaves. He also served as the director of National Institute of Technology, Rourkela, from 2016 to 2021.

Index

R

S

Electromagnetic Diffraction Modeling and Simulation with MATLAB®, Gökhan Apaydin and Levent Sevgi

Engineering Applications of the Modulated Scatterer Technique, Jean-Charles Bolomey and Fred E. Gardiol

Far-Field Wireless Power Transfer and Energy Harvesting, Naoki Shinohara and Jiafeng Zhou, editors

Fast and Efficient Algorithms in Computational Electromagnetics, Weng Cho Chew, et al., editors

FDTD Modelling of Metamaterials: Theory and Applications, Yang Hao and Raj Mittra

Generalized Multipole Technique, Christian Hafner

Grid Computing for Electromagnetics, Luciano Tarricone and Alessandra Esposito

High Frequency Electromagnetic Dosimetry, David A. Sánchez-Hernández, editor

High-Power Electromagnetic Effects on Electronic Systems, D. V. Giri, Richard Hoad, and Frank Sabath

Integral Equation Methods for Electromagnetics, Joseph R. Mautz and Nagayoshi Morita

Intersystem EMC Analysis, Interference, and Solutions, Uri Vered

Introduction to Electromagnetic Wave Propagation, Paul Rohan

Introduction to the Uniform Geometrical Theory of Diffraction, Derek A. McNamara, C. W. I. Pistorius, and Jan Malherbe

Iterative and Self-Adaptive Finite-Elements in Electromagnetic Modeling, Magdalena Salazar-Palma, et al.

Machine Learning Applications in Electromagnetics and Antenna Array Processing, Manel Martínez-Ramón, Arjun Gupta, José Luis Rojo-Álvarez, and Christos Christodoulou

New Foundations for Applied Electromagnetics: The Spatial Structure of Fields, Said Mikki and Yahia Antar

Numerical Analysis for Electromagnetic Integral Equations, Karl F. Warnick

Parallel Finite-Difference Time-Domain Method, Wenhua Yu, et al.

Practical Applications of Asymptotic Techniques in Electromagnetics, Francisco Saez de Adana, et al.

A Practical Guide to EMC Engineering, Levent Sevgi

Quick Finite Elements for Electromagnetic Waves, Second Edition, Giuseppe Pelosi, Roberto Coccioli, and Stefano Selleri

The Scattering of Electromagnetic Waves from Rough Surfaces, Peter Beckman and Andre Spizzichino

Shipboard Electromagnetics, Preston Law

Substrate Integrated Waveguides, Soumava Mukherjee, Prasun Chongder, and Animesh Biswas

Understanding Electromagnetic Scattering Using the Moment Method: A Practical Approach, Randy Bancroft

Wavelet Applications in Engineering Electromagnetics, Tapan K. Sarkar, Magdalena Salazar-Palma, and Michael C. Wicks

For further information on these and other Artech House titles, including previously considered out-of-print books now available through our In-Print-Forever® (IPF®) program, contact:

Artech House
685 Canton Street
Norwood, MA 02062
Phone: 781-769-9750
Fax: 781-769-6334
e-mail: artech@artechhouse.com

Artech House
16 Sussex Street
London SW1V 4RW UK
Phone: +44 (0)20 7596 8750
Fax: +44 (0)20 7630 0166
e-mail: artech-uk@artechhouse.com

Find us on the World Wide Web at: www.artechhouse.com